Modern Software Tools for Scientific Computing

Erlend Arge
Are Magnus Bruaset
Hans Petter Langtangen

Editors

Springer Science+Business Media, LLC

Erlend Arge
Are Magnus Bruaset
SINTEF Applied Mathematics
Blindern
N-0314 Oslo
Norway

Hans Petter Langtangen
University of Oslo
N-0316 Oslo
Norway

Library of Congress Cataloging-in-Publication Data

Modern software tools for scientific computing / Erlend Arge, Are Magnus Bruaset, Hans Petter Langtangen, editors.
p. cm.
Includes bibliographical references and index.
ISBN 978-1-4612-7368-4 ISBN 978-1-4612-1986-6 (eBook)
DOI 10.1007/978-1-4612-1986-6
1. Computer software--Development. 2. Object-oriented programming (Computer science) 3. Science--Data processing. I. Arge, Erlend, 1960- . II. Bruaset, A. M. (Are Magnus) III. Langtangen, Hans Petter, 1962- .
QA76.76.D47M63 1997
502.85'5--dc21 97-6613
CIP

Printed on acid-free paper

Originally published by Birkhäuser Boston in 1997
Softcover reprint of the hardcover 1st edition 1997

ISBN 978-1-4612-7368-4

Camera-ready text provided by the editors in La-TEX.

9 8 7 6 5 4 3 2 1

Contents

Preface

Looking back at the years that have passed since the realization of the very first electronic, multi-purpose computers, one observes a tremendous growth in hardware and software performance. Today, researchers and engineers have access to computing power and software that can solve numerical problems which are not fully understood in terms of existing mathematical theory. Thus, computational sciences must in many respects be viewed as experimental disciplines. As a consequence, there is a demand for high-quality, flexible software that allows, and even encourages, experimentation with alternative numerical strategies and mathematical models. Extensibility is then a key issue; the software must provide an efficient environment for incorporation of new methods and models that will be required in future problem scenarios. The development of such kind of flexible software is a challenging and expensive task. One way to achieve these goals is to invest much work in the design and implementation of generic *software tools* which can be used in a wide range of application fields.

In order to provide a forum where researchers could present and discuss their contributions to the described development, an *International Workshop on Modern Software Tools for Scientific Computing* was arranged in Oslo, Norway, September 16–18, 1996. This workshop, informally referred to as *SciTools'96*, was a collaboration between SINTEF Applied Mathematics and the Departments of Informatics and Mathematics at the University of Oslo. In total, the workshop was visited by 76 attendees from 14 countries world-wide. During the workshop, one introductory, four invited, and 33 contributed talks were given. From this comprehensive program, 17 carefully selected and refereed research papers are presented in this book. The editors want to thank all the authors, as well as the other speakers at SciTools'96, for their high-quality contributions that made the workshop a successful meeting.

The papers found in this volume emphasize the design of large software codes, computational efficiency, object-oriented programming in scientific computing, reliability of numerical software, and parallel computing. The scientific topics covered are parallel CFD software, multilevel methods for PDEs, advanced ODE solvers, computational steering, computational finance, scattered data approximation, splines and hierarchical surfaces, high-level algorithmic specification and automatic code generation, debugging tools for large codes, quality measures for numerical software libraries, migration of real-world FORTRAN 77 codes to C++, as well as modern and flexible implementations of finite element and difference methods for

PDEs. Several of the papers give extensive examples on the use of C++ and object-oriented design for numerical applications.

All chapters in this book have been peer reviewed by at least two referees. As editors, we want to thank all the referees who generously have spent their valuable time reading the submitted manuscripts. Through numerous corrections and suggestions they have provided invaluable assistance in increasing the quality of this book.

As organizers of the SciTools'96 workshop, we want to thank the Norwegian Research Council of Norway for their financial support through the research program *Numerical Computations in Applied Mathematics*[1] (grant no. 110673/420). This program and its predecessor, *Toolkits in Industrial Mathematics*, have prepared the scientific basis needed to take on this arrangement. We want to also express our gratitude to the invited speakers; Professor Richard Bartels (University of Waterloo, Canada), Professor Christopher R. Johnson (University of Utah, USA), Professor Ulrich Rüde (University of Augsburg, Germany), and Dr. Barry F. Smith (Argonne National Laboratory, USA) for their outstanding contributions. Moreover, we would like to thank Professor Aslak Tveito (University of Oslo, Norway) for his keynote address, and Director Kari Kveseth (Research Council of Norway) for conducting the opening of the workshop.

In the production of the final manuscript, we have also benefitted from the LaTeXnical assistance provided by Mr. Fredrik Tyvand, and the linguistic expertise of Mrs. Janet Skallerud. Finally, we are grateful to Mr. Wayne Yuhasz, who as editor for Birkhäuser's books in Computational Sciences and Engineering suggested the publication of this volume.

Oslo, January 1997

Erlend Arge *Are Magnus Bruaset* *Hans Petter Langtangen*

For further information on the SciTools'96 workshop, including abstracts of the talks not presented here, we refer to the World Wide Web page

`http://www.oslo.sintef.no/SciTools96`

In addition, Birkhäuser Boston will host the web page

`http://www.birkhauser.com/book/isbn/0-8176-3974-8`

which will have pointers to relevant research material and software made available by the contributors to this book.

[1]See `http://www.oslo.sintef.no/NAM` for more information on this program and its associated software projects.

Modern Software Tools for Scientific Computing

Part I

Computational Differential Equations

1

The SCIRun Computational Steering Software System

Steven G. Parker
David M. Weinstein
Christopher R. Johnson[1]

ABSTRACT
We present the design, implementation and application of SCIRun, a scientific programming environment that allows the interactive construction, debugging, and steering of large-scale scientific computations. Using this "computational workbench," a scientist can design and modify simulations interactively via a dataflow programming model. SCIRun enables scientists to design and modify model geometry, interactively change simulation parameters and boundary conditions, and interactively visualize geometric models and simulation results. We discuss the ubiquitous roles SCIRun plays as a computational tool (e.g. resource manager, thread scheduler, development environment), and how we have applied an object oriented design (implemented in C++) to the scientific computing process. Finally, we demonstrate the application of SCIRun to large scale problems in computational medicine.

1.1 Introduction

1.1.1 Visual Computing and Interactive Steering

In recent years, the scientific computing community has experienced an explosive growth in both the possible size and the possible complexity of numeric computations. One of the significant benefits of this increased computing power is the ability to perform complex three-dimensional simulations. However, such simulations present new challenges for computational scientists. How does one effectively analyze and visualize complex 3D data? How does one solve the problems of working with very large datasets often consisting of tens to hundreds of gigabytes? How does one provide tools that address these computational problems while serving the needs of scientific users?

---

[1]Department of Computer Science, University of Utah, Salt Lake City, Utah, 84112, USA. Email:`{sparker,dweinste,crj}@cs.utah.edu`.

Scientific visualization clearly plays a central role in the analysis of data generated by scientific simulations. Unfortunately, though visualization may in itself be more computationally intensive than the original simulation, it is often performed only as a mystical post-processing step after a large-scale computational batch job is run. For this reason, errors invalidating the results of the entire simulation may be discovered only during post-processing. What is more, the decoupling of simulation and visualization presents serious scientific obstacles to the researcher. A visualization package may provide only a limited data analysis capability and may be poorly matched to the underlying physical models used in the simulation code. As a result, the researcher may expend significant effort trying to use a data analysis package only to walk away frustrated.

In 1987, the Visualization in Scientific Computing (ViSC) workshop reported [5]:

> Scientists not only want to analyze data that results from super-computations; they also want to interpret what is happening to the data during super-computations. Researchers want to *steer* calculations in close-to-real-time; they want to be able to change parameters, resolution or representation, and see the effects. They want to drive the scientific discovery process; they want to *interact* with their data.
>
> The most common mode of visualization today at national supercomputer centers is batch. *Batch processing* defines a sequential process: compute, generate images and plots, and then record on paper, videotape or film.
>
> *Interactive visual computing* is a process whereby scientists communicate with data by manipulating its visual representation during processing. The more sophisticated process of *navigation* allows scientists to *steer*, or dynamically modify computations while they are occurring. These processes are invaluable tools for scientific discovery.

Although these thoughts were reported close to ten years ago, they express a very simple and still current idea: scientists want more interaction than is currently present in most simulation codes. While the scientific computing community is still trying to find better ways to address these needs, we feel that the problems encountered by computational scientists encompass a wider range of issues, including but not restricted to scientific visualization. Our efforts, therefore, include a diverse range of techniques, including, among others, the use of scripting languages, existing software, visual dataflow programming, and a sophisticated system designed exclusively for computational steering. In this chapter, we focus on the latter, the SCIRun[2] [30] computational steering software system.

SCIRun is a scientific programming environment that allows the interactive construction, debugging and steering of large-scale scientific computations [26]. SCIRun can be envisioned as a "computational workbench," in which a scientist can design and modify simulations interactively via a dataflow programming model. SCIRun enables scientists to modify geometric models and interactively change numerical parameters and boundary conditions, as well as to modify the level of mesh adaptation needed for an accurate numerical solution. As opposed to the typical "off-line" simulation mode - in which the scientist manually sets input parameters, computes results, visualizes the results via a separate visualization package, then starts again at the beginning - SCIRun "closes the loop" and allows interactive steering of the design, computation, and visualization phases of a simulation.

The dataflow programming paradigm has proven useful in many applications. In the scientific community, it has been successfully applied in several scientific visualization packages, including AVS from Advanced Visual Systems Inc., and Iris Explorer from SGI. We have extended the use of the dataflow programming model into the computational pieces of the simulation. To make the dataflow programming paradigm applicable to large scientific problems, we have identified ways to avoid the excessive memory use inherent in standard dataflow implementations, and we have implemented fine-grained dataflow in order to further promote computational efficiency.

1.1.2 An Iterative Environment for Scientific Computing

Currently, the typical process of constructing a computational model consists of the following steps:

1. Create and/or modify a discretized geometric model;
2. Create and/or modify initial conditions and/or boundary conditions;
3. Compute numerical approximations to the governing equation(s), storing results on disk;
4. Visualize and/or analyze results using a separate visualization package;
5. Make appropriate changes to the model; and
6. Repeat.

---

[2]SCIRun is pronounced "ski-run" and derives its name from the Scientific Computing and Imaging (SCI) research group which is pronounced "ski" as in "ski Utah."

The "art" of obtaining valuable results from a model has up until now required a scientist to execute this process time and time again. Changes made to the model, input parameters, or computational processes are typically made using rudimentary tools (text editors being the most common). Although the experienced scientist will instill some degree of automation, the process is still time consuming and inefficient. Ideally, scientists and engineers would be provided with a system in which all these computational components were linked, so that all aspects of the modeling and simulation process could be controlled graphically within the context of a single application program. While this would be the preferred *modus operandi* for most computational scientists, it is not the current standard of scientific computing because the creation of such a program is a difficult task.

Difficulties in creating such a program arise from the need to integrate a wide range of disparate computing disciplines (such as user interface technology, 3D graphics, parallel computing, programming languages, and numerical analysis) with a wide range of equally disparate application disciplines (such as medicine, meteorology, fluid dynamics, geology, physics, and chemistry). Our approach to overcoming these difficulties is to separate the components of the problem. SCIRun's dataflow model employs "modules" that can be tailored for each application or computing discipline. Although this method is proving successful at partitioning many of the complexities, we have found that some complexities remain, such as the burdens of parallel computing and user interfaces. Much work goes into simplifying the programming interfaces to these features so that they will be used, rather than ignored, by module implementors.

1.1.3 Steering

The primary purpose of SCIRun is to enable the user to interactively control scientific simulations while the computation is in progress [9, 33]. This control allows the user to vary boundary conditions, model geometries, or various computational parameters during simulation. Currently, many debugging systems provide this capability in a very raw, low-level form. SCIRun is designed to provide high-level control over parameters in an efficient and intuitive way, through graphical user interfaces and scientific visualization. These methods permit the scientist or engineer to "close the loop" and use the visualization to steer phases of the computation.

The ability to steer a large scale simulation provides many advantages to the scientific programmer. As changes in parameters become more instantaneous, the cause-effect relationships within the simulation become more evident, allowing the scientist to develop more intuition about the effect of problem parameters, to detect program bugs, to develop insight into the operation of an algorithm, or to deepen an understanding of the physics of the problem(s) being studied.

The scientific investigation process relies heavily on answers to a range of "What if?" questions. Computational steering allows these questions to be answered more efficiently and therefore to guide the investigation as it occurs.

1.2 Requirements of SCIRun as a Computational Steering System

Initially we designed SCIRun to solve specific problems in Computational Medicine [13–16], but we have made extensive efforts to make SCIRun applicable in other computational science and engineering problem domains. In attacking the specific problems, we found that there were a wide range of disparate demands placed on such a system. Each of these demands reveals a different facet of what we call SCIRun.

1.2.1 SCIRun the Operating System

In a sophisticated simulation, each of the individual components (modeling, mesh generation, nonlinear/linear solvers, visualization, etc.) typically consumes a large amount of memory and CPU resources. When all of these pieces are connected into a single program, the potential computational load is enormous. In order to use the resources effectively, SCIRun adopts a role similar to an operating system in managing these resources. SCIRun manages scheduling and prioritization of threads, mapping of threads to processors, inter-thread communication, thread stack growth, memory allocation policies, and memory exception signals (such as segmentation violations).

1.2.2 SCIRun the Scientific Library

SCIRun uses a visual programming interface to allow the scientist to construct simulations through powerful computational components. While the visual programming environment is the central focus of SCIRun, it requires a powerful set of computational tools. In the first stage of SCIRun, we have concentrated on integrating the computational components that we have used to solve our own computational problems. We have recently expanded focus and are now in the process of integrating popular libraries and tools, such as *Diffpack* [3,17], *SparseLib++* [7], and *PETSc* [1,8] into the SCIRun environment.

1.2.3 SCIRun the Development Environment

Perhaps the most powerful facet of SCIRun is the ability to use it in the development phases of a simulation. SCIRun augments the development environment by providing convenient access to a powerful set of computational components. However, these components could never be comprehensive, so SCIRun also provides an environment whereby new modules can be developed efficiently. If a module triggers a segmentation violation, bus error or failed assertion, SCIRun stops the module at the point of error, thus allowing the developer to attach a debugger to the program at the point of failure. This avoids the frustrating experience of trying to reproduce these errors in the debugger. In addition, SCIRun provides simple instrumentation of module performance (CPU times printed out interactively), feedback execution states (waiting for data, percent completed, etc.), and visualization of memory usage. SCIRun employs dynamic shared libraries to allow the user to recompile only a specific module without the expense of a complete re-link. Another SCIRun window contains an interactive prompt which gives the user access to a Tcl shell that can be used to interactively query and change parameters in the simulation.

1.2.4 Requirements of the Application

SCIRun is not magic – it is simply a powerful, expressive environment for constructing steerable applications, either from existing applications or starting from the ground-up. The application programmer must assume the responsibility of breaking up an application into suitable components. In practice, this modularization is already present inside most codes, since "modular programming" has been preached by software engineers as a sensible programming style for years.

More importantly, it is the responsibility of the application programmer to ensure that parameter changes make sense with regard to the underlying physics of the problem. In a CFD simulation, forexample, it is not physically possible for a boundary to move within a single time-step without a dramatic impact on the flow. The application programmer may be better off allowing the user to apply forces to a boundary that would move the boundary in a physically coherent manner. Alternatively, the user could be warned that moving a boundary in a non-physical manner would cause gross errors in the transient solution.

1.3 Components of SCIRun

In order to implement the requirements described above, we have broken down SCIRun into a layered set of libraries. These libraries are organized as shown in Figure 1.1.

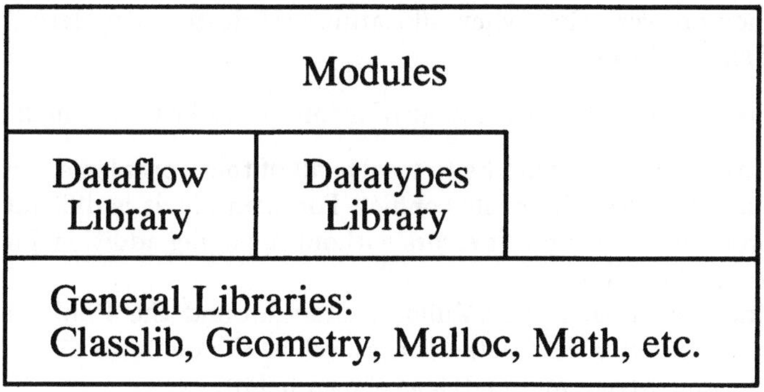

FIGURE 1.1. SCIRun library organization.

SCIRun uses an object oriented design; however, it should be stressed that we have paid careful attention to avoid over-using the object oriented paradigm to a point that efficiency suffers.

In implementing the SCIRun kernel and modules, we leverage off of a powerful toolbox of C++ classes that have been tuned for scientific computing and operation in a multi-threaded environment. We discuss these classes below, starting with the lowest level library and proceeding to more complex libraries. We describe each of the toolbox components here, starting with the lowest layer. In discussing higher layers, we describe how features of the lower layers are leveraged to facilitate implementation.

1.3.1 `Malloc, operator new`: *libMalloc*

We have encountered several problems with the implementations of `malloc/free` and `new/delete` that are available on current Unix systems. Difficulties with the current implementations of `malloc` and `new` include:

1. They are not robust against erroneous behavior. This is particularly confusing when the user's program crashes in `malloc`, while the actual error resulted from freeing a bad pointer in a previous call. A multithreaded environment further exacerbates this problem, allowing errors in one thread to cause another thread to crash.

2. They are not thread-safe (reentrant) on many systems. This is typically the case on systems without a native implementation of threads. Accessing `malloc` and `free` in such an environment can cause frequent non-deterministic crashes.

3. They do not reveal statistics about their operation.

4. They do not return memory to the operating system when it is no longer being used.

5. They are very slow when allocating and deallocating large numbers of small objects.

6. They have a large percentage of memory overhead for small objects.

Of course, the goal would be to resolve all of these problems, but we find that many of the requirements conflict. For example, it is difficult to have bullet-proof behavior against errors without incurring additional overhead, even for small objects.

The implementation of libMalloc centers around the **Pool** class. **Pool** defines a constructor and destructor, as well as the methods `alloc`, `free`, `realloc`, `get_stats` and `audit` as shown below.

```
class Pool {
   Pool();
   ~Pool();
   Mutex lock;
   void* alloc(size_t size, char* ctag, int itag);
   void free(void* p);
   void* realloc(void* p, size_t size);
   void audit();
   void get_stats(size_t statbuf[18]);
   int nbins();
   void get_bin_stats(int bin, size_t statbuf[6]);
   ...
};
```

Pool represents a pool of memory. At startup, there is a single pool, `default_pool`, from which requests from `malloc` and `new` are granted. The implementations of `malloc` and the `new` operator simply call the `alloc` method of the default pool. Subsequently, the `free` and operator `delete` methods call the `free` method of the default pool. The default `malloc` and operator `new` provide generic information as the two tags for the allocation, but there are alternate interfaces that automatically provide the file and line number for these tags.

The `alloc` method uses three slightly different memory allocation algorithms for small, medium and large objects. Based on heuristics from current applications, small objects are those less than 512 bytes, medium objects range from 513 bytes-64k bytes, and large objects are those over 64k bytes. These ranges are configurable at compile time.

Small and medium objects both use an algorithm based on bins. A bin contains a list of free objects. When free space is requested, `alloc` figures out which bin contains objects of the appropriate size, and the first one from the list is removed. Sentinels are placed at the beginning and at the end of the actual allocation. Small and medium bins differ in how the bins are refilled when they become empty. Small bins use an aggressive fill scheme, where 64k worth of objects are placed in the bin's free list in order to minimize the number of refills. Medium objects, on the other hand, use a less aggressive scheme - objects are allocated from a larger pool one at a time. Large objects are allocated with independent `mmap` calls to the operating system. This allows the large objects to be returned to the operating system when they are no longer needed. In order to avoid releasing and re-requesting memory, these large chunks are returned to the operating system (unmapped) in a lazy fashion. It is possible for this policy to fragment the address space of the program, but in practice this has not been a problem, and will never be a problem for 64 bit programs.

The algorithms for the three different allocation ranges are based on the philosophy that bigger objects can afford to use more CPU cycles in trying to be efficient, since large objects will be allocated less frequently and used for a longer period of time. It is also more valuable to minimize waste for large objects than for small allocations.

In order to make the pool thread safe, each of the methods acquires the mutex before accessing or modifying any data in the Pool, and releases the mutex when these operations are complete. The **`alloc`** and **`release`** methods attempt to minimize the time that the pool is locked by performing most operations (tag/header manipulation, verification, etc.) without holding the lock.

This implementation resolves all of the problems that we described above, except for items five and six. The memory overhead (item six) is approximately the same as current implementations, and the time overhead for small objects (item five) is considerably smaller, but still too large. In the next section, we will see a mechanism that may be layered on top of libMalloc to resolve these problems.

This memory allocator can also reveal statistics about its operation. Figure 1.2 shows these statistics displayed by a running program.

1.3.2 The Multitask Library: libMultitask

SCIRun derives much of its flexibility from its internal use of threads [31]. Threads allow multiple concurrent execution paths in a single program. SCIRun uses threads to facilitate parallel execution, to allow user interaction while computation is in progress, and to allow the system to change variables without interrupting a simulation. However, standards for implementing threads are only starting to appear, and the standards that are appearing are, thus far, cumbersome. libMultitask is a layer that provides a simple, clean C++ interface to threads and provides abstraction from the actual standard used to implement them.

1.3.2.1 Tasks

The Multitask library provides a class **Task**, which encapsulates a thread. The **Task** constructor requires a name for the **Task** and a priority. A new

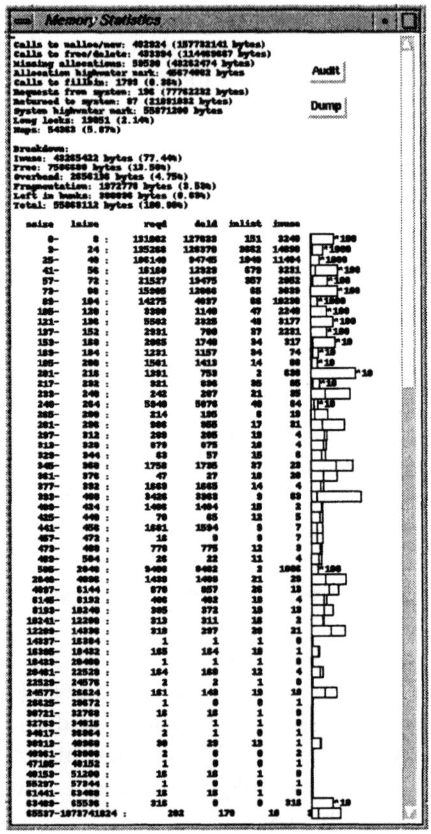

FIGURE 1.2. Statistics of the custom allocator, showing bytes allocated and freed, high water marks, and spinlock statistics. Some of these statistics are also displayed for each bin. To the right of each bin a small graph shows the objects in the freelist and the objects in use for each range of sizes.

thread is created when the creator calls the `activate` method of the **Task** class, which will cause the `body` method to be started in a separate thread. `Activate` will return immediately and will not wait for `body` to complete - thus triggering concurrent execution in the program, similar to a `fork()` operation. However, all of the threads (**Tasks**) will share access to a common heap - unlike the traditional `fork()` function. **Task** is an abstract base class because it does not actually provide a `body` function. Other classes inherit from **Task**, providing a `body` function to do the actual work of the thread. The thread continues until the `body` function returns, or until `Task::exit` is called.

Task also provides static functions to return the number of processors available on the system, to start-up multiple threads for a function, and to cause all threads to exit.

1.3.2.2 Intertask Communication

The Multitask library also provides a number of convenient synchronization primitives for these tasks to communicate with each other – Inter-Task Communication (ITC). ITC primitives are:

- **Mutex** provides `lock, try_lock` and `unlock` methods.
- **Semaphore** is a counting semaphore, providing `down, try_down` and `up` methods.
- **Barrier** provides a single `wait(int nwait)` method to allow a group of threads to stop executing at the barrier until all `nwait` threads arrive.
- **ConditionVariable** provides `wait(Mutex& lock), cond_signal` and `cond_broadcast` methods.
- **CrowdMonitor** a multiple-reader, singler-writer access control primitive, provides `read_lock, read_trylock, read_unlock, write_lock, write_trylock` and `write_unlock methods`.
- **Mailbox** a fixed-length, thread-safe FIFO (First-In, First-Out communication pipe), allows multiple senders and multiple receivers. This is a template class that provides `send, try_send, receive` and `try_receive` methods. The mailbox allows multiple threads to send tokens to the mailbox and an arbitrary number of threads to receive tokens from the mailbox. These tokens are typically pointers to a message structure. Using this primitive, one can implement threads which behave like a small server.
- Other structures, including **AsyncReply** which provides a single pigeon hole rendezvous point, and classes to perform reduction operations.

The Task and ITC methods have been implemented in four different environments: SGI IRIX (using `sproc` and `us*` primitives), and with Posix threads (aka pthreads), with Solaris threads, and with `setjmp/longjmp`.

1.3.3 Generic Tools: libClasslib

This is a collection of various tools that are valuable in constructing SCIRun's kernel and computational modules. Some of these data structures overlap those available in the Standard Template Library (STL) [24], but our implementation predates common acceptance of STL. In implementing these, we have not tried to make extravagant general-use interfaces. Rather, we have designed our tools to be simple, easy to use, and efficient. We have also designed these interfaces to perform the operations that we require, avoiding the temptation to over-engineer them. Data structures that we have implemented include an unbounded queue, a bounded stack, a dynamic hashtable, and a dynamic array class. These structures use templates to make them usable as containers for any type.

1.3.3.1 TrivialAllocator

Another useful tool is the **TrivialAllocator** class. This class is designed to increase the efficiency of the `new/delete` operator for small objects that are allocated and deleted frequently. The **TrivialAllocator** simply keeps a list of free objects that are ready to be used. Using the **TrivialAllocator** for a particular class simply requires redefining the operator `new` and operator `delete` methods to use the `alloc` and `free` methods of the **TrivialAllocator**. Using `Small_alloc.alloc` is significantly more efficient than using the general operator `new`, because most of the time it simply returns the first item off of the free list. Objects are allocated in groups, using the second parameter to the constructor as the number of objects to be allocated at a time. `Small_alloc.free` always just puts the object back on the free list. The free list is accessed in a last-in/first-out manner to maximize cache locality. Since these will be used in a multithreaded environment, `alloc` and `free` both require the acquisition and release of a mutex. However, this is a separate mutex from the global allocator, so it will not be subject to the same contention.

This tool allows us to work around the per-object overhead and allocation time required for small, high-use objects. However, it does so at the expense of the overrun detection and consistency checks that our implementation of `new/delete` provides. A future implementation will provide a mechanism by which trivial allocators can be disabled through an environment variable - reducing run time performance, but allowing the consistency checks to be made.

1.3.3.2 Handles and LockingHandles

Handles are a "smart pointer" mechanism for automatically maintaining reference counts in objects [32]. SCIRun uses these to facilitate the sharing of large data structures between modules. The last **Handle** to "let go" is responsible for deleting the object. Reference counting provides a simple form of garbage collection, without the high overhead and unpredictability associated with a full garbage collection system. The largest weakness is that reference counting can fail to destroy objects which contain circular references (including circular lists, self references, and back pointers). However, it does provide the advantage that objects are destroyed immediately when the last handle releases the object. This feature is essential for large scale scientific computing where the memory resources held by such a handle need to be carefully controlled.

A **handle** contains a single data member `rep`, which contains a pointer to the actual representation. It also defines constructors, a destructor and access methods that increment and decrement a reference count in the object. The objects used in a handle must provide a member called `ref_cnt`, which is initialized to zero at construction time. In addition, objects which support the `detach` operation must support a `clone` method which will duplicate the object. Since template syntax is sometimes rather clumsy, it is often convenient to typedef a "smart pointer" type for different object types, such as:

typedef LockingHandle<Object> `ObjectHandle;`

A **LockingHandle** is similar to **Handle**, but with each object also providing a mutex that is locked under all of the `ref_cnt` operations. For more details, see the appendix.

1.3.3.3 Timers

Two very convenient classes are the **CPUTimer** and the **WallClockTimer**. These provide a stopwatch interface to acquire time information. Methods include `start`, `stop`, `clear`, and `time`. The `time` method simply returns the total accumulated time between `start` and `stop` pairs. Typically, these timers access the Unix timer functions appropriate for the operating system, but they can also utilize a high resolution user-space mappable timer on systems that support it.

1.3.3.4 Persistent Objects

Modern scientific computing codes use modern data structures that are generally more complex than a large array of numbers. We needed a facility to save complex data structures on disk. In order to solve this problem, we employed the idea of "persistent objects," where complex data structures can be flattened into a data file and then reconstructed into their original form. Since C++ does not support persistent objects automatically, we designed a set of utilities to make these input and output routines in a simple manner.

There are three different levels on which the persistent I/O routines operate:

1. **Primitive:** There is an overloaded global function, `Pio`, for each of the basic types (float, double, int, etc.), which can both read and write these primitive types.

2. **Simple Structure:** For basic structures without class derivations, the user can overload a global `Pio` function to serialize these objects. Usually, this function just calls `Pio` on each of the individual data elements.

3. **Complex Structure:** For a complicated class hierarchy, the user writes a virtual `io` function which emits a class name and version number, then the `io` method for the superclass, and then calls other `Pio` functions on the individual components.

The basic stream I/O routines support both binary files and somewhat humanly readable text files. Binary input and output uses the Sun Microsystem's XDR library for portable binary I/O, which allows us to move files between different architectures without the need to convert to ASCII text. Binary files are sometimes smaller, and are always a factor of 3-5 faster in reading and writing. In addition, we support versioning of objects so that we can change the data structures in the code without requiring the conversion of all old data files.

To illustrate the simplicity of reading and writing objects in this manner, consider an example of a 3D tetrahedral mesh:

```
class Mesh {
    Array1<Node*> nodes;
    Array1<Element*> elems;
    ...
};

#define MESH_VERSION 1

void  Mesh::io(Piostream& stream)
{
    stream.begin_class("Mesh", MESH_VERSION);  // identify type & version
    Pio(stream, nodes);                        // read/write the nodes
    Pio(stream, elems);                        // read/write the elements
    stream.end_class();
}
```

The **Array1** classes know how to read and write themselves:

```
template<class T>
void Pio(Piostream& stream, Array1<int>& array)
{
    stream.begin_class("Array1", ARRAY1_VERSION); // id the type & version
    int size=array._size;               // grab the size of the array...
    Pio(stream, size);                  // ...and write it
    if(stream.reading())                // if we're reading...
        array.resize(size);             // ...allocate space for objects
    for(int i=0;i<size;i++)
        Pio(stream, array.objs[i]);     // read/write all of the objects
    stream.end_class();
}
```

and the **Node/Element** classes know how to read and write themselves:

```
void Pio(Piostream& stream, Node*& node)
{
    stream.begin_cheap_delim();
    if(stream.reading())              // if we're reading...
        node=new Node();              // ...allocate a new node
    Pio(stream, node.pt);             // read/write the node's location
    stream.end_cheap_delim();
}

void Pio(Piostream& stream, Element* elem)
{
    if(stream.reading())              // if we're reading...
        elem=new Element();           // ...allocate a new element
    stream.begin_cheap_delim();
    Pio(stream, elem->n[0]);          // read/write all of the
    Pio(stream, elem->n[1]);          //    indices for the four
    Pio(stream, elem->n[2]);          //    nodes composing the
    Pio(stream, elem->n[3]);          //    tetrahedral element
    stream.end_cheap_delim();
}
```

It is important to remember that these small functions will both read and write the mesh for both binary files and text files. This feature virtually eliminates the potential for incompatibilities between the reading code and the writing code. However, the real power comes when we emit a scalar field based on these meshes:

```
class ScalarFieldUG : public ScalarField {
    MeshHandle mesh;
    Array1<double> data;
    ...;
};

void ScalarFieldUG::io( Piostream& stream)
{
    stream.begin_class("ScalarFieldUG", SCALARFIELDUG_VERSION);
    ScalarField::io(stream); // This serializes the base class
    Pio(stream, mesh);          // read/write the mesh
    Pio(stream, data);          // read/write the data
}
```

Then, we could emit multiple scalar fields:

```
    Pio(stream, field1);        // read/write field1
    Pio(stream, field2);        // read/write field2
    ...
```

In this example, the fields might share a common mesh. In this case, the mesh would be written into the file only once. We have omitted many of the details of how this is implemented internally, but the `Pio` routines do not need to be concerned with this mechanism – it is handled internally by the **Piostream**.

We have found that this mechanism is a very powerful way to implement I/O for complex data structures. The versioning system allows us to use data files that were written years ago without converting them. Using the same code for reading and writing drastically reduces the number of errors in the input/output routines. In addition, this mechanism has increased the utility of binary files by avoiding the need to write a separate I/O function.

1.3.4 Geometry Library

libGeometry provides **Point**, **Vector**, **Plane**, **Ray**, **Transform**, and **BoundingBox** classes for convenient computation of 3D geometry. The addition, subtraction, and multiplication operators have all been implemented to allow these components to be used in a convenient fashion.

We have chosen to separate the concept of a **Point** from the concept of a **Vector** [6]. For the sake of efficiency, these are both specialized for 3 dimensions, with an `x`, y, and `z` component. A **Point** differs from a **Vector** only in the operations that can be performed on it. A **Point** indicates position in 3D space, and a **Vector** indicates direction. A **Point** subtracted from another **Point** produces a **Vector**, a **Vector** added to another **Vector** produces a **Vector**, and so forth. A cross product is defined only for **Vectors**, since they do not make geometric sense for **Points**. This has proven to be a useful way to help the programmer reason about geometric transformations and to write correct code for geometric manipulations.

1.3.5 The Math Library

libMath provides a somewhat eclectic collection of various core numerical functions, such as `Min`, `Max`, and `Abs`, which have all been overloaded for various numerical types. libMath also contains several core linear algebra loops, such as dot products, vector-vector multiply, etc. These loops have all been highly tuned to take maximum advantage of various architectures.

1.4 The Datatypes Library

The libraries described above are the building blocks from which the SCIRun kernel and SCIRun modules are created. The next layer consists of the Datatypes library. These datatypes are the data structures that get passed from one SCIRun module to another. Reflecting our work in Finite Element Analysis, the Datatypes library has four main parts:

1. **Meshes:** Unstructured tetrahedral meshes.
2. **Fields:** Scalar and Vector fields representing scalar and vector valued functions of space. These are currently defined on regular grids or with per-element or per-node values on an unstructured mesh.
3. **Surfaces:** Triangular mesh and parametric surface boundaries, optionally tagged with boundary condition information.
4. **Matrices:** Dense, Compressed row storage (symmetric and non-symmetric), Tridiagonal, and other matrix formats.

The Datatypes library is a powerful set of data structures for scientific computing, but it also provides a powerful method of extending SCIRun. A typical dataflow oriented program can be extended by adding new modules to implement new algorithms. However, SCIRun can be extended by extending the abstract interfaces in the Datatypes library to operate on other data formats. For example, one could make a new **Field** datatype which implements the **ScalarField** interface for some type of spectral method. Then most downstream modules would be able to operate on the data without modifying those modules and without converting and duplicating the data. Note that we are careful to say "most of the time." There are times when we violate these abstract interfaces for efficiency purposes. There will be an example of this shown when we discuss the **IsoSurface** module.

1.4.1 The Mesh Class

The **Mesh** class is not an abstract interface but is meant to be a powerful class for operating on unstructured grids. A **Mesh** consists of a set of **Nodes** and a set of **Elements**. A **Node** contains its **Point** in 3D space, and an **Element** contains a pointer to four different **Nodes**. This data completely specifies the mesh, but we also maintain other information for making mesh operations efficient. The full version of the **Node** data structure also contains a list of the elements to which it is connected. The **Element** data structure contains the face neighbors - the elements which neighbor its four faces. In addition, the **Element** data structure can optionally contain the element basis function and element volume. The element basis function is a large amount of data, increasing the size of the element data structure from 40 bytes to 176 bytes. This can cause memory limitations for large problems, but it avoids repeated recalculation of these basis functions and so increases the speed of finite element matrix assembly, element interpolation, and many other operations. The user may select at compile time whether or not to maintain the element basis functions.

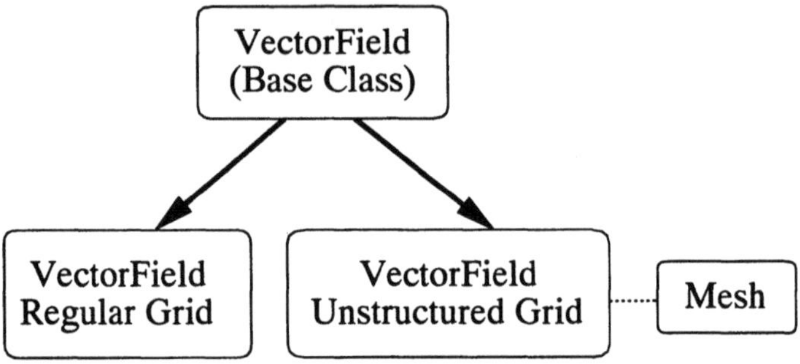

FIGURE 1.3. VectorField class hierarchy showing how the unstructured grid and regular grid versions are derived from a common base class. The dotted line to the **Mesh** shows that the Unstructured Grid version contains a 3D tetrahedral **Mesh**. In addition, the field contains a Vector value for each node or element in the mesh.

We could have chosen to make meshes of arbitrary dimension, but we opted for the simplicity and efficiency afforded by hard coding the dimension to three.

1.4.2 Fields Data Structures

The fields data structures provide two base classes: **ScalarField** and **VectorField**. We chose to separate these two types in order to clarify which operations make sense on which type. **ScalarField** provides interfaces for interpolating the value at a point (`interpolate`), determining the minimum and maximum scalar values (`minmax`), and for determining the spatial bounds of the field (`get_bounds`). There are currently two different implementations of the **ScalarField**: **ScalarFieldRG**, which defines a scalar field using a regularly sampled grid, and **ScalarFieldUG**, which contains a **Mesh**, and values for each node or element. **ScalarFieldRG** and **ScalarFieldUG** implement the `interpolate`, `minmax`, and `get_bounds` methods described above. Most modules use this abstract interface to access **ScalarField** data.

The **VectorField** class is similar to the **ScalarField** class, but does not provide a `get_minmax` interface. Figure 1.3 shows the class structure for the **VectorField** classes. The **ScalarField** classes are structured similarly.

1.4.3 Surfaces Data Structures

There is also another class hierarchy that describes a surface in three-dimensional space. For our work, the most important surface is a triangulated surface defined as a collection of 3D points and the corresponding triangles that connect them. These surfaces can be tagged with boundary condition information for later integration with finite element problems. A few other surfaces are provided such as cylinders, spheres, and points (a degenerate surface).

1.4.4 The Matrix Class

We implement a base class called **Matrix**, which defines `multiply` and `transpose multiply` abstract methods. Instances of this class may be implemented with *SparseLib++* [7], or other sparse matrix packages. SCIRun also implements a compressed-row storage matrix which has been highly tuned for the architectures which we use most often.

The discussion of the **SolveMatrix** module will display how these abstract interfaces are used to implement the conjugate gradient algorithm without regard for the actual layout of the matrix. Other operations, such as matrix-matrix multiply, are better implemented with code that checks the actual type of the matrices and uses the most efficient multiplication algorithm and resultant matrix format.

1.5 Dataflow

To this point, we have described a powerful set of computational elements that may be composed by a variety of means. SCIRun composes computational algorithms with these data elements using a dataflow style "boxes and wires" approach. An example of a dataflow network is shown in Figure 1.4.

1. A module, drawn as a box in the network, represents an algorithm or operation. A set of input ports and a set of output ports define its external parameters.

2. A port provides a connecting point for routing data to different stages of the computation. Ports are typed: each datatype has a different color, and datatypes cannot be mixed. In SCIRun, ports can be added to and removed from a module dynamically. Input ports are represented on the top of the module icon, and output ports are on the bottom. Output ports can cache datasets to avoid recomputation. The user can select which datasets should be cached and which should not.

3. A datatype represents a concept behind the numbers. Datatypes are quantities such as scalar fields or matrices, but are not the specific representations, such as regular grids, unstructured grids, banded matrices, sparse matrices, etc.

4. A connection connects two modules together: the output port of one module is connected to the input port of another module. These connections control where the data is sent to be computed. Output ports can be connected to multiple input ports, but input ports accept only a single connection.

5. A network consists of a set of modules and the connections between them. This represents a complete dataflow "program."

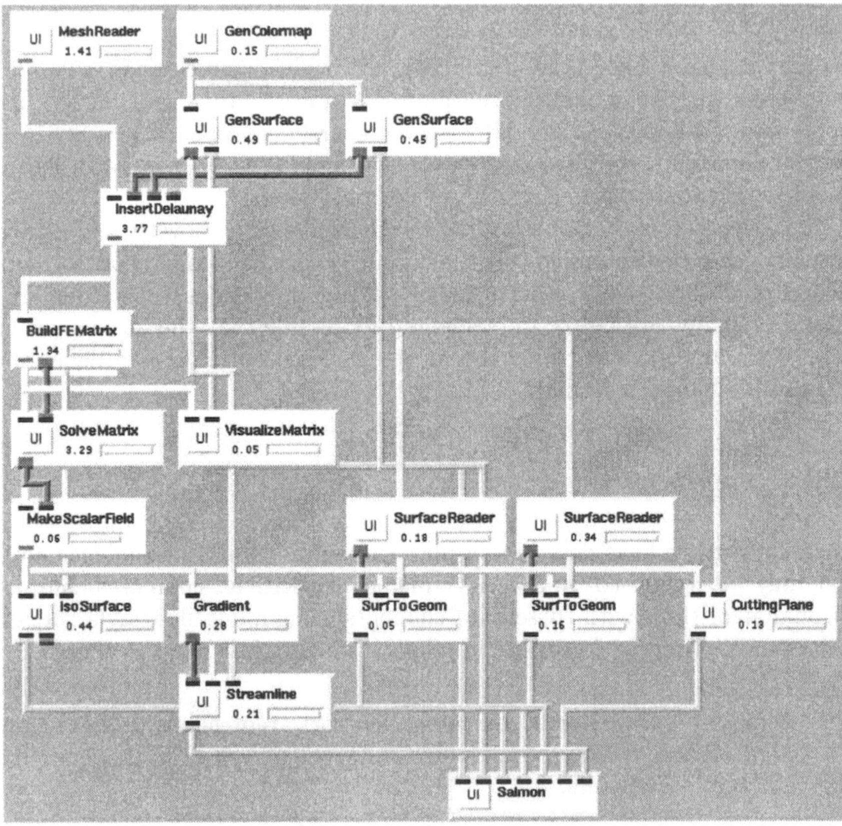

FIGURE 1.4. An example of a fairly complex dataflow network, showing the modules (the boxes), the connections (the wires between them), and the input/output ports (the points on the modules that the wires connect). On a color monitor, the colors of the ports and connections indicate the type of data that flows through them.

The dataflow library is responsible for deciding which modules need to be executed and when. A module is typically re-executed when the user changes a module parameter, when new data is available at an input port, or when data is required at its output port.

When the user moves a user interface component, such as a 2D slider or a 3D widget, the module sends a message to the scheduler requesting execution. The scheduler analyzes the dataflow graph to determine what other modules will also need to be executed. The dataflow scheduler uses the following algorithm to determine which modules need to be executed:

```
execute_list = modules that requested execution;
resend_list = empty;
foreach module in the execute_list {
  foreach module connected to an output {
    if the connected module is not in the execute list,
      add it
  }
  foreach module connected to an input {
    if the connected output port has a cached dataset,
      add it to the resend list
    else add it to the execute_list
  }
}
cull all ports from the resend_list whose modules appear
      in the execute list
send resend messages to the ports in the resend list
send execute messages to the modules in the execute list
```

Note that all of the modules are executed at the same time. The modules actually communicate directly with each other using a threadsafe FIFO for the dataset handoffs. Each thread will wait for data to appear on the input ports. A module does not have to gather data from the inputs in order, and it may interleave computation with receiving the data. However, in order to satisfy the requirements for determinacy in a dataflow program, the module is not able to request datasets in a first-come, first served or any other non-deterministic order. It is important to note that modules do not send actual data, but rather a handle to the data (see 1.12 for a discussion of handles).

Sending a dataset to the `matrix_out` port looks like this:

```
MatrixHandle matrix = new DenseMatrix(...);
.. build the matrix ..
matrix_out->send(matrix);
```

Receiving a dataset from the `matrix_in` port looks like:

```
MatrixHandle matrix;
if(!matrix_in->receive(matrix))
     return; // returns false if the dataset is not
             // available - this is usually due to
             // an error upstream
.. do cool stuff with the matrix ..
```

For each input port, the module must do exactly one receive, and for each output port, the module must do exactly one send. If the module wishes to send multiple datasets (as intermediate results, or for feedback loops), it can use a special method called `send_multi`, which also arranges for the modules downstream to be executed again.

The `send/receive` pairs shown above are a simple atomic protocol where the entire dataset is transferred at once. However, it often makes more sense to use "fine grain" dataflow where appropriate. Fine-grain protocols are tuned to the specific layout of the data, such as for receiving slabs of a regular grid for isosurfacing or scanlines for image processing. In such cases the module receives scanlines and sends scanlines for each scanline in the image, but it semantically sends or receives the full dataset exactly once.

We chose to implement a centralized scheduler in order to avoid redundant module re-execution in a branching situation. Since we leave the central scheduler out of the loop for individual dataset handoffs, it does not become a bottleneck in the execution of the program.

The Dataflow library also contains a base class from which all modules are derived. This class contains the data structures that are required to implement the dataflow structures; it also contains various utility functions, such as `update_progress`, a function that the module writer can call periodically to update the graph on the module icon that indicates the approximate percentage of work the module has completed.

1.6 Steering in a Dataflow System

All of the pieces described above have been designed to support steering of large scale scientific simulations. SCIRun uses three different methods to implement steering in this dataflow oriented system:

1. Direct lightweight parameter changes. The **SolveMatrix** module allows the user to change the target error even while the module is executing. The parameter change does not pass a new token through the dataflow network, but simply changes the internal state of the **SolveMatrix** module, effectively changing the definition of the operator rather than triggering a new dataflow event.

2. Cancellation. When parameters are changed, the module can choose to cancel the current operation. For example, if boundary conditions are changed, it may make sense to cancel the computation in order to focus on the new solution. This makes the most sense when solving elliptic problems, since the solution does not depend on any previous solution.

3. Feedback loops in the dataflow program. For a time varying problem, the program usually goes through a time stepping loop with several major operations inside. The boundary conditions are integrated in one or more of these operations. If this loop is implemented in the

dataflow system, then the user can make changes in those operators which will be integrated on the next trip through the loop.

These three methods provide the mechanisms whereby computational parameters can be changed during the execution of the program.

1.7 Modules

We have thus far described a handy set of tools and a dataflow system for imposing control structure. However, the real value in SCIRun comes from how these components are leveraged in actual algorithms, or in SCIRun - actual modules. This section covers how a module is constructed and then describes some of the modules in SCIRun, showing how they use the features that we have described above.

1.7.1 Writing a Module

The process of writing a new module involves writing a new C++ class. The constructor for this class creates the input and output ports for the module and defines parameters which the user interface may control. A single virtual function, `execute`, is overloaded to perform the actual work of the module. The `execute` function typically receives data on the input ports, performs some computation and then sends data on the output ports. Other callback functions can provide input from user interface components and 3D widgets. Adding a user interface to a module involves writing a small Tcl script which sets up the components of the interface.

Existing code may be integrated into SCIRun by writing a small wrapper module which passes data into and out of an existing C, C++ or Fortran program. The wrapper module may be required to translate data structures before passing them to the existing code and before sending them down the pipe. In order to avoid this translation, existing code can also be incorporated by extending the class hierarchy that flows through the dataflow network. For example, instead of translating to a specific **ScalarField** class the module could send a new subclass of the **ScalarField** that would provide `interpolate` and other methods for use by downstream modules.

SCIRun was originally designed as an environment for developing new simulations and computational components. We are currently working on ways to more automatically incorporating existing packages into the SCIRun visual programming environment.

1.7.2 FEM and Matrix Modules

The **BuildFEMatrix** module takes a tetrahedral mesh and builds a stiffness matrix for a finite element approximation. The current version of this module approximates the generalized 3D Poisson's equation (or its homogeneous counterpart, Laplace's equation) in the discretized physical domain Ω:

$$\nabla \cdot \sigma \nabla u = f \tag{1.1}$$

where f is a vector of sources and σ is 3×3 tensor corresponding to the materials within the domain. The scalar field u is subject to the boundary conditions:

$$u = u_0 \text{ on } \Gamma_1 \text{ and } \sigma \nabla u \cdot \mathbf{n} = 0 \text{ on } \Gamma_2 \tag{1.2}$$

The mesh elements have been tagged with a corresponding tensor property and boundaries have been tagged with the appropriate boundary condition. The module makes two passes - first to build the sparse structure of the matrix using a compressed row storage scheme, and second to fill in the resulting sparse matrix. Building the matrix is done in parallel in order to take full advantage of multiple processors. The module also builds the right hand side (load) vector. Users who wish to use other governing equations can extend this module to build the stiffness and load matrices appropriately.

Usually, these matrices are passed into the **SolveMatrix** module, which uses direct or iterative schemes to find the solution, x to the matrix equation:

$$Ax = b \tag{1.3}$$

The **SolveMatrix** module graphs the convergence of the residual as the algorithm iterates. The algorithm stops when the residual is less than some target level (i.e., the solution has converged). The user may move the target residual up or down while the solution is still in progress, thus steering the computation in a simple manner. Figure 1.5 shows the interface for this module in operation. **SolveMatrix** uses the `mult` and `mult_transpose` virtual methods in the **Matrix** base class to perform each iteration. As a result, the user could implement a new matrix type which **SolveMatrix** could use without even recompiling the module. This new matrix type does not even need to explicitly store the matrix - it could build the matrix dynamically or implicitly.

SolveMatrix also has an option to use the previous solution as the initial guess for the iterative solver. When small changes are made in the boundary conditions (a common case in an engineering design scenario), the system converges rapidly. This is one instance where an integrated system can actually be more efficient than implementing separate programs for each of the phases.

The **VisualizeMatrix** module draws a figure representing the non-zeros in a matrix. Non-zero entries are draw with small red dots, and zero entries are left black. This gives the user a quick representation of the sparse

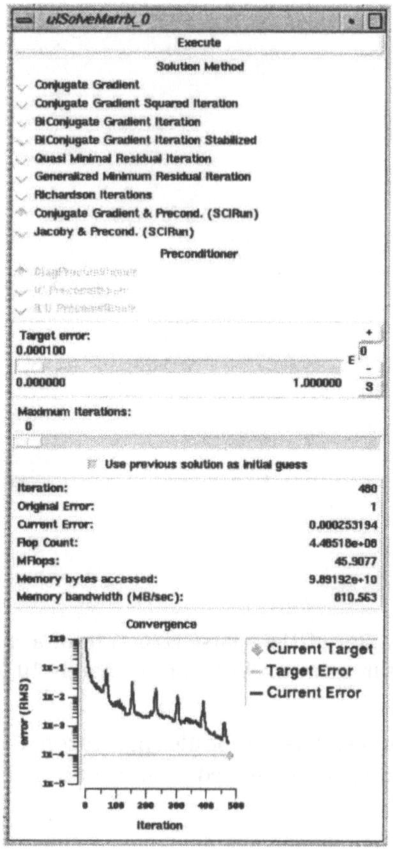

FIGURE 1.5. The user interface for the **SolveMatrix** module, showing the iterative methods available, the graph of the convergence and the target residual, which may be changed while the computation is in progress.

structure of the matrix, as shown in Figure 1.6. A magnifying glass can reveal the actual numbers in a portion of the matrix by clicking the mouse in the desired region. **VisualizeMatrix** uses two additional virtual functions in the **Matrix** base class: `getRowNonZeros`, which returns the nonzeros in a particular row, and `get`, which returns the number in a specific row and column of the matrix.

1.7.3 Readers and Writers

The **Reader** and **Writer** modules are very straightforward. They simply call the Persistent object `io` routines which were described above in Classlib (1.3.3.4). The readers read in a text or binary persistent object file and send the resulting object downstream, and the writers receive an object and then write it out to disk. Since the support for these modules has been provided by the lower layers, these modules do nothing more than provide user in-

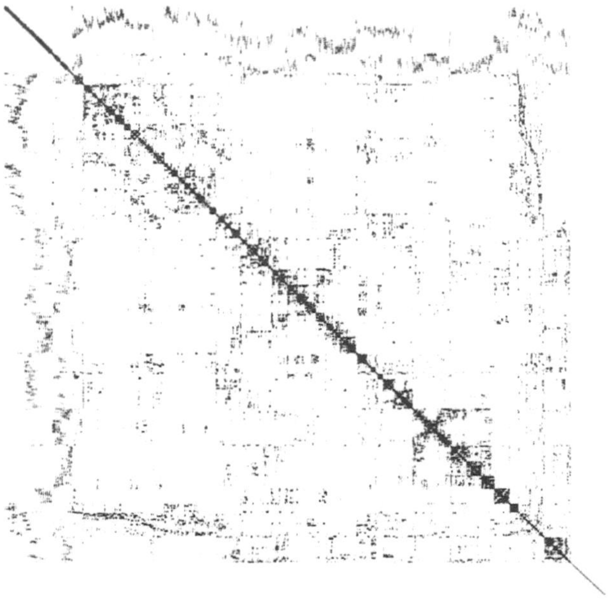

FIGURE 1.6. Visualization of the sparse structure of a matrix. The black dots represent the non-zeros in a matrix with approximately 10,000 rows and columns.

terfaces for the filenames. These modules are automatically generated, but provide hooks for adding user-defined reader/writer functions.

1.7.4 Visualization Modules

In implementing SCIRun's datatypes, we considered what type of operations SCIRun modules would require. One example of such a consideration was the implementation of the vector and scalar field datatypes. Each of these fields comes in a variety of flavors, corresponding to the internal representation of the data - implicit, explicit, parametric - and the topology of the field - structured (implicit) or unstructured (explicit). But in designing the field datatype, we chose several generic operators to allow module writers to access information from the field without needing to know how the field is internally represented. As discussed above (see Section 1.4.2), these operators can query the field for minimum and maximum scalar values or geometric bounds, or for the field's value at an arbitrary point in space. This last operation, retrieving the value of the field at any location, is implemented by an interface called `interpolate`. In the following subsection, we will discuss how we have exploited this generic operator in several of our modules, and we will provide an example of when we chose to side-step this abstraction in order to write a more efficient algorithm based on the specific internal representation of the field.

1.7.4.1 Interpolation

Many of SCIRun's modules exploit the field interface `interpolate`. The `interpolate` method takes a point as an argument and calculates the value of the field at that specific location. It accomplishes this by determining which element of the field contains the point, and linearly interpolating the values at the element's vertices.

One example of a module that calls the `interpolate` method is the **Streamline** module. The **Streamline** module is used for vector field visualization: by tracing particles advecting through the vector field, the user can examine local flow phenomena around critical points (such as vortices and turbulence) while also gaining a global sense of the field's flow. Figure 1.7 shows an example of the electrical current flow near the heart as computed by the **Streamline** module. We compute the paths of particles through the vector field as discrete line integrals, each corresponding to a streamline. We have several implementations of this integration; one of these is a fourth order Runge-Kutta method. Given a point `p`, we integrate along the path to find the next point along the streamline, `pNew`. The following piece of code accomplishes this:

```
Vector f1, f2, f3, f4;
Point p1, p2, p3, pNew;

vfield->interpolate(p, f1);
p1 = p + (f1 * 0.5);
vfield->interpolate(p1, f2);
p2 = p + (f2 * 0.5);
vfield->interpolate(p2, f3);
p3 = p + f3;
vfield->interpolate(p3, f4);
pNew = p + (F1 + F2 * 2.0 + F3 * 2.0 + F4) / 6.0;
```

The principal idea of this algorithm is that we find the vectors corresponding to particular points near `p` and take a weighted average of these vectors to determine our next location along the streamline. Note that this algorithm will work for any type of vector field that implements the `interpolate` interface. By designing a generic interface to the `interpolate` method, we have abstracted away the details of how interpolation is implemented for different field types, and consequently the module writer needs not be concerned with what *type* of vector field is generating the streamlines.

Another module that uses the `interpolate` interface is **SurfToGeom**. This module takes a surface, a scalar field, and a color map as input, and outputs colored geometric primitives that can be passed to a renderer. The module computes these primitives in the following way: it queries the field for scalar values corresponding to points on the input surface; these points are assigned material properties by indexing the resultant scalar values in the color table; and finally, these colored points are grouped into geometric

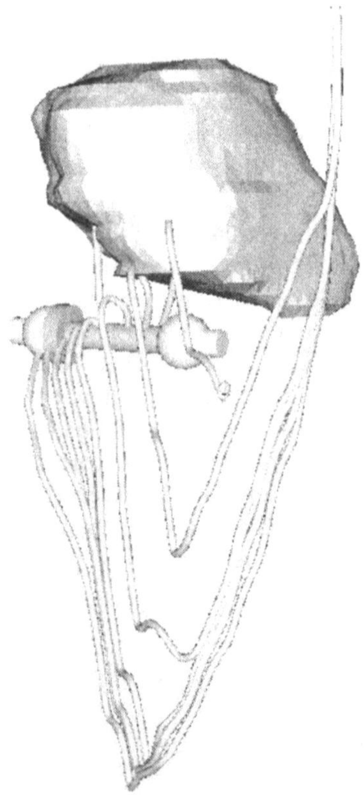

FIGURE 1.7. Visualization of a defibrillator design simulation, showing an electrode, the surface of the heart (epicardium), a 3D widget (rake), and electrical current lines (streamlines). The other electrode in the simulation is obscured by the heart. Much of the current leaving the visible electrode travels away from the heart, due to the high conductivity of blood in a nearby vessel.

primitives (such as triangles or tri-strips). Here again the **SurfToGeom** module utilizes the `interpolate` interface in order to find the scalar values of the points on the surface.

The user programming the **SurfToGeom** module does not ever have to worry about the type of the incoming scalar field - structured, unstructured, or spectral. As long as the `interpolate` interface is provided, the **SurfToGeom** module will work. Furthermore, as new scalar field types are implemented in the future, they will automatically work with the **SurfToGeom** module as long as the `interpolate` method is implemented, without requiring code to be rewritten or even recompiled.

In contrast to the **Streamline** and **SurfToGeom** modules, the **IsoSurface** extraction module does not use the `interpolate` method to find values at points within the field. While the isosurfacing algorithm could have been implemented this way, we were able to dramatically improve the speed of our algorithm by taking into account (and exploiting) the underlying structure of the data. For example, we have implemented Lorensen and Cline's Marching Cubes algorithm [19] as one option for isosurface extraction. For the case of both structured and unstructured grids, it is much faster to solve the "forward problem" - finding the intersection of the field with each element than it is to solve the "inverse problem" - using various calls to `interpolate` to track down the isovalue points within the field. In this case, it would have been highly inefficient to use the generic `interpolate` operator, so we wrote data-structure dependent code to solve the problem. The **IsoSurface** module lacks the abstraction (and as a result, the cleanliness) of the **Streamline** and **SurfToGeom** modules. As other field types are implemented, we will have to write new isosurfacing code to handle each field type. In this case, there was considerably more to be gained by using data-structure specific information in our implementation than there was to be gained by using the generic accesses permitted by abstracting this information away.

1.7.5 Salmon Module

One of SCIRun's key modules is the graphical viewer called **Salmon**. **Salmon** was named for its ability to spawn multiple views (Roe), and its ability to send messages upstream in the dataflow network. **Salmon** collects the geometric primitives from any number of modules and presents them in a single 3D view. The user can rotate, scale, and translate the objects, as well as manipulate lighting, camera parameters and rendering method, in order to obtain the desired view. Other views can be spawned to separate windows in order to simultaneously display the objects from other viewpoints or with different subsets of the objects.

Geometric primitives are passed from the modules to **Salmon** as a subset of a scenegraph. These scenegraphs are a tree-based display list that define geometric primitives, colors, and other rendering parameters. Drawing the scene involves traversing the graph and emitting OpenGL commands at each node. Scenegraphs can be composed, stored to disk via the persistent object mechanism, and then read back for later display. Parameters in the scenegraph can be changed dynamically by the module using the **CrowdMonitor** (multiple reader, single writer) lock described above in libMultitask (1.3.2).

In addition to using the **Salmon** module for visual output, we can also use it for 3D input by allowing the user to interact with specific objects in the scene. These objects, called Widgets [28], allow the user to intuitively augment parameters directly in the 3D scene. For example, in order to provide the starting point for a streamline advection, the user simply drags a **SphereWidget** around in the scene. This interaction is generally more intuitive to a user then typing in numbers or manipulating 2D sliders. This interaction becomes even more powerful when a rake [4,10] is used to specify a line of starting points for streamline advection.

1.7.6 Other Modules

We have talked about several of the Modules in SCIRun, but there are many more:

- **FEMError** - computes the upper and lower error bounds of a finite element simulation [34, 35].
- **MeshRefiner** - uses the Error fields described above to decide where to add new nodes and remove old nodes in order to refine and derefine the mesh.
- **Gradient** - computes a vector field which is the gradient of the given scalar field.
- **Magnitude** - computes a scalar field which is the magnitude of the given vector field.
- **FFTImage / IFFTImage** - takes the FFT (inverse FFT) of an image, producing another image.
- **FilterImage** - multiplies two images together to perform a filter operation in frequency space.

Even this is not an exhaustive list. Despite the number of modules available, the user will always need something different than what is supplied. In such cases, the user can simply create a new module (by writing a small C++ program). User modules are dynamically linked with SCIRun, thus avoiding the need to relink the SCIRun executable.

1.8 Applications of SCIRun in Computational Medicine

Here we address the application of SCIRun to two bioelectric field problems in medicine: simulation of cardiac defibrillation and simulation of temporal lobe epilepsy.

Every year, a large number of people die suddenly because of abnormalities in their hearts' electrical system (cardiac arrhythmia) and/or from coronary artery disease. While external defibrillation units have been in use for some time, their use is limited because it takes such a short time for a heart attack victim to die from insufficient oxygen to the brain. Lately, research has been initiated to find a practical way of implanting electrodes within the body of patients with recurring and life-threatening arrhythmia to defibrillate a person automatically upon onset of cardiac fibrillation. Because of the complex geometry and inhomogeneous nature of the human thorax and the lack of sophisticated thorax models, most past design work

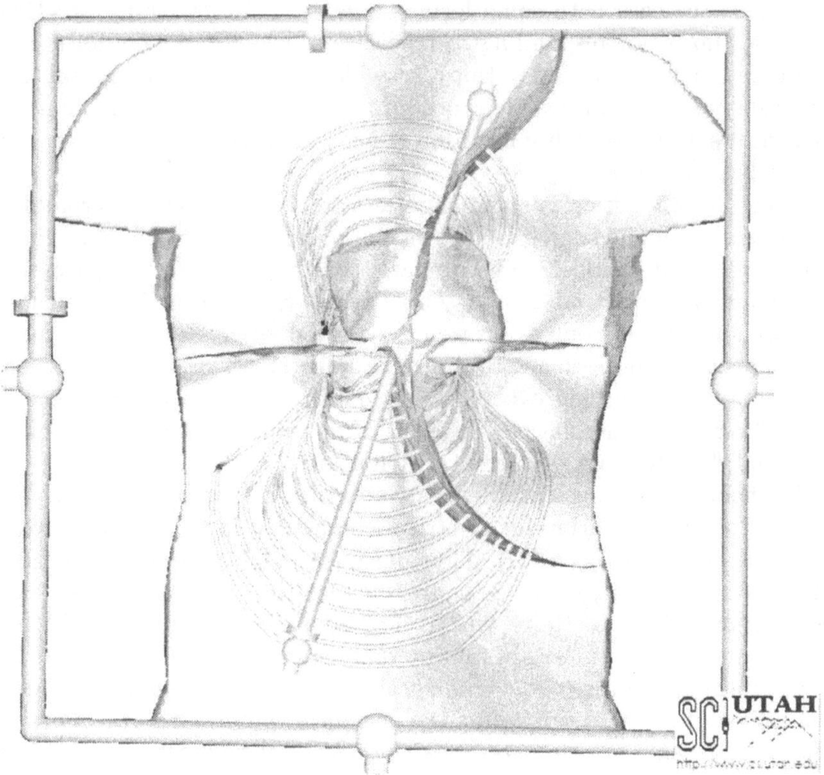

FIGURE 1.8. Visualization of the electrical current lines and an isovoltage surface from a simulation of a cardiac defibrillator design simulation.

on defibrillation devices has relied on animal studies. We have constructed a large scale model of the human thorax, the Utah Torso Model [13,14,20,21], for simulating both the endogenous fields of the heart and applied current sources (defibrillation devices). Using these computer models, we are also able to simulate the multitude of electrode configurations, electrode sizes, and magnitudes of defibrillation shocks. Figure 1.8 shows the results of such a simulation. Given the large number of possible external and internal electrode sites, magnitudes, and configurations, it is a daunting problem to computationally test and verify various configurations. For each new configuration tested, geometries, mesh discretization levels, and a number of other parameters must be changed.

Excitation currents in the brain produce an electrical field that can be detected as small voltages on the scalp. By measuring changes in the patterns of the scalp's electrical activity, physicians can detect some forms of neurological disorders. Electroencephalograms, EEGs, measure these voltages; however, they provide physicians with only a snapshot of brain activity. These glimpses help doctors spot disorders but are sometimes insufficient for diagnosing them. For the latter, doctors turn to other techniques; in rare cases, they rely on investigative surgery.

Such is the case with some forms of epilepsy. To determine whether a patient who is not responding to medication has an operable form of the disorder, known as epilepsy (most of which occur in the temporal lobe), neurosurgeons use an inverse procedure to identify whether the abnormal electrical activity is highly localized (thus operable) or diffused over much of the brain.

Using SCIRun, scientists and engineers are able to design internal defibrillation devices and source models for the epileptic foci, place them directly into the computer model, and automatically change parameters (size, shape and number of electrodes) and source terms (position and magnitude of voltage and current sources) as well as the mesh discretization level needed for an accurate finite element solution. Furthermore, engineers can use the interactive visualization capabilities to visually gauge the effectiveness of their designs and simulations in terms of distribution of electrical current flow and density maps of current distribution.

Mathematically, these problems are governed by variations of the generalized Poisson's equation for electrical conduction in the physical domain Ω of the thorax or head [27]. Thus one solves (1.1) where $u = \Phi$ are the electrostatic voltages, $f = -I_V$ are the electrical current sources and σ is an electrical conductivity tensor. The boundary conditions in (1.2) are such that Γ_1 is the surface of the internal defibrillator electrodes and Γ_2 is the surface of the torso. $u_0 = \Phi_0$ specifies a Dirichlet boundary of known voltages and $\sigma\nabla\Phi \cdot \mathbf{n}$ represents the current flow normal to the surface of the torso (scalp), which is zero for the insulated boundary of the thorax (or head).

Once the electrostatic potentials are known, one can calculate the current density $\mathbf{J}$ according to:

$$\mathbf{J} = -\sigma\nabla\Phi. \tag{1.4}$$

For the defibrillation problem, electrodes are either implanted internally or applied directly to the chest in order to deliver sufficient electric energy to stop the irregular heart rhythms that signify a fibrillating heart [18, 23]. Mathematically, this can be posed as solving equations (1.1-1.4) with the voltage boundary condition applied on a portion of the torso boundary $\Sigma \subseteq \Gamma_2$ for external defibrillation or from the surface of the defibrillation electrode(s) within the volume of the thorax for internal defibrillation.

Past (and much current) practice regarding the placement of the electrodes for either type of defibrillator has been determined by clinical trial and error. One of our goals is to allow engineers to use SCIRun to assist in determining the optimum electrode placement, size, shape, and strength of shock to terminate fibrillation by solving equations (1.1-1.4) within a detailed model of the human thorax [2, 11, 12, 22, 29].

For the neuroscience problem, the epileptic foci are represented as a set of idealized dipole sources situated in the brain. Using a model of the human

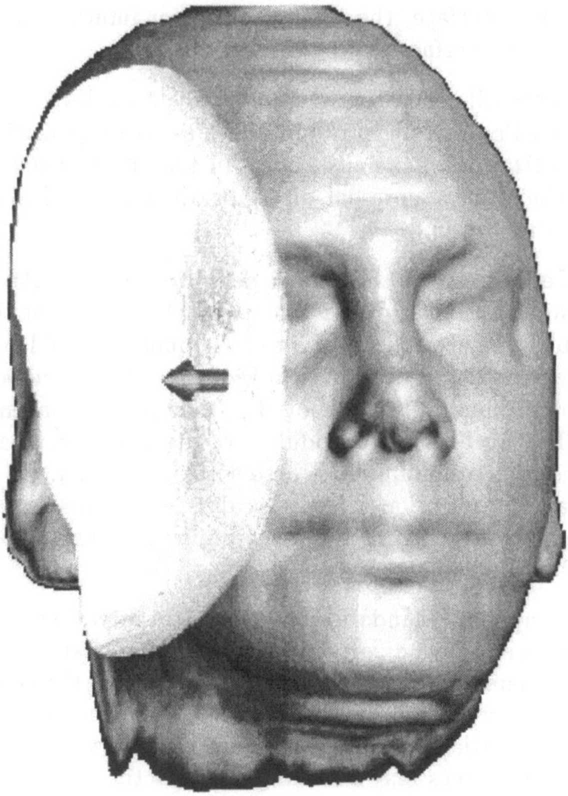

FIGURE 1.9. Visualization of a computational neuroscience simulation, showing an epileptic focus (a dipole source, indicated by the arrow) and an isovoltage surface in the resulting field.

skull and brain, the direct EEG problem is posed by solving equations (1.1-1.4) for the voltage and current distribution within the brain and upon the surface of the scalp. For the inverse EEG problem, measured scalp voltages are used as the inputs and equation (1.1) is solved for the source currents I_V. We are currently using SCIRun to investigate the direct and inverse EEG problems as well as using SCIRun as an interactive modeling and visualization tool.

A SCIRun System for Bioelectric Field Problems

A network that can be used to model cardiac defibrillation is shown in Figure 1.4. A similar network is used for a forward solution in the neuroscience application. The network consists of the following modules:

- **SurfaceReader** reads a triangulated surface definition from a file. One of these modules will read the torso boundary (body surface) geometry, and the other will read the epicardium (heart surface) geometry. In the neuroscience application, there may be several surfaces - including the scalp surface, the scalp-skull interface, the skull-cerebrospinal fluid (CSF) interface, the CSF-grey matter interface, and the grey-white matter interface.

- **GenSurface** will generate two cylindrical electrodes for the defibrillation study. Parameters in the interface allow the scientist to control the discretization of the electrodes. In the neuroscience application, these surfaces may represent dipole or other types of source configurations in the brain.

- **SurfToGeom** converts the surface definitions into displayable geometry. A flag in the user interface controls whether or not the geometry is movable. The epicardium and torso boundary should not be moved, since they correspond to physical geometry. The electrode cylinders, on the other hand, must be moved so that various placements can be tested. The **SurfToGeom** module provides 3D widget handles which allow the user to manipulate these surfaces directly. An optional input parameter selection will map scalar field values onto the surface. This input is attached to the epicardium and shows the voltages on the surface of the heart.

- **ApplyBC** applies boundary conditions to the various surfaces. The torso and scalp have a zero-flux Neumann boundary. In the defibrillation problem, the two electrode cylinders have Dirichlet boundary conditions corresponding to their respective voltage. In the neuroscience application, there are discrete current sources. The voltage and current sources may be changed interactively.

- **GenerateMesh** discretizes the volume defined on the outside by the torso (scalp) and on the inside by the two voltage source electrodes. This discretization occurs after the application of the boundary conditions so that the mesh generator may optimize the mesh for the particular boundary conditions.

- **BuildFEMatrix** uses the mesh structure, the boundary conditions, and finite element theory to construct a matrix that describes the user-specified configuration. Utilizing controls on the user interface, the user may instruct the module to create a dense matrix, a band diagonal matrix or a compressed sparse-row matrix.

- **SolveMatrix** uses direct or iterative algorithms to find the solution to the matrix equation. For this problem, we use a preconditioned conjugate gradient algorithm for iterative solutions. The scientist controls convergence parameters through the graphical user interface.

- **MakeScalarField** combines the solution of the finite element matrix to the volume mesh generated by **GenerateMesh**. This mesh/solution combination provides a representation of the solution in terms of a scalar field of voltage values.

- **IsoSurface** allows interactive extraction of isosurfaces in the voltage field. A small sphere controls the starting point of the isosurface algorithm, and an attached 3D arrow shows the direction of the gradient. The sphere and arrow widget may be moved using the mouse to allow interactive exploration of the voltage field. Dragging on the body of the arrow moves the widget along the line defined by the gradient; dragging on the sphere allows unconstrained movement of the seed point.
- **Gradient** computes a vector field from the scalar voltage field according to equation (1.4). This yields another form of the solution in terms of electric current density.
- **Streamline** produces vector field lines that reveal the flow of electrical currents within the torso and brain. These field lines are analogous to massless particle traces in fluid flow fields. The streamlines are advected using a 4th order Runge-Kutta technique. The user may choose between a single streamline or a row of streamlines. Adaptation parameters and step sizes are controlled via the 2D user interface, while the positions of the particle sources are controlled with 3D widgets [10].
- **Salmon** provides the underlying structure for viewing geometry and 3D user interaction for both viewpoint control and control of the 3D widgets described above. As the **Streamline** module computes streamlines, or as the **Isosurface** module computes isosurfaces, it will send geometry (lines, triangles or other primitives) to **Salmon**. **Salmon** displays these objects geometry in a rendering window.

Each of these modules is simple enough to be managed easily, but when they are joined together, they accomplish a very complex task. Sample visualizations from this type of network are shown in Figures 1.7 and 1.9.

Pressing the "UI" (user interface) button on any of the modules' icons summons a pop-up window which allows the user to change various parameters for that module. For example, the **Streamline** module's pop-up allows the user to specify integration parameters, display modes, 3D widget types, and other information. As an example user interface, Figure 1.10 shows the pop-up windows for the **Streamline** module.

1.9 Summary

We have presented an overview of the SCIRun software architecture. SCIRun provides support for computational steering by providing powerful libraries and tools at a variety of levels, ranging from operating system level to a high-level visual user interface. SCIRun allows computational components to be efficiently assembled using a dataflow programming paradigm.

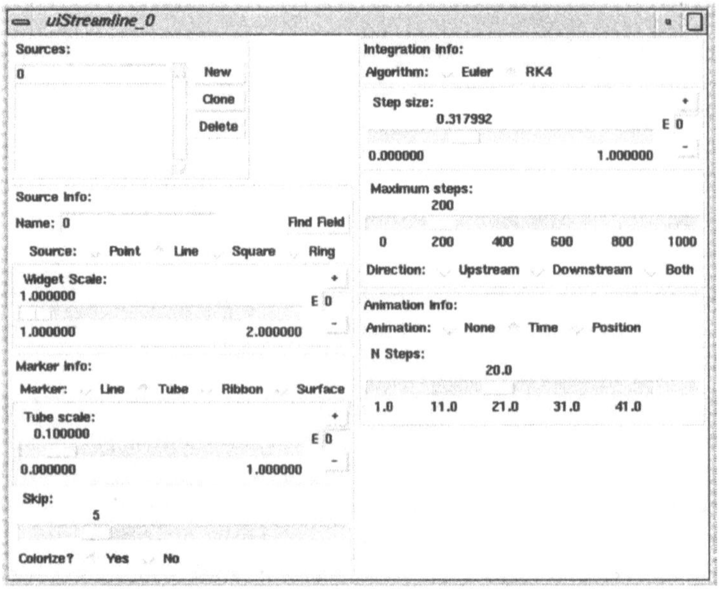

FIGURE 1.10. Pop-up window for the **Streamline** module.

The real power of SCIRun comes from its modules. The set of modules can be extended by the user in order to provide custom tools and interfaces to existing programs.

1.10 Future Work

SCIRun is continually evolving. We look forward to the challenges of integrating SCIRun components with other computational paradigms (alternatives to dataflow), and also to the converse: connecting SCIRun's visual programming environment to other computational components. In addition, we are continually trying to stretch SCIRun in new directions by applying it to new applications in a range of disciplines.

In addition to porting SCIRun to new shared memory architectures [25], we have ideas of how to use SCIRun in a distributed environment. SCIRun's role as a miniature operating system would be greatly expanded in such a scenario.

Acknowledgments: This work was supported in part by the National Science Foundation, the National Institutes of Health, the Department of Energy through a Computational Science Graduate Fellowship, and the Utah State Centers of Excellence Program. The authors would like to thank K.

Coles and P. Sloan for their helpful comments and suggestions. Furthermore, we appreciate access to facilities that are part of the NSF STC for Computer Graphics and Scientific Visualization.

1.11 References

[1] S. Balay, W. Gropp, L. C. McInnes, and B. F. Smith. Efficient management of parallelism in object-oriented numerical software libraries. In E. Arge, A. M. Bruaset, and H. P. Langtangen, editors, *Modern Software Tools for Scientific Computing*, pages 163–202. Birkhäuser, 1997.

[2] D. Blilie, J. Fahy, C. Chan, M. Ahmed, and Y. Kim. Efficient solution of three-dimensional finite element models for defibrillation and pacing applications. In *Proceedings of the 13th Annual Conference of the IEEE Engineering in Medicine and Biology Society*, volume 13, pages 772–773, 1991.

[3] A. M. Bruaset and H. P. Langtangen. A comprehensive set of tools for solving partial differential equations: Diffpack. In M. Daehlen and A. Tveito, editors, *Numerical Methods and Software Tools in Industrial Mathematics*. Birkhauser, Oslo, 1997.

[4] D. B. Conner, S. S. Snibbe, K. P. Herndon, D. C. Robbins, R. C. Zeleznik, and A. van Dam. Three-dimensional widgets. In *Computer Graphics (Proceedings of the 1992 Symposium on Interactive 3D Graphics)*, pages 183–188. ACM, 1992.

[5] T. De Fanti et al. Special issue on visualization in scientific computing. *Computer Graphics*, 21(6), November 1987.

[6] T. D. DeRose. A coordinate-free approach to geometric programming. 1989.

[7] J. Dongarra, A. Lumsdaine, R. Pozo, and K. Remington. A sparse matrix library in C++ for high performance architectures. In *Proceedings of the Second Object Oriented Numerics Conference*, pages 214–218, 1994.

[8] W. Gropp and B. Smith. Scalable, extensible, and portable numerical libraries. In *Proceedings of the Scalable Parallel Libraries Conference*, pages 87–93. IEEE Computer Society Press, 1994.

[9] W. Gu, J. Vetter, and K. Schwan. An annotated bibliography of interactive program steering. *Georgia Institute of Technology Technical Report*, 1994.

[10] K. P. Herndon and T. Meyer. 3D widgets for exploratory scientific visualization. In *Proceedings of UIST '94, ACM Siggraph*, pages 69–70. ACM, November 1994.

[11] S. A. Hutchinson, S. Gao, L. Ai, K. T. Ng, O. C. Deale, P. T. Cahill, and B. B. Lerman. Three-dimensional modeling of electrical defibrillation on a massively parallel computer. In *Computers in Cardiology*, pages 343–346. IEEE Computer Society Press, 1992.

[12] S. A. Hutchinson, K. T. Ng, J. N. Shadid, and A. Nadeem. Electrical defibrillation optimization – an automated, iterative parallel finite-element approach. *IEEE Trans. Biomed. Eng.*, 1995.

[13] C. R. Johnson, R. S. MacLeod, and P. R. Ershler. A computer model for the study of electrical current flow in the human thorax. *Computers in Biology and Medicine*, 22(3):305–323, 1992.

[14] C. R. Johnson, R. S. MacLeod, and M. A. Matheson. Computational medicine: Bioelectric field problems. *IEEE COMPUTER*, pages 59–67, Oct., 1993.

[15] C. R. Johnson and S. G. Parker. A computational steering model for problems in medicine. In *Supercomputing '94*, pages 540–549. IEEE Press, 1994.

[16] C. R. Johnson and S. G. Parker. Applications in computational medicine using SCIRun: A computational steering programming environment. In *Supercomputer '95*, pages 2–19. Springer-Verlag, 1995.

[17] H. P. Langtangen. Diffpack: Software for partial differential equations. In *Proceedings of the 2nd Annual Conference on Object Oriented Numerics (OON-SKI)*, 1994.

[18] B. Lerman and O. Deale. Relation between transcardiac and transthoracic current during defibrillation in humans. *Circulation Research*, 67:1420–1426, 1990.

[19] W. E. Lorensen and H. E. Cline. Marching cubes: A high resolution 3d surface construction algorithm. *Computer Graphics*, 21(4):163–169, 1987.

[20] R. S. MacLeod, C. R. Johnson, and M. A. Matheson. Visualization tools for computational electrocardiography. In *Visualization in Biomedical Computing*, pages 433–444, 1992.

[21] R. S. MacLeod, C. R. Johnson, and M. A. Matheson. Visualization of cardiac bioelectricity — a case study. In *IEEE Visualization '92*, pages 411–418, 1992.

[22] X. Min, L. Wang, M. Hill, J. Lee, and R. Mehra. Detailed human thorax fem model for the cardiac defibrillation study. In *Proceedings of the 15th Annual International Conference of the IEEE Engineering in Medicine and Biology Society*, pages 838–839, Piscataway, New Jersey, 1993. IEEE Press.

[23] M. Mirowski. The automatic implantable cardioverter-defibrillation: An overview. *Journal of the American College of Cardiology*, 6:461–466, 1985.

[24] D. Musser and A. Saini. *C++ Programming with the Standard Template Library*. Addison-Wesley, California, 1996.

[25] ORIGIN2000 and ONYX2 Technical Report. Silicon Graphics, Revision 6.0, October 4, 1996.

[26] S. G. Parker and C. R. Johnson. SCIRun: A scientific programming environment for computational steering. In *Supercomputing '95*. IEEE Press, 1995.
http://www.supercomp.org/sc95/proceedings/499_spar/sc95.htm.

[27] R. Plonsey and R. C. Barr. *Bioelectricity*. Plenum Press, New York, 1988.

[28] J. T. Purciful. Three-dimensional widgets for scientific visualization and animation. Master's thesis, University of Utah, 1995.

[29] J. A. Schmidt and C. R. Johnson. Defibsim: An interactive defibrillation device design tool. In *IEEE Engineering in Medicine and Biology Society 17th Annual International Conference*. IEEE Press, 1995.

[30] SCIRun. http://www.cs.utah.edu/~sci/.

[31] A. Silberschatz, J. Peterson, and P. Galvin. *Operating Systems Concepts*. Addison-Wesley, California, 1991.

[32] B. Stroustrup. *The C++ Programming Language*. Addison-Wesley, Reading, 1991.

[33] J. Vetter, G. Eisenhauer, W. Gu, T. Kindler, K. Schwan, and D. Silva. Opportunities and tools for highly interactive distributed and parallel computing. *Proceedings of the Workshop On Debugging and Tuning for Parallel Computing Systems*, October 1994.

[34] F. Yu and C. R. Johnson. An automatic adaptive refinement and derefinement method for elliptic problems. *Proceedings of IMACS*, 3:1555–1557, 1994.

[35] F. Yu, Y. Livnat, and C. R. Johnson. An automatic adaptive refinement and derefinement method for 3D elliptic problems. *Applied Numerical Mathematics*, 1997 (to appear).

1.12 Software Appendix

1.12.1 Handles

A **handle** contains a single data member `rep`, which contains a pointer to the actual representation. It also defines ten methods:

1. `Handle()` - initializes `rep` to `NULL`.
2. `Handle(Type* p)` - sets `rep` to `p`, and increments `p->ref_cnt`.
3. `Handle(Handle<Type>& copy)` - copies `rep` from another handle, and increments the reference count.
4. `~Handle()` - decrements `rep->ref_cnt`. If it is now zero, delete the object pointed to by `rep`.
5. `Handle<Type>& operator = (Type* p)` - decrements `rep->ref_cnt`, deleting when it reaches zero; sets `rep` to `p` and increments `p->ref_cnt`.
6. `Handle<Type>& operator = (Handle<Type>& copy)` - decrements `rep->ref_cnt`, deleting if it reaches zero; copies `rep` from `copy` and increments `copy->ref_cnt`.
7. `Type* operator -> () const` - this simply returns `rep`, but it allows the handle to be dereferenced using `->` like a normal C++ pointer.
8. `Type* get_rep() const` - returns the actual pointer `rep`. This is useful for some code, but is also dangerous since the handle must continue to exist as long as the pointer is being used - otherwise the object could be deleted while it is still being accessed.
9. `bool isNull() const` - returns true if `rep==0`.
10. `void detach()` - if `rep->ref_cnt` is greater than one, then `void detach` clones the object, thus obtaining an exclusive copy of the object. If the value of `rep->ref_cnt` equals one, then we already own an exclusive copy of the object. This allows subsequent code to make changes to the referred object without sharing those changes with other handle owners.

2

Object-Oriented Solvers for Initial Value Problems

Hans Olsson[1]

ABSTRACT
The Godess (Generic Ordinary Differential Equations Solving System) Project aims at development of an object-oriented solver for initial value problems. In this project, the use of object-oriented techniques has been instrumental in development, testing, evaluation and verification of methods and control strategies. Based on comparisons with traditional solvers, we find that Godess is competitive with respect to accuracy and computational speed. These experiments demonstrate that this very generic solver is of real practical value.

2.1 Introduction

In this article, we present the object-oriented design features of the Godess Project, a development platform designed to advance the state-of-the-art methodology used in the numerical solution of initial value problems. This project is a Scandainavian joint effort between the Danish Technical University, the Norwegian University of Science and Technology and Lund University in Sweden. The main focus of this presentation is on the actual software, which is written in C++. Those who are interested in details about the algorithms should consult [5,9,11,16,20].

The advantage of the object-oriented design will be shown by what it has allowed us to accomplish. The details of the design are mostly suppressed because they are lengthy and would require a more detailed presentation of the algorithms. The overall design will be discussed, and we will start by examining the problem to be solved. A general observation is that we primarily have classes that encapsulate algorithms and not data.

[1]Department of Computer Science, Lund University, P.O. Box 118, S-221 00 Lund, Sweden. Email: `Hans.Olsson@dna.lth.se`.

2.1.1 Initial Value Problems

We view an initial value problem (IVP) as consisting of the following components:

Initial value $y(t_0) = \ldots$

Differential equation Any one of the following types:

> **ODE** Ordinary Differential Equation, $dy(t)/dt = \dot{y} = f(t, y, p)$
>
> **Semi Explicit** $B(p)\dot{y} = f(t, y, p)$
>
> **DAE** Differential Algebraic Equation, $r(t, y, \dot{y}, p) = 0$.
>
> Note that the semi explicit form is a special case of DAEs, which is easier to analyze and solve, [14]. Here p denotes a parameter vector whose values are not governed by the differential equation(s).

State Condition An inequality $g(t, y, \dot{y}, p) \geq 0$ and the specification of what to do when g becomes negative. Normally, if g becomes negative, one switches to a new parameter vector p, which results in a new definition for f or r.

A solution $y(t)$ is sought.

There are also some assumptions that we have not made explicit in this description, such as:

- The differential equations should be well posed and sufficiently continuous. If the mathematical problem has discontinuities we will have to use state conditions g to switch between different models, each of which is identified by a particular choice of the parameter vector p.
- The state vectors are finite dimensional. For instance, they might be the result of a partial differential equation (PDE) discretized by the method of lines.
- The B-matrix is square or similarly r has the same dimension as y. This excludes overdetermined systems of equations, [4, p. 477], [14].

If we want to model this problem in an object-oriented fashion we immediately see two abstractions. First the "state", which we represent by an instance of the class DynState, composed of:

t The independent variable ("time").

y The state vector governed by differential equations.

$\dot{y}$ The time derivative of the state vector.

p Parameters and states not governed by differential equations.

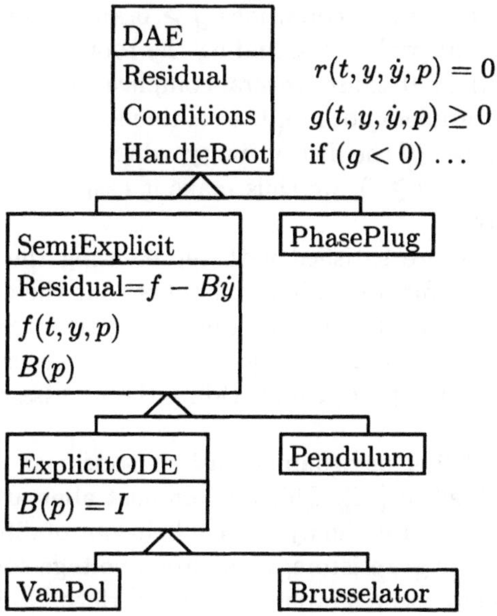

FIGURE 2.1. The hierarchy of dynamical systems expressed in Godess. The base class is DAE, where the subclasses restrict the differential equations even further. The triangle below DAE indicate that SemiExplicit and PhasePlug are subclasses of DAE.

Additional information. This has expanded over time, and we first added the DAE index of the individual components of y to help DAE solvers. We have later introduced facilities for delay equations. By having the state in a separate class it has been easy to extend the state by additional information without having to rewrite existing code.

We also see that the differential equations and the state conditions build up a hierarchy of dynamical systems, see Figure 2.1. We here use the notation of [23], but have reduced the number of operations to make it more readable. All attributes have been omitted, as they are in general used internally in the classes. The residual formulation of DAE, $r(t, y, \dot{y}, p) = 0$, is further restricted in SemiExplicit, $B\dot{y} = f(t, y, p)$, and even more in ExplicitODE, $\dot{y} = f(t, y, p)$. The actual problems are instantiated from one of the concrete problem classes, such as Pendulum or PhasePlug.

The difference from traditional solvers is mainly the modeling of the discontinuities. Traditional solvers that handles discontinuities, [13], stop the integration and return to the caller instead of calling a routine such as HandleRoot which is responsible for treating the roots of $g = 0$. The caller is in those cases responsible for determining the cause, handling it (i.e. calling HandleRoot), and restarting the solver. We believe this to be wrong from a modeling point of view because the problem becomes entangled with the solution process.

Another difference is that state conditions $g \geq 0$ are traditionally represented by a more general problem of finding the roots of g. The added generality gains very little, but causes several complications if $g = 0$ at the starting point. This is a problem both for the routine localizing the roots of g and for the actual handler of the discontinuity, which is often left for the user to write. By requiring $g \geq 0$, we thus make it easier both to specify the problem and to solve it.

One might ask why we are satisfied with representing the differential equations as mathematical functions. The answer is that the construction of differential equations should be done in a modeling environment suited for the particular problem domain. Instead of writing such environments we construct a computational platform providing the necessary numerical support.

An example of a modeling environment using Godess as a computational platform is described in [15]. This environment aims at electronics and it allows the user to describe the dynamical behavior of different types of electronic components (e.g. resistors, transistors, voltage sources). The components are connected via a graphical user interface, which directly corresponds to traditional circuit diagrams in electronics. This example indicates that we could model the differential equations by having every component know about its behavior, and to whom it is connected. The differential equations are distributed to the individual components. The system of differential equations is, however, only one aspect of the components; suggesting that every modeling environment should use our conventions is unrealistic.

The modeling environment should also preprocess the differential equations by e.g. solving small linear systems, simplifying symbolic expressions, replacing discontinuities by state conditions, and eliminating redundant variables. We found it beyond the scope of the project to perform such an analysis. The analysis might give more information than what is provided by the differential equations, e.g. on DAE indices of variables, bandedness of the system, analytical Jacobians, and the complexity of the system. In Godess we can take advantage of these aspects to ease the task of the solvers, and we will try to include more problem-dependent information in the future. The problem to be solved is, however, still given by mathematical functions.

2.1.2 Timestepping Methods

We will here give a very short introduction to timestepping methods for IVPs exemplified by how ODEs are solved using Runge-Kutta methods, [13,14].

From the starting point (t_n, y_n) we use intermediate stage values Y_i and stage derivatives $\dot{Y}_i$ in order to take a step with step size h_n:

$$\begin{aligned} \dot{Y}_i &= f(t_n + h_n c_i, Y_i) && i = 1, \ldots, s \\ Y_i &= y_n + \sum_{j=1}^{s} \dot{Y}_j A_{i,j} h_n && i = 1, \ldots, s \\ y_{n+1} &= y_n + \sum_{j=1}^{s} \dot{Y}_j b_j h_n \\ \text{error}_n &= \sum_{j=1}^{s} \dot{Y}_i (b_j - \hat{b}_j) h_n. \end{aligned}$$

The endpoint of the step is (t_{n+1}, y_{n+1}) where $t_{n+1} = t_n + h_n$. The step size is chosen such that the error estimate, error_n, is acceptable. The coefficients of this method are A, b, $\hat{b}$ and c.

For explicit Runge-Kutta methods we can sequentially evaluate $\dot{Y}_i$, but for implicit ones we have to solve for $\dot{Y}_i$ in some way.

2.1.3 Motivation for Godess

There have existed many successful solvers for IVPs for quite some time. The motivation for the Godess Project was that even if the solvers were successful, substantial improvements were possible.

The first area of improvement was the supporting algorithms that are necessary to turn a method for solving IVPs into a working implementation. They handle such decisions as selecting step size, recomputing Jacobians, and obtaining initial values for iterations. These have traditionally been based on simple asymptotic analysis (for step sizes), *ad hoc* strategies with some "fudge factor" (when to recompute the Jacobians), or just using whatever data available (initial values for iterations). We will in the following sections discuss these examples in more detail.

This situation was unsatisfactory, because without analysis we cannot know whether the algorithms can be used for a new solver — or even whether they are suited for existing ones. If the algorithms fail to work as expected, we need analysis to explain the problem and to find good alternative strategies.

Another problem is that constructing, developing and verifying a new code is a very time-consuming undertaking if we have to select, code and tune the control structures. It can only be justified if there are strong reasons to believe that a new method will have significant advantages in practical computations.

Godess has addressed these problems by implementing a generic solver instead of a particular method or a collection of such. Thus, in order to implement a new method, all that needs to be specified is the coefficients of the method, i.e. A, b, $\hat{b}$, c for a Runge-Kutta method. This is not the same as taking an old code and just plugging in the new coefficients, because specific details of the algorithms depend on method properties. A generic solver must therefore ensure that the supporting algorithms are not tuned to some specific method but adapt well to the method in question. This can only be done correctly after analyzing these algorithms, and finding the dependencies between the method and the supporting algorithms.

This generic nature also allows us to directly compare the performance of *methods*, instead of the traditional comparison of *codes*. This allow us to draw firm conclusions about the methods, provided that we know that the solver is correctly adapted to the methods.

Another point is that users do not use codes solely because of the methods they implement. Other factors such as familiarity with the code and additional functionality in the code are often more important.

As an example of the latter we consider a problem with state conditions $g \geq 0$, and consider a typical situation from the user's point of view. Suppose that one *method* is clearly better suited for the differential equations, but unfortunately the *code* implementing that method cannot handle state conditions. We may also have a code that handles state conditions, but implements an inappropriate method for the problem. Which code will the user choose?

The user will in this case often use the inferior method just to be able solve the problem, see e.g. [3]. A system of generic solvers would offer the user complete functionality in *all* solvers. The user then has every possibility to select the right method independently of other requirements. This also makes it easier to turn a newly developed method into a code, because there is no need to rewrite its supporting algorithms.

2.1.4 Goals for Godess

The previous section explained why we were unsatisfied with the state-of-the-art, and we thus set the following goals for the Godess program:

Provide good solvers. The first goal of the Godess program is that we want good solvers for IVPs implemented in a good way. This is almost self-evident, but we must remember that object-oriented solvers will probably not be much used unless they are competitive with existing solvers. Due to the use of virtual functions and stronger typing, object-oriented codes can run slower than traditional codes. However, it is widely believed that the slight loss in efficiency suffered by object-oriented codes is offset by the superiority of these codes in other areas. In the Godess Project, we demonstrate that our code is superior to traditional codes in many important areas, and that in many cases, we in fact get improved run-time efficiency.

Good control strategies. The control strategies and supporting algorithms should be consistent and based on analysis rather than simple heuristics (possibly with an exception for error recovery).

Support for method development. A major goal of the overall Godess Project is to develop and evaluate new methods, and to do this we must make it straightforward to add a new method. This means that dependencies on the methods in the code should be clearly specified (and preferably computed on the fly). It should also be easy to specify a new method in a natural way.

Usable in packages. We consider it natural that Godess should be used in modeling environments or other environments and not for solving individual IVPs. This makes it even more crucial that all solvers should behave in a coherent way and have the same calling sequence. We also want the solver to have the broad functionality required by modern modeling environments and simulation tools (delays, discontinuity handling etc., cf. [13]).

These requirements naturally suggest an object-oriented design. The original program was presented at the first OON-SKI conference by Kjell Gustafsson, [7] (also in [8]). We have since redesigned the system, concentrated on investigating algorithmic features of IVP solvers and extending the capabilities of the solvers in several directions. For some explanations of the theoretical developments see [20], and for more object diagrams than in this article see [21].

2.2 Overview of the Code

Some experimentation was required in order to find the organization of the solver. Since we are interested in both the timestepping and the supporting algorithms, we concluded that the best way of organizing the solver was to have one class for each task related to solving the problem, see Figure 2.2.

The separation of actual computation and control of it has enabled us to add new methods, while reusing the existing control strategies. We have also been able to develop new control strategies, and other supporting items as will be discussed in the following sections. The basic idea is thus to separate the solver into its parts, and then analyze and improve the parts on an individual basis.

The overall result is that we have a strong, consistent solver where the components are improved over time and new components add new capabilities.

2.2.1 Case Studies: Code Organization

We will here discuss how the generic and object-oriented nature of the Godess program has helped in organizing the code. As a first example we

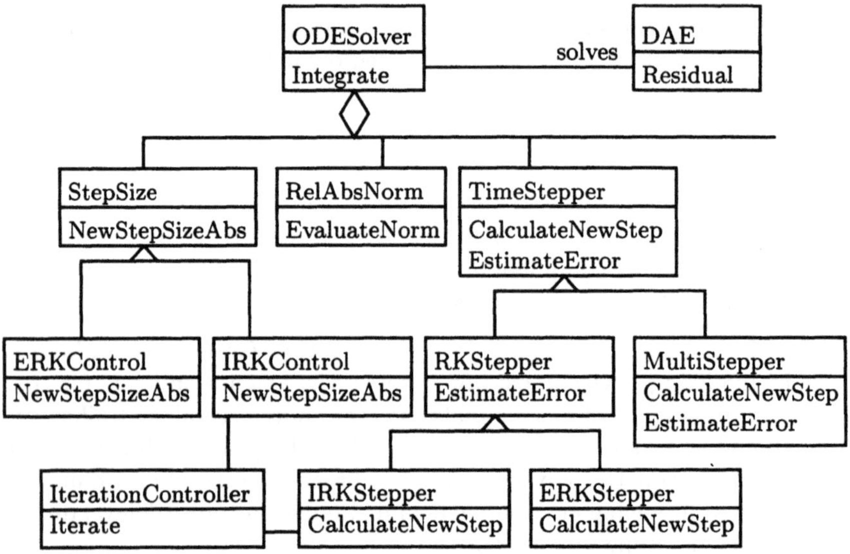

FIGURE 2.2. Overview of the Godess program. The notation indicates that the ODESolver consists of (=diamond) step size control, norm, timestepper etc. The actual implementation of the timestepper is in the subclasses below it. The ODE-Solver solves the DAE by obtaining a step size from StepSize, it then "Calculates a New Step" in the TimeStepper, and "Estimates the Error". The norm of this error estimate is then evaluated by the RelAbsNorm object, and this information is fed back to the StepSize control. For IRK methods the IterationController is responsible for controlling the iterations necessary to solve the implicit equations. Several other class hierarchies have been suppressed.

consider the differential equations, and by examining traditional solvers we see that they are usually built for a single problem class, cf. Table 2.1. By having the systems in an object-oriented class hierarchy we ensure that an explicit ODE can be directly plugged into a DAE solver and used without changing the ODE problem or the DAE solver.

By having the differential equations as a common interface it has been easy to add features such as handling state conditions, $g \geq 0$, and delay equations with very little change to the solvers.

2.2.2 Jacobians

In order to use stiff methods (e.g. implicit Runge-Kutta (IRK) methods, or backward differentiation formulas (BDFs), [14]), we have to solve nonlinear equations. We solve them using modified Newton iterations and thus need the Jacobian.

The Jacobian classes provide a common interface such that different solvers can ask to solve the nonlinear equation without caring about details such as:

Code	Ref.	Solves	State Cond.	Jacobians	Norm
DASx	[17]	$r(t,y,\dot{y})=0$	Yes	d/b	rms
DP12	[17]	$\dot{y}=f(t,y)$	No	d	rms
RKF45	[17]	$\dot{y}=f(t,y)$	No	-	max
RKC	[17]	$\dot{y}=f(t,y)$	No	-	rms
ODE	[17]	$\dot{y}=f(t,y)$	No	-	2
RKSuite	[17]	$\dot{y}=f(t,y)$	No	-	max
LSODAx	[18]	$\dot{y}=f(t,y)$	Yes	d/b	max
LSODE	[18]	$\dot{y}=f(t,y)$	No	d/b	rms
LSODES	[18]	$\dot{y}=f(t,y)$	No	s	rms
LSODI	[18]	$A(t,y)\dot{y}=f(t,y)$	No	d/b	rms
LSODIBT	[18]	$A(t,y)\dot{y}=f(t,y)$	No	bt	rms
SDIRK	[14]	$A\dot{y}=f(t,y)$	No	d/b	rms
Godess		All of the above	Yes	d/b/s	rms

TABLE 2.1. An overview of some available codes. More versions may be available. The reference gives information about where to obtain the codes. For the Jacobians we have dense (d), banded (b), sparse (s) and block tridiagonal (bt). For the Norm we have given the default, and some solvers (including Godess) allow the user to change this. The rms-norm is the root-mean-square norm, i.e. the 2-norm divided by $\sqrt{\text{number of equations}}$.

Structure of Jacobian matrix
Dense, banded, sparse.

Computation of Jacobian
Analytic, numerical.

Structure of the system

- DAE, $r(t,y,\dot{y},p)=0$.
- Semi Explicit, $B\dot{y}=f(t,y,p)$.
- ODE, $\dot{y}=f(t,y,p)$.

Behind this common interface we have also ensured that the numerical differentiation routine can be reused for different structures of the systems, without incurring the penalty of treating all systems as DAEs.

The differential equation can on the other hand specify the Jacobian without caring about how it will be used. For BDF we have to factor a matrix of the type $r_y - h\gamma r_{\dot{y}}$, whereas most IRKs use a matrix of the type $I\otimes r_y - hA\otimes r_{\dot{y}}$. To gain additional performance for IRKs it is possible to decompose A and partition the Jacobian appropriately (the decomposition of A can be performed automatically in Godess).

2.2.3 Time Dependent Discontinuities

When we introduced state conditions as a way of handling discontinuities we neglected to mention that often discontinuities only depend on time. Such discontinuities can be caused by sampling, or by a step response experiment.

Often there is no direct support for time discontinuities in solvers. The unstated reason is that the user can "easily" integrate up to the discontinuity, change the system and then restart the integration. This is simpler to handle than state conditions because we do not have to search for the time of the discontinuity. How to change the system and restart the solver is left to user; the preferred way differs from one solver to another.

If we on the other hand take a look at test suites for IVPs, [13] and [1,2], we find systems of this form without any note about proper treatment of the discontinuity. The solver will without information about the discontinuity spend a lot of effort in finding it, and if it is not determined accurately enough we will lose precision. These test suites were developed by writers of IVP solvers, and to expect that users of codes will handle time discontinuities better than writers of the codes seems a bit optimistic.

The reason for this is that handling time discontinuities is only "almost" trivial. We therefore include the ability to give the points in time (either as a few points or as repeated ones) in Godess. The solver should furthermore not restart from scratch because some initializations are redundant.

We take the `medakzo` example from [1,2], which is a PDE discretized by the methods of lines. It models a chemical reaction with two reactants, and the reactants also diffuse in the medium. One of the reactants is present from the start and the other is inserted after a certain time. The onset of this inflow causes a discontinuity in the equations, and is a typical step response experiment. As originally formulated this discontinuity was hidden in the equations. Specifying the sole discontinuity requires about three extra lines of code, but could decrease the CPU time significantly for the same precision compared to integrating past it with brute force, Figure 2.3. The CPU time vs. precision graph is also more regular, which makes it easier to draw firm conclusions about the performance of the method, when comparing this method to another or investigating different control algorithms.

2.3 Case Studies: New algorithms

Godess has proven to be an invaluable tool for developing and testing new algorithms. It has forced us to reconsider the algorithms in use, and thus clearly showed the problems with the traditional algorithms. We will give examples where the traditional asymptotic analysis has been deficient (step size selection, predictors), as well as another case of tuned heuristics which can be replaced by strategies based on control theory (step size selection),

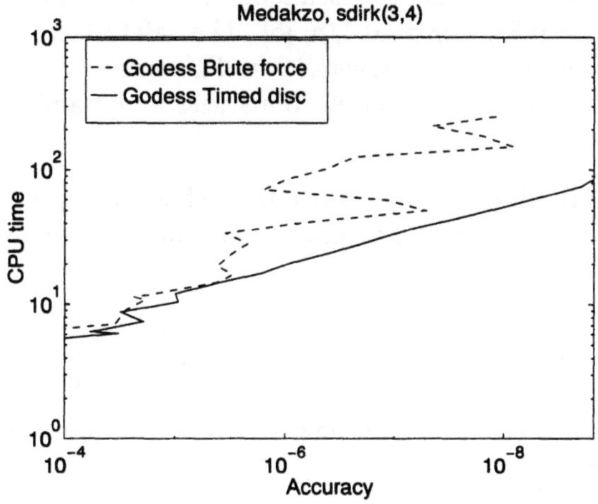

FIGURE 2.3. Impact of a timed discontinuity.

optimization (recomputing Jacobians) and more appropriate asymptotic analysis (predictors). Significant improvements in robustness, performance and general coherence result from these efforts. We have made use of Godess as a testbed for the new algorithms, and for comparisons with traditional ones.

2.3.1 Step Size Selection

Modern timestepping methods choose the step size adaptively to maximize efficiency. With few exceptions, the error is represented by a simple asymptotic model

$$\text{error} = \phi h^p,$$

see [13, p.168], where p is the order and the principal error function ϕ is assumed constant. Based on this model, the next step size is computed as

$$h_{n+1} = h_n \left(\frac{0.8 \cdot \text{tol}}{\text{error}_n} \right)^{1/p}.$$

The factor 0.8 (or some similar number) is used to prevent step size rejections and to compensate for the shortcoming of the oversimplified model assumptions.

Two new improved control algorithms have been developed, tested and verified within Godess:

- For nonstiff methods, an PI control law (cf. [24, p.222-232]) was developed [5, 10] and [14, p. 28-31]. It uses the formula

$$h_{n+1} = h_n \left(\frac{\Theta \cdot \text{tol}}{\text{error}_n} \right)^{k_I} \left(\frac{\text{error}_{n-1}}{\text{error}_n} \right)^{k_P}.$$

Analysis of the controller's closed loop stability led to the recommendations $k_I = 0.3/p$, $k_P = 0.4/p$, [5]. The safety factor $\Theta = 0.8$ was found appropriate [5, 20]. Overall, the PI controller yields more regular step size sequences, added robustness and stability at no extra cost.

- For stiff methods improved control was achieved by providing an estimator for ϕ, which in real computations may vary significantly. This yielded the control law

$$h_{n+1} = h_n \frac{h_n}{h_{n-1}} \left(\frac{\Theta \cdot \mathrm{tol}}{\mathrm{error}_n}\right)^{1/p} \left(\frac{\mathrm{error}_{n-1}}{\mathrm{error}_n}\right)^{1/p}$$

  based on an analysis of the how to observe ϕ and the controller's stability [6, 9].

The benefits of the new controllers have been widely recognized, [14].

2.3.2 Predictors

In order to start the iterations in implicit methods we need some guess of the initial values. We exemplify with a semiexplicit DAE where the iterations for the stage derivatives, $\dot{Y}$, are computed as:

$$\dot{Y}^{k+1} = J^{-1}(B\dot{Y}^k - f(t, y_n + hA \otimes \dot{Y}^k)), \quad k = 1, 2, \ldots$$

where $\dot{Y}^0$ is the predicted value and J is some approximate Jacobian. We stop this iterative process when the difference $\dot{Y}^k - \dot{Y}^\infty$ is deemed acceptable and take $\dot{Y} = \dot{Y}^k$; the stopping criteria for this is discussed in [12, 20], [14, p. 121]. The task of determining the initial value, $\dot{Y}^0$, is often accomplished by simply extending the continuous solution from the previous step into the current one.

This is a typical example of using available data instead of proper analysis and design — the motivation is to avoid extra work that could impede performance. A closer examination, [20], shows that this view is not valid in general. For Singly Diagonally Implicit Runge Kutta, SDIRK methods, we get a first order prediction error although the continuous extension can be more accurate in tracking the actual solution. This is illustrated in Figure 2.4.

We used a fourth order method from [14, p.100], with a second order continuous extension from [20] to verify this. The original continuous extension, [14, p. 100], was not used because it produced even worse results. The explanation why SDIRK methods have these problems can be interpreted in terms of how close the individual stages are to the local solution, i.e. the stage order [13, p. 210], [14, p. 226].

By constructing new 2nd order predictors at negligible extra cost, we gain up to 10% raw CPU performance for normal solvers, Figure 2.5. This

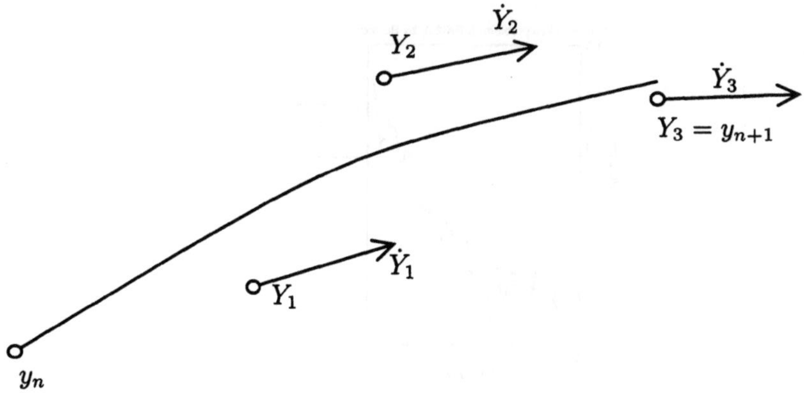

FIGURE 2.4. An SDIRK method. The solid line represents the local solution starting at y_n. The starting point of the arrows represents the stage values, Y, and the direction the stage derivatives, $\dot{Y}$. Note that the intermediate stages, Y_1, Y_2 are less accurate than the endpoint of the step y_{n+1}.

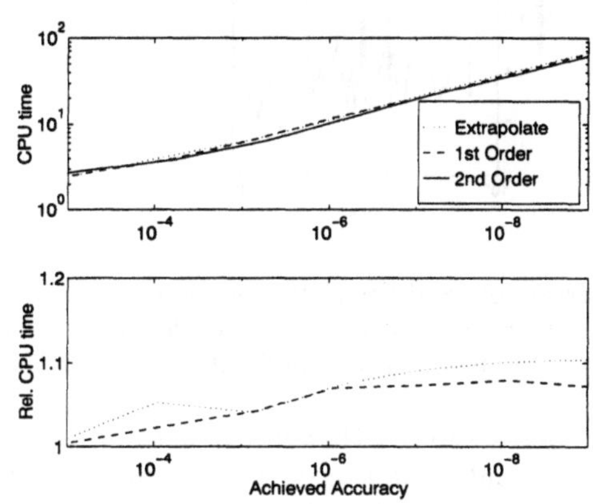

FIGURE 2.5. Comparison of different predictors. We have used our 2nd order predictor as baseline in the second diagram. This shows that the new 2nd order predictor is up to 10% faster for the same accuracy.

is attributed both to a reduction in function evaluations and in convergence failures that would require new Jacobians. For systems with more costly Jacobians we expect the reduction in CPU time to be even greater.

2.3.3 Recomputing Jacobians

In conventional strategies, the Jacobian is recomputed when the convergence rate is "slow". The threshold value can vary between 0.001 [14, p.121] and 0.8 [19]. Only rough guidelines are given [14, p. 573] for how to choose this value. At best it is a tuning parameter.

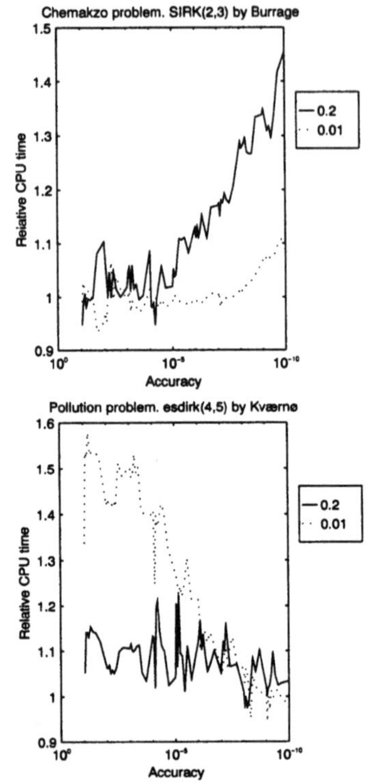

FIGURE 2.6. Comparison of fixed threshold for the rate of convergence with our adaptive value. Our adaptive value is relative performance of one, such that a relative CPU time of 1.3 indicates that the fixed value is 30% slower.

Analysis within Godess showed that in the case of fixed-point iterations the convergence rate should not exceed e^{-1}, with 0.2 as a recommended upper limit [6,11]. Viewing modified Newton iterations as a preconditioned fixed-point iteration led to the same conclusions. In addition it provided a recomputation/refactorization strategy which prevents the convergence rate from slowing down while at the same time being compatible with the new step size control law, [11]. Further analysis, [20,22] also studies the *complexity* in order to choose an adaptive (sharper) upper limit. Based on the cost of computing and factoring a Jacobian compared to solving a system this adaptive strategy can yield considerable gains in CPU time compared to a fixed-value strategy, see Figure 2.6. Since this strategy yields a value that depends on the method in several ways, it is also apparent that it should be automatically set in a generic solver.

We expect further development in this area, in particular with respect to DAEs and the construction of consistent overall strategies (step size, iteration) that try to bring the techniques towards globally optimal performance. This is a challenging task, as control decisions are made locally and based on locally observable quantities.

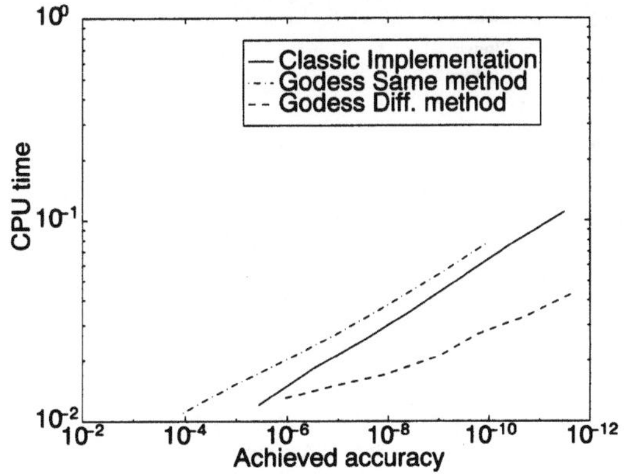

FIGURE 2.7. Comparing Godess to an explicit solver.

2.4 Comparison with Classical Solvers

The first goal of the Godess Project is to produce good solvers, which means that we are not interested in trading too much efficiency to get object-oriented solvers. We will here present these comparisons as CPU time vs. achieved accuracy diagrams following [13, 14]. This is crucial when comparing different methods, e.g. a Runge-Kutta method and BDF, because a Runge-Kutta method will use more function evaluations per step whereas the BDF method will instead use information from several preceding steps.

We thus have a tradeoff between the cost of function evaluations and using extra information from previous steps, and this is best modeled by using the total CPU time for realistic problems. The term realistic depends crucially on the applications we are interested in, because the relative cost of a function evaluation depends upon the complexity of the system. Traditional PDE discretizations often lead to simple systems, whereas mechanical systems often have complex structure, cf. the examples of [14, p. 530-536].

This does not mean that the diagrams are ideal for comparing different solvers, but they are the best we can offer at the present state-of-the-art. The problem with comparing different solvers is that we compare differently optimized programs implementing different supporting algorithms for different methods. We have reduced the number of differences by using the same method, and we have tacitly assumed that the classical solvers written in Fortran have the benefit of a better optimizer. Our only hope should in this case be our improved algorithms.

For a classical explicit Runge-Kutta method and small test problems we normally get results like the ones presented in Figure 2.7. Thus, we may lose some CPU time for the same achieved accuracy. This is caused by the overhead of the generic and object-oriented nature, and partially by the different languages and compilers.

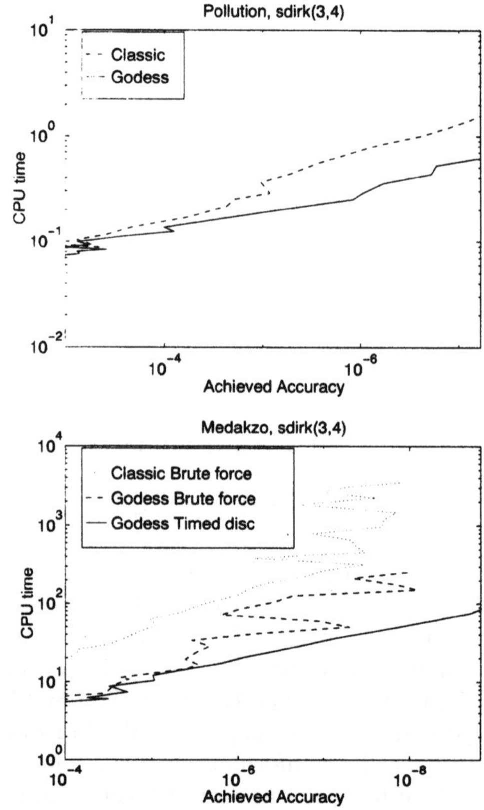

FIGURE 2.8. Comparing Godess to an implicit solver.

The important part is that the loss is modest and acceptable. It is compensated by the ability to switch to better suited methods and other features of the Godess program, cf. Figure 2.7.

In contrast to explicit methods, implicit ones need more supporting algorithms, but we expect our improvements to pay off. In Figure 2.8 we compare the SDIRK code from [14] and use examples from [1,2]. Here, our code is actually ahead in terms of performance. We also see that the ability to treat discontinuities gives Godess an extra advantage. The `pollution` problem is a small chemical system and the `medakzo` problem is a PDE discretized by the method of lines.

The important part is, however, not to show that we are faster but to demonstrate that a generic object-oriented solver can compete with a classical solver written in Fortran. We have thus gained flexibility without sacrificing efficiency. We can also break down the numbers and see that the difference is due to the improvements in the predictors and a better understanding of when to recompute the Jacobian. This means that we do not only have a good generic solver, but we can also explain why it is good.

2.5 Conclusions

We have shown that our object-oriented solver compares favorably to classical IVP solvers. It was also shown that the object-oriented design makes it is easy to add new methods and new capabilities to the system, a task that is difficult to do with classical IVP solvers. There are still methods not included and algorithms which need more analysis, but our experience indicates that Godess will be a useful framework for this.

Acknowledgments: I would like to thank the organizers for inviting me to contribute to this book, and the referees and Professor G. Söderlind for valuable comments.

2.6 References

[1] J. J. B. de Swart. Test set for IVP solver. http://www.cwi.nl/cwi/projects/IVPtestset.html, 1995.

[2] J. J. B. de Swart, W. A. van der Veen, and B. P. Sommeijer. Test set for IVPODE and IVPDAE solvers. Technical report, CWI, 1995.

[3] J. Eborn. Modelling and simulation of an industrial control loop with friction. Master's thesis, Lund Institute of Technology, March 1994.

[4] C. Führer and B. J. Leimkuhler. Numerical solution of differential-algebraic equations for constrained mechanical motion. *Numerische Mathematik*, 59:55–69, 1991.

[5] K. Gustafsson. Control theoretic techniques for stepsize selection in explicit Runge-Kutta methods. *ACM TOMS*, 17(4):533–554, December 1991.

[6] K. Gustafsson. *Control of Error and Convergence in ODE Solvers.* PhD thesis, Lund Institute of Technology, 1992.

[7] K. Gustafsson. Object oriented implementation of software for solving ordinary differential equations. In *OON-SKI'93*, P.O Box 2328, Corvallis, OR 97339, 1993. Rogue Wave Software.

[8] K. Gustafsson. Object-oriented implementation of software for solving ordinary differential equations. *Scientific Programming*, 2:217–225, 1993.

[9] K. Gustafsson. Control theoretic techniques for stepsize selection in implicit Runge-Kutta methods. *ACM TOMS*, 21(4):496–517, December 1994.

[10] K. Gustafsson, M. Lundh, and G. Söderlind. A PI stepsize control for the numerical solution of ordinary differential equations. *BIT*, 28(2):270–287, 1988.

[11] K. Gustafsson and G. Söderlind. Control strategies for the iterative solution of nonlinear equations in ODE solvers. *SIAM Journal on Scientific and Statistical Computing*, Jan 1997.

[12] N. Houbak, S. P. Nørsett, and P. G. Thomsen. Displacement or residual test in the application of implicit methods for stiff problems. *IMA Journal of Numerical Analysis*, 5:297–305, 1985.

[13] E. Hairer, S.P. Nørsett, and G. Wanner. *Solving Ordinary Differential Equations I.* Springer Verlag, 2nd edition, 1993.

[14] E. Hairer and G. Wanner. *Solving Ordinary Differential Equations II.* Spring Verlag, Berlin, 2nd edition, 1996.

[15] A. D. Jensen. An object-oriented platform for dynamic simulation. Master's thesis, Denmark Technical University, Institute of Mathematical Modelling, DK-2800 Lyngby Denmark, 1994. IMM-EKS-1994-47.

[16] A. Kværnø. More, and to be hoped, better DIRK methods for the solution of stiff ODEs. Technical report, Mathematical Sciences Div., Norwegian Institute of Technology, Trondheim, 1992.

[17] Netlib ODE solvers. http://www.netlib.org/ode/.

[18] Netlib ODEPACK. http://www.netlib.org/odepack/.

[19] S. P. Nørsett and P. G. Thomsen. Switching between Newton and fix-point iteration for implicit ODE-solvers. *BIT*, 26:339–348, 1986.

[20] H. Olsson. Practical implementation of Runge-Kutta methods for initial value problems. Licentiate Thesis, Lund Institute of Technology, December 1995.

[21] H. Olsson. Object oriented solvers for initial value problems. In *OONSCI'96*, 1996. To appear.

[22] H. Olsson and G. Söderlind. Stage value predictors and efficient Newton iterations in implicit Runge-Kutta methods. Technical Report LU-CS-TR:96-182, Department of Computer Science, Lund University, P.O. Box 118, S-221 00 Lund, Sweden, 1996. Also submitted for publication.

[23] J. Rumbaugh, M. Blaha, F. Eddy W. Premerlani, and W. Lorensen. *Object-Oriented Modeling and Design.* Prentice Hall, 1993.

[24] K. J. Åström and B. Wittenmark. *Computer-Controlled Systems.* Prentice-Hall, 2nd edition, 1990.

3

SPRINT2D Software for Convection Dominated PDEs

M. Berzins
S. V. Pennington[1]
P. R. Pratt[2]
J. M. Ware[3]

ABSTRACT
SPRINT2D is a set of software tools for solving time-dependent partial differential equations in two space variables. The software uses unstructured triangular meshes and adaptive error control in both space and time. This chapter describes the software and shows how the adaptive techniques may be used to increase the reliability of the solution procedure for a challenging combustion problem. The recent construction of a problem solving environment (PSE) has partially automated the use of SPRINT2D. This PSE consists of tools such as a visual domain specification tool, which helps ease the input of complex geometries, and a visual problem specification tool. After describing these components an evaluation will be made of SPRINT2D and its associated PSE.

3.1 Introduction

Two important trends in the development of numerical software for partial differential equations are to make numerical methods more reliable through the use of adaptive error control, and to make the software easier to use and its results easier to understand by embedding it in an interface layer. The combination of this layer and the core numerical software is classified as a Problem Solving Environment (PSE). The form of such a PSE is by no means clear; PSEs have been defined as capable of solving problems by

---

[1]Computational PDEs Unit, School of Computer Studies,University of Leeds, Woodhouse Lane, LS2 9JT Leeds, United Kingdom. Email: {martin,vp}@scs.leeds.ac.uk.
[2]PA Data Design, The Bishop's Mannor, Howden, East Yorkshire, DN14 7BL, UK. Email: paulp@padd.press.net.
[3]Renaissance, 22/25 Finsbury Square, London, UK. Email: justin@rs.com.While doing the work described in this paper, the latter two authors were associated with the University of Leeds.

communicating in the user's own terms, [6]. Stetter, [13], looks at the wide variety of scientific tools currently available and advocates the utilisation and integration of such tools to form PSEs devoid of explicit programming. A PSE can therefore be viewed as a collection of tools that provide a bridge between the problem the user wishes to solve and scientific software. The aim of a PSE is that it should enhance the solution process for example by increasing the reliability of the solution or decreasing the time spent from specification to solution. Adjerid and Flaherty et al. [1] identify eight components of a successful PSE:

- A computer algebra interface to describe the PDE and data.
- A geometric modelling system to describe the domain.
- A mesh generator to create a computational mesh.
- A numerical solution procedure to solve the PDE.
- Error estimation procedures to give accuracy measures.
- Adaptive strategies to improve solution resolution when needed.
- Parallel solution capabilities for increased performance when needed.
- Visualisation tools to analyse and interpret results.

The aim of this chapter is to address the twin themes of mesh adaptation and PSEs [10] in connection with the SPRINT2D software for convection-dominated PDEs [15]. Sections 3.2 and 3.3 will consider the SPRINT2D numerical PDE code. Section 3.4 looks at the software tools that are part of the PSE surrounding the numerical code. Section 3.5 contains two case studies, including an engineering example from combustion modelling, to show how the software functions. Finally Section 3.6 evaluates the success of the approach taken.

3.2 The SPRINT2D Software

The SPRINT2D software [15] solves time-dependent partial differential equations by using the method of lines to discretise in space thus reducing the PDEs to a system of ODEs (Ordinary Differential Equations) which can then be integrated using existing software packages. This separation of space and time and the use of ODE software makes it possible to combine different combinations of spatial and temporal discretisation as required. SPRINT2D uses a cell-centred finite volume method in which the PDE is integrated over an element and the divergence theorem applied to replace the area integral for the fluxes by a line integral around the edge of the element. The flux functions in the PDE are then used to calculate the numerical flux between adjoining elements. Although the finite volume method may use any form of spatial elements, the use of triangular elements allows complex domains to be modelled, and when used in conjunction with spatial adaptivity provides a powerful modelling environment, [4]. This is particularly true when temporal local error control and spatial error estimation and control are used. The PDEs solved by SPRINT2D are:

$$\beta \frac{\partial \underline{U}}{\partial t} + \frac{\partial}{\partial x} \underline{f}^x + \frac{\partial}{\partial y} \underline{f}^y = \frac{\partial}{\partial x} \underline{g}^x \left(\frac{\partial \underline{U}}{\partial x}, \frac{\partial \underline{U}}{\partial y} \right) + \frac{\partial}{\partial y} \underline{g}^y \left(\frac{\partial \underline{U}}{\partial x}, \frac{\partial \underline{U}}{\partial y} \right) + \underline{S} \tag{3.1}$$

where all the functions $\beta, \underline{f}^x, \underline{f}^y, \underline{g}^x, \underline{g}^y$ and $\underline{S}$ are allowed to depend on $\underline{U}$, x and t. For steady problems β is set to zero. The convective fluxes $\underline{f}^x$ and $\underline{f}^y$ may give rise to wave-like structures in the solution $\underline{U}$, and the terms $\underline{g}^x$ and $\underline{g}^y$ define the diffusive fluxes. The source term $\underline{S}$ can be used to add other processes such as reaction terms including chemical kinetics. The three types of boundary conditions allowed by the package are Dirichlet, Neumann and flux conditions in which the solution, normal derivatives and fluxes are specified. Examples of the types of PDE problems that SPRINT2D has been applied to are given in [3] and include atmospheric dispersion problems, shallow water equations, combustion problems and gas jets. A taxing combustion modelling problem is considered in Section 3.5.

3.2.1 An Overview of SPRINT2D

The SPRINT2D software has the layered structured shown in Figure 3.1. The top layer of the software consists of the PSE type interface tools—the Visual Display Tool (VDS) and the Visual Problem Specification Tool (VPS). An important part of the specification process for solution of two space dimensional PDEs is the definition of the region over which the problem is to be solved. Once this is done, either by hand or by the VDS tool, this region can then be meshed to provide a suitable triangulation of the domain for the numerical solution process. There are currently two mesh generation software packages that can be used by SPRINT2D; the KSLA mesh generator [5] and the GEOMPACK mesh generator [7]. Once the mesh has been generated the user must either use the VPS tool to create a C driver for SPRINT2D or write a driver program by hand. The SPRINT2D package is implemented on top of two existing numerical packages: SPRINT and NAESOL - all the codes being written in C. After applying spatial discretisation to time-dependent problems, the resulting system of ODEs is integrated in time by the SPRINT integrators.

Spatially discretising steady problems results in a system of non-linear equations which are solved by the non-linear solver package NAESOL, [12], but are not considered further here. The modular nature of the software

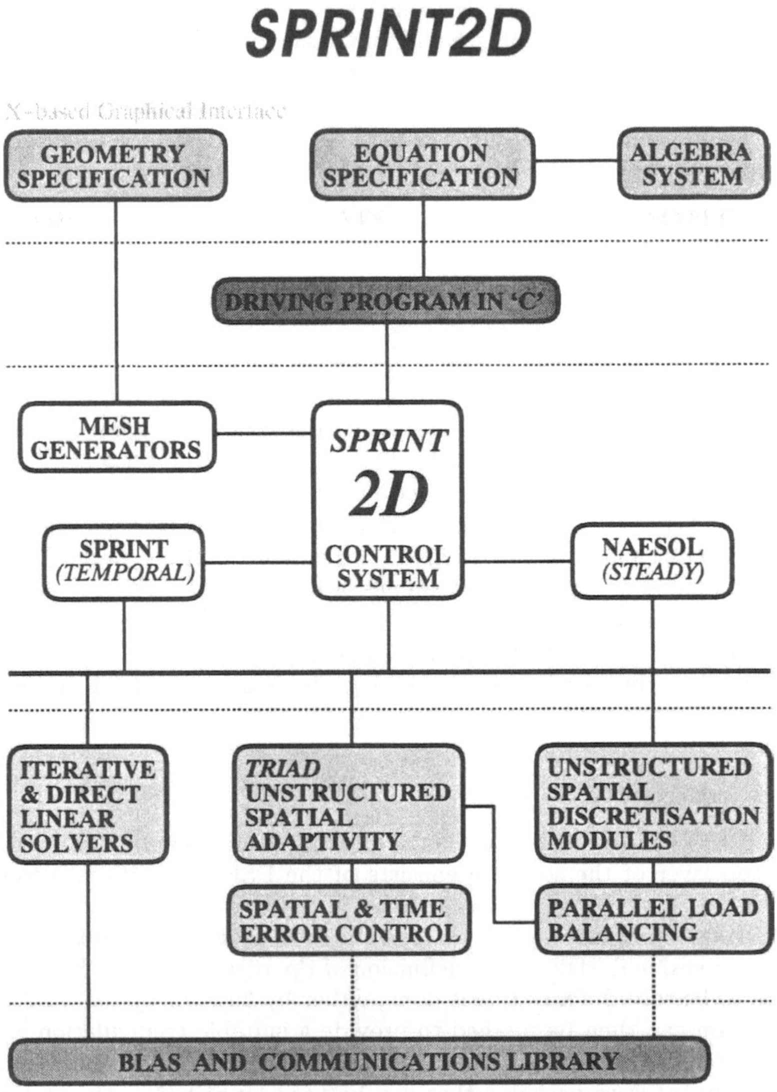

FIGURE 3.1. Outline of Sprint Software.

allows additional solution modules to be added to the package. The main options are: dense sparse and iterative linear algebra modules; an operator splitting module; and Theta and Backward Differentiation Formula time integration methods. The TRIAD package provides the routines to perform any spatial adaptivity using h-refinement.

The C driving program for SPRINT2D specifies the PDEs and the solution techniques to be used. The first part of the driver program must include the relevant header files for the SPRINT2D package and modules that are to be used. The user, in the driver program, needs to specify the following information: a file containing a specification of the physical domain; relative and absolute tolerances for the adaptivity routines; a Riemann solver for the advective fluxes $\underline{f}^x$, $\underline{f}^y$ (see Section 3.2.2); a flux function for $\underline{g}^x$ and $\underline{g}^y$; a source term function $\underline{S}$; boundary conditions; and initial conditions. These functions have to follow a fixed interface in returning values to SPRINT2D.

3.2.2 Riemann Solvers and Boundary Conditions

For convection-dominated PDEs, correct specification of the flux and careful space discretisation are essential to avoid unphysical oscillations in the discrete solution. A standard approach to ensure a stable solution is to place more emphasis on the information coming from the direction of the flow (the upwind values) in discretising the advective parts of the PDE. In order to maintain stable solution values upwinding is combined with a non-linear scheme that changes order by limiting the numerical flux that passes between cells, see [4, 9]. For simple problems, e.g. linear advection, the choice of upwinding direction is obvious. However, for complex systems, the direction may alternate, or a combination of both left and right values maybe needed for a system of PDEs, see [9]. In this case an approximate (sometimes exact) Riemann solver is used to calculate the advective flux in the code using a combination of knowledge about the PDE and left and right solution values, [9]. For example in the case of the knock problem defined in Section 3.5.2 below, the fluxes $\underline{f}^x$ and $\underline{f}^y$ must be calculated given left $\underline{U}_L = (\rho, \rho u, \rho v, E, \rho z)_L^T$ and right $\underline{U}_R = (\rho, \rho u, \rho v, E, \rho z)_R^T$ solution values at the midpoint of each edge. This calculation is a nontrivial task, see [4]. The Riemann solver is used to implement flux or derivative boundary conditions. For example, the reflective boundary conditions in Section 5.2 are imposed by setting the exterior 'normal' velocity to be the opposite sign to the normal velocity at the boundary from the interior. The values of all the other variables on the 'exterior' being the same as the interior values. All other 'outside' solution values are the same as the interior values. Although it may be possible to construct approximate Riemann solvers automatically, there are many situations in which the user must specify the Riemann solver.

3.3 Mesh Generation and Adaptivity

The main attraction of unstructured triangular meshes is that they can approximate arbitrary domains more easily than quadrilateral based meshes. The initial meshes used in SPRINT2D are created from a geometry description using the KSLA [5] or GEOMPACK [7] mesh generators. GE-

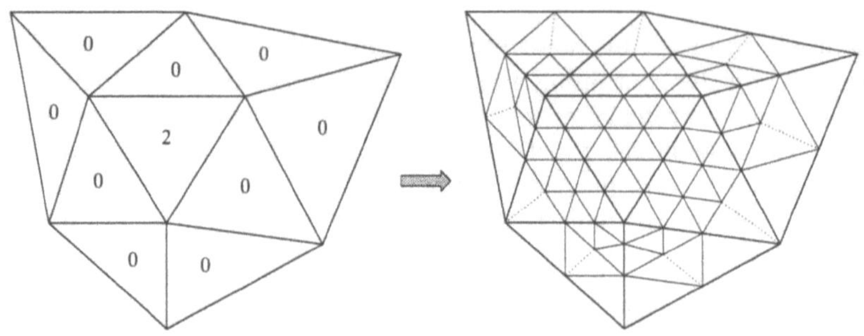

FIGURE 3.2. Regular and Green Refinement.

OMPACK constructs the mesh by decomposing the input geometry into simpler polygons and then meshing these polygons. As a semi-automatic mesh generator GEOMPACK requires additional information at the beginning of the specification file to accomplish this. This information provides the user with the ability to control various aspects of the final mesh such as desired number of triangles, mesh smoothness and the way in which the geometry is decomposed into simpler polygons. These meshes are then refined and coarsened by the TRIAD [15] adaptivity module which uses data structures to enable efficient mesh adaptation. For the ith PDE component on the jth triangle, a local error estimate $e_{i,j}(t)$ is calculated from the difference between the solution using a first order method and that using a second order method, see [4] for details. For time dependent PDEs this estimate shows how the spatial error grows locally over a timestep. A refinement indicator for the jth triangle is defined by an average scaled error ($serr_j$) measurement over all *npde* PDEs

$$serr_j = \sum_{i=1}^{npde} \frac{e_{i,j}(t)}{atol_i / A_j + rtol_i \times u_{i,j}}, \qquad (3.2)$$

where *atol* and *rtol* are the user-supplied absolute and relative error tolerances. This formulation for the scaled error provides a flexible way to weight the refinement towards any PDE error. An integer refinement level indicator is calculated from this scaled error to give the number of times the triangle should be refined or derefined.

In the refinement case, all the neighbouring triangles which share an edge with a refined triangle are refined towards that shared edge. Similarly, all triangles with a vertex in common with the original triangle are refined towards that vertex. Finally Bank's green rule is applied to ensure the mesh is conforming, [1]. This is illustrated in Figure 3.2 in which a level 2 refinement is applied to the central triangle where dashed lines represent the bisecting edges of green triangles.

De-refinement is a reversal of the refinement process, that is, the four children created through regular subdivision can be deleted, leaving the parent. Only one level of de-refinement is allowed at any one remeshing time and, in addition, all four children must be marked for deletion. Derefinement will not be allowed if a triangle in the initial mesh, produced by the mesh generator, is specified. The triangles created as a result of application of the green rule may be of poor quality and so are removed before any further mesh refinement takes place.

3.3.1 Time Integration

Although in many time dependent PDE codes a CFL stability condition is used to control the timestep, the SPRINT2D Theta or Backward Differentiation Formula codes with functional, Newton-Krylov or operator splitting methods allow automatic control of the local error. Efficient time integration requires that the spatial and temporal errors are roughly the same order of magnitude. The need for spatial error estimates unpolluted by temporal error requires that the spatial error is the larger of the two errors. The SPRINT2D software also has an option to use the strategy of Berzins, see [4], which controls the local time error to be a fraction of the growth in the spatial discretisation error over a timestep.

3.3.2 Visualisation

The driver program also allows the user to extract information about the numerical solution each time it changes or is updated. This is achieved by the user providing a *monitor* routine which SPRINT2D calls at regular intervals with a large amount of solution information. For example, each triangle has a solution value, a spatial error value and, for time-dependent problems, a temporal error value. The code can also provide a large quantity of spatial information about the unstructured mesh such as areas of triangles, lengths of edges, unit normals to edges etc. This information is used by the visualisation package which complements the SPRINT2D solver. This visualisation package is developed in IRIS GL and runs on a local host whilst SPRINT2D runs on a computationally intensive platform elsewhere. Solution frames are sent across the network to the visualisation package within which the user can interrogate the solution whilst the next frame is being calculated. An example of this is the frame shown in Figure 3.5 for the knock problem described above. The visualisation package displays the solution values for each triangle in the spatial mesh or error estimates in space and time. This information is not used to steer the calculation directly, but has proved to be invaluable for users learning how to apply adaptivity to their applications.

In displaying the numerical solution values for convection-dominated problems great care must be taken to avoid introducing physically unreal values not already present in the numerical solution. For example physical values of density should always be positive. The solution to convection-dominated PDE problems may have shocks and discontinuities present.

Numerical PDE solvers take great care to preserve, say, the positivity of the solution. However, such discontinuities may lead to numerical undershoot and overshoot if standard interpolation techniques are used. This can mislead the user. In [11] a triangular based interpolant is described which achieves the desired properties by bounding the values it produces to be between the maximum and minimum values used to define it. This interpolant provides a more reliable and natural way for the user to view the solution.

3.4 A PSE for SPRINT2D

In designing a visually based PSE, the need for portability makes it desirable to use tools that are either industry standard or are as de-facto standards. Such tools are the X Window System and associated Widget sets which are high level X toolkits. A widget is defined as an X window with associated manipulation procedures for the window and data structures. One such set is the The Open Software Foundation (OSF) Motif Widget Set which has a distinctive look and style, and was used to construct the X interfaces for the PSE described here. It is also necessary to convert the user's information into C functions; this is done by using the Maple system which was chosen because of its wide availability, robustness, and because it provides C and TEX output.

3.4.1 A Visual Domain Specification Tool

The Visual Domain Specification (VDS) tool aids the key task of specifying the initial domain so that it can be meshed, thus reducing the time spent on the problem specification process. The tool must provide a convenient way for the user to specify and manipulate the geometry and allow for the user to visualise a coarse mesh defined over the domain.

The VDS tool uses an internal data structure to construct the geometry. This information is then transferred to mesh generation software via a postprocessing routine. This intermediate step allows the separation of the visual specification process and the creation of the numerical domain specification file, although these two processes are closely linked.

Initially an interface was constructed to the fully automatic KSLA mesh generator [5], as used in SPRINT2D. In addition to this, an option to construct output files for the GEOMPACK and PLTMG mesh generators [2, 7] was added to demonstrate the generic nature of the tool. Overall the VDS tool is split into three main components: a drawing canvas where the user can specify the geometry using the mouse; a display canvas that shows the mesh generated from the user specified geometry; and a control panel containing buttons and labels which allows the user to control the tool.

As well as the visual components of the tool a suitable internal data structure needs to be defined. The problem faced is to design a data structure that fits around the requirements for final output format and allows flexibility for the input requirements. The description of the geometry adopted is hierarchical in that each level is built up from the lower levels, composed of vertices, lines or arcs, and regions respectively. The ability of the data structure to store geometries in a tree structure, where more complex elements are constructed from simpler elements, is one which has proved successful as it allows top-down and bottom-up manipulation of the information stored in the data structure. The structure made it possible to provide geometry output in a form suitable for GEOMPACK.

3.4.2 A Visual Problem Specification System

The creation of the SPRINT2D driver program can be a lengthy process. The user must first define the PDE problem and the spatial domain. The SPRINT2D modules to be used must be initialised by calling the appropriate command. The VPS system aims to provide an easy and natural way for the user to visually specify the information required by SPRINT2D. The aim is to decrease the time taken to create a valid driver program, guide the user to provide all the information needed by SPRINT2D and to avoid the need for explicit programming wherever possible.

The information needed to specify the problem may be split into: the mesh information (including boundary and initial conditions); the equation specification and finally the error control information. The VPS system uses this information to create a suitable driver program for the numerical software via the postprocessing step. At the end of each stage of this user specification process, the information supplied is stored so that the postprocessing subsystem can use it to create the driver program. For continuity, this information is used as the new default values for the user interfaces when next invoked. Although the time saved by this visual interface is important, the interface also ensures that all information is input, that sensible defaults are chosen, and that user errors are reduced.

The user defines the mesh in two stages; first, information concerning the number of boundary conditions and the mesh file is supplied. An outline of the boundary is then displayed and the boundary edges are named with unique integers. The second stage of the interface extracts and reproduces the geometry to allow the specification of the boundary conditions. The boundary conditions (Neumann, Dirichlet or Flux) are specified using Maple and mapped onto edges by placing conditions on consecutive lists of edges. The PDE functions in Equation (3.1) and the initial conditions are also specified using Maple syntax which is then converted into the functions required by the driver program. Figure 3.3 shows these functions for the case study of Section 3.5.1. The adaptivity interface allows the user to select values for the maximum number of triangles, and to set the adaptivity and visual states to either on or off. If the adaptivity is set to on the user is prompted for the absolute and relative spatial tolerance values. Finally the user is asked to specify the integrator module (BDF or Theta) and the linear algebra module to be used (sparse, iterative or operator splitting).

3.4.3 Construction of the Driver Program

The information provided by the users and encoded by the interfaces is passed on to the postprocessing subsystem responsible for creating the SPRINT2D driver program. The approach used is that of fitting information into a template program. The postprocessing subsystem first provides a visual summary of the information so that the user may easily validate the problem definition and then trigger the creation of the driver program. The final output of the postprocessing subsystem is a valid C driver program which can be compiled and linked to run the numerical software.

The driver program starts by including the relevant header files for the solution modules used. The initial conditions, the boundary conditions and the appropriate functions required for the finite volume method are then defined. The user may provide a monitor routine which provides a means of examining the numerical solution. This is then followed by a set of routines to instruct the numerical code where to find the previously defined functions it requires and which software packages to use. The driver program then starts the solution process by calling SPRINT2D.

The driver program, as well as having the ability to execute the numerical software, must also be easy to understand, well-structured and well-documented to allow possible user modifications. One example where this is important is the Riemann solver function required by the finite volume method. The default solver supplied by the system uses the average of the 'left' and 'right' values in the flux calculation, see Section 3.2.2. The user must specify the approximate Riemann solver if flow directionality has to be taken into account, see Section 3.5.1 for example.

3.5 Case Studies

This section will demonstrate the use of the tools by two case studies involving time dependent PDEs; other problems are in Pratt [10].

3.5.1 Convection-Dominated PDE – Burgers' Equation

The first problem is a Burgers' equation given by

$$\frac{\partial u}{\partial t} + \frac{\partial}{\partial x}\left(\frac{u^2}{2}\right) + \frac{\partial}{\partial y}\left(\frac{u^2}{2}\right) = p\frac{\partial^2 u}{\partial x^2} + p\frac{\partial^2 u}{\partial y^2}$$

where p is a constant defined as 0.01. The Dirichlet boundary conditions and the initial conditions on the square domain are

$$u(x,t) = (1 + e^{\frac{x+y-t}{p}})^{-1}.$$

This time dependent problem is solved with the Theta integration module and iterative linear algebra. The start time is 0.15 with 15 output points and a step between output points of 0.10. Spatial adaptivity is used with absolute and relative spatial tolerance of 0.05 and a maximum number of 10,000 triangles. This example demonstrates the construction of the approximate Riemann solver by the postprocessing subsystem for the driver program. The routine from the driver program for the averaging Riemann solver is given below; its use results in negative solution values close to the wave front.

```
void problem_rs(TRIAD_Line *line, int npde, double x,
                double y, double t, int sub_name,
                double norm_x, double norm_y, double u_l[],
                double u_r[], void *users_data, double nf[])
{ /*  Burgers eqn: crude averaging Riemann solver */
    double u = ( u_l[0] + u_r[0] ) / 2.0 ;
    double f_x, f_y;
    f_x = 0.5*u*u;
    f_y = 0.5*u*u;
    nf[0] = f_x * norm_x + f_y * norm_y;
} /* Riemann solver */
```

The negative values vanish when Roe's Riemann solver, see [9] is implemented by inserting the code:

```
/* Burgers eqn: Roe fix to averaging Riemann solver */
if u > 0.0 u = u_l[0];
else       u = u_r[0];
```

before the assignment to `f_x`. This example shows that although the VPS tool does not know how to produce a correct Riemann solver for a general problem, the code it does produce may be easily modified by the user.

3.5.2 Combustion Knock-Modelling Problem

A challenging test problem is a combustion model related to the modelling of 'knock' in car engines, see [8]. The model is used to investigate the effects of autoignition in end gases in an idealised car engine cylinder. The onset of 'knock' is seen when large pressure pulses interact with the edges of the cylinder. Mathematically the problem is specified by a system of five PDEs representing conservation of mass, momentum and energy together with a species equation. The functions in equation (3.1) are defined by

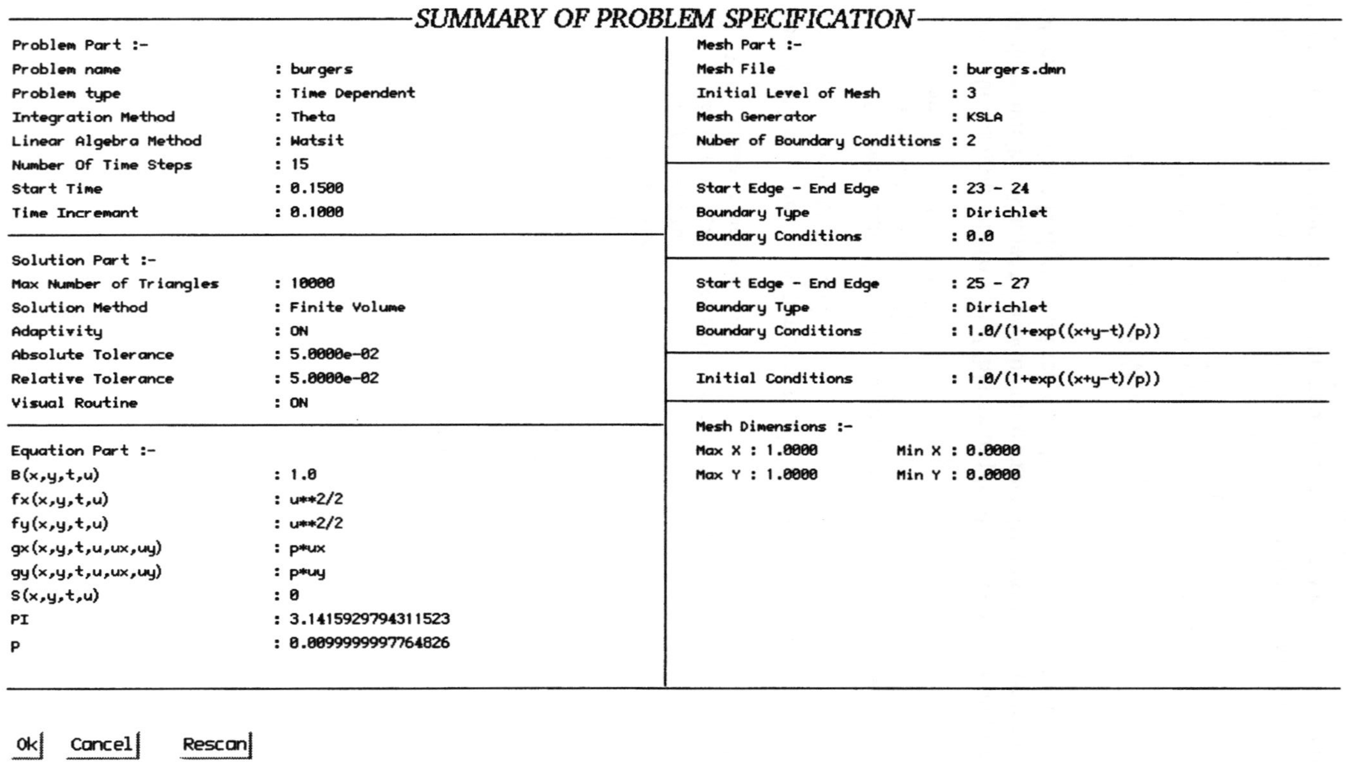

FIGURE 3.3. Summary Window for Burgers' Problem.

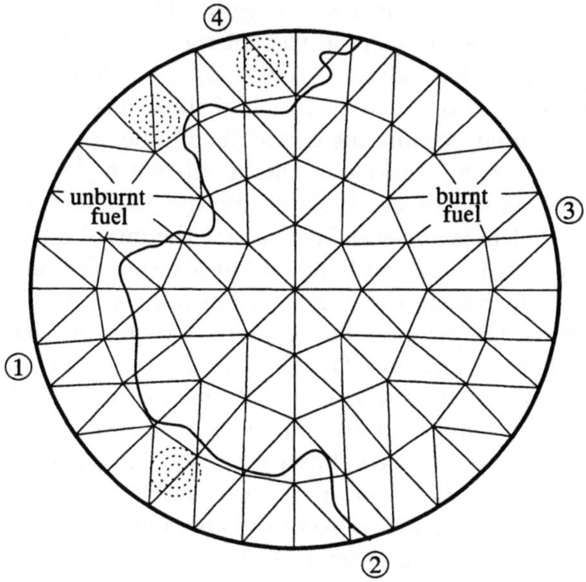

FIGURE 3.4. Diagram of Knock Model.

$\underline{U} = (\rho, \rho u, \rho v, E, \rho z)^T$, and

$$\underline{f^x} = \begin{pmatrix} \rho u \\ \rho u^2 + p \\ \rho uv \\ u[E+p] \\ \rho uz \end{pmatrix}, \underline{f^y} = \begin{pmatrix} \rho v \\ \rho uv \\ \rho v^2 + p \\ v[E+p] \\ \rho vz \end{pmatrix}, \underline{S} = \begin{pmatrix} 0 \\ 0 \\ 0 \\ 0 \\ -\rho z\ k(T) \end{pmatrix}$$

and $k(T) = \exp(\hat{\beta}\,(1 - \frac{1}{T}))$, $T = P/\rho$ and $\hat{\beta} = 20.0$. The variables ρ, u, v, p are the density, the velocities in the x and y dimensions and the pressure, respectively. The variable z represents the scaled fuel concentration. The energy E is defined by the equation of state

$$E = \frac{p}{(\gamma - 1)} + \frac{\rho u^2 + \rho v^2}{2} + \alpha \rho z$$

where $\gamma = 1.2$ and $\alpha = 8.0$. The geometry of the problem is shown in Figure 3.4 in which the irregular solid line represents the initial position of the flame front, as taken from experimental data. The area to the left of this front contains unburnt fuel while that to the right is one in which the fuel has burnt. The dotted concentric circles indicate temperature hot spots which will lead to autoignition and pressure pulses travelling across the cylinder to cause 'knock'. Points numbered 1 to 4 are the four pressure transducers at which experimental time histories of pressure are available.

The initial conditions are as follows. The initial velocities u and v are zero; the pressure has the value $p = 1$; the fuel concentration z is zero in the burnt region and one in the unburnt region. The scaled temperature t is 0.75 in the unburnt region except at the hot spots where it rises to one and in the burnt region it has value $1 + \alpha(\gamma - 1)/\gamma$, $\rho = p/T$. The quantities E, γ and α are defined by above equation of state. The implementation of the reflective boundary conditions is described in Section 3.2.2.

Although fixed mesh solutions to this problem are given in [4], the focus here will be on the effect of adaptivity and the PSE on the solution process. The problem is non-standard in that the initial conditions cannot be specified by a mathematical function but are specified on a triangle by triangle basis from camera data and read in from a data file. Soon after integration starts, the complex flow patterns for this problem mean that heavy mesh refinement occurs. Thus it is useful to let the user refine the mesh a priori. To allow this, the adaptivity module was modified to allow user specified mesh refinement around a specific location.

The SPRINT2D code was employed with the Theta and operator splitting options as in [4]. Runs were performed with fixed triangular meshes with 2048 and 8192 elements respectively, and adaptive meshing with the maximum number of triangles set to 8192 and 32768 respectively. These modes are referred to in the results table as FIXED and ADAPT respectively. The SPRINT2D code was used with standard local error control with absolute tolerances of 10^{-4} for all the PDE variables except the species concentration z for which 10^{-5} is used. A maximum step size of 5.0×10^{-4} was imposed during the initial combustion phase in order to prevent unphysical solution values being passed into the Riemann solver. The geometry of the problem and the solution shortly after the start of integration are shown in Figure 3.5, which shows the output the user sees from SPRINT2D.

In contrast, the fixed timestep regular square mesh code LUMAD, [4], uses an ad-hoc Riemann solver approach to determine the flux values. Timestepping is done using the forward Euler method with only a CFL condition to control the timestep. In order to obtain results consistent with SPRINT2D and to resolve the reaction transients, LUMAD must use a square regular mesh with 100x100 mesh points and 40,000 timesteps giving a CFL number of 0.01.

The entries marked TIME show the time of the peak pressure pulse at pressure transducer 1. PEAK indicates the values of this peak. The physical significance of the PEAK value is that it indicates the strength of the pressure pulse that causes 'knock' while the TIME value indicates when this occurs. Correct computation of these values is thus important if the mathematical model is to be validated against experiments. The results in Table 3.1 show a consistent trend with those of [4]. Moreover, by the time the adaptive run has encountered the maximum pressure spike the mesh has about 24,000 triangles, with the finest mesh in the region of the pressure spikes. In this case the spatial refinement tolerances are 10^{-4} for density and 10^{-2} for the other PDE variables and hence refinement is biased towards the density errors. It is worth stating that a poor choice of tolerances can lead to inappropriate refinement.

Code	*Mesh*	MODE	TIME	PEAK
LUMAD	10000	CFL	32.08	5.87
SPRINT2D	2048	FIXED	41.09	4.44
SPRINT2D	8192	FIXED	28.59	5.61
SPRINT2D	8192	ADAPT	28.63	5.59
SPRINT2D	32768	ADAPT	27.12	6.45

TABLE 3.1. Transducer 1 Pressure Spike.

One of the challenging aspects of this problem is that of writing a good physically realistic Riemann solver [4]. For this problem the Riemann solver function within the driver program is approximately 650 lines of code and comments and took an expert user of SPRINT2D about 5 days to write and debug.

The table shows that unless great care is taken with the choice of time step and spatial mesh, over-large pressure pulses at incorrect times may be recorded. In particular the use of adaptivity enables the mesh to be concentrated where it is needed.

3.6 Conclusions

The aim in this paper has been to show how SPRINT2D and its associated PSE together form a powerful and semi-automatic way of solving time-dependent PDEs. The software fits into the PSE framework described in Section 3.1 by using Maple to describe the PDEs, the VDS tool to model the geometry and GEOMPACK to generate the mesh. The SPRINT2D finite volume scheme and associated error indicators are used to compute the solution and adapt the mesh, with the visualisation module and IRIS Explorer being used to display the results. Although not described here a prototype distributed parallel version of the code exists, [14]. The overall approach has proved successful for a broad range of convection-dominated problems with complex geometries needing adaptivity. The modularity of the software does make it possible to devise efficient components for particularly important and/or difficult problems; one example of this being the operator splitting iterative scheme used in [4].

The combination of end-users and the developers of SPRINT2D and its PSE has helped to construct a package with numerical reliability, has eased the solution process, reduced the time spent and provided a more natural and convenient way to solve the PDEs. The users were enthusiastic about the VDS and VPS tools and about how easy it was to generate working code. The benefit of this however must be balanced against the many months of effort spent on the knock problem experimenting with different meshes, tolerances, Riemann solvers and initial conditions.

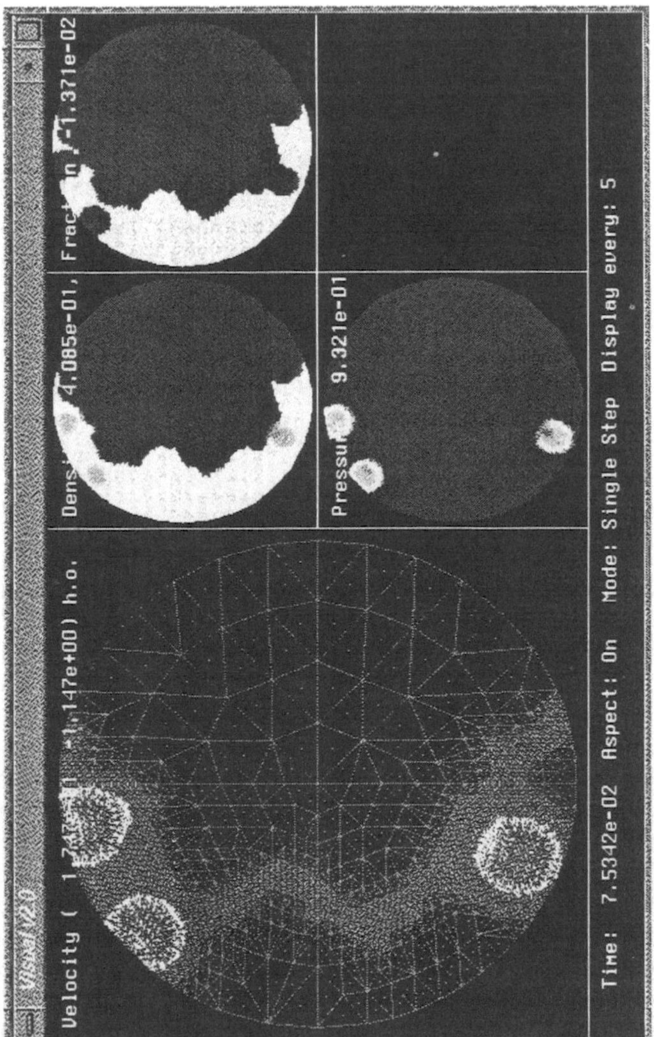

FIGURE 3.5. SPRINT2D Solving the Knock Problem.

Perhaps the largest problem faced with automatically producing a driver program capable of giving a valid solution is that of the Riemann solver. In the case of problems with source terms the best Riemann solver may not even be known. However, as understanding in this area increases, PSEs can develop alongside the numerical code.

This work has shown that it is possible to utilise current scientific computing technology to build software tools and packages that when combined form an easy-to-use layer surrounding complex computational code. This layer can help both novice and experienced users to better utilise their time, efforts and knowledge, even if the layer provides only partial help for difficult problems and specialised modules may still need to be written.

Acknowledgments: Thanks are due to L. E. Scales of Shell Research and J. G. Pan and C. G. W. Sheppard of the Mechanical Engineering Department at Leeds. This research was supported by EPSRC and Shell Research UK.

3.7 References

[1] S. Adjerid, J. E. Flaherty, P. K. Moore, and Y. J. Wang. High-order adaptive methods for parabolic equations. In J. M. Hyman, editor, *Experimental Mathematics: Computational Issues in Non-Linear Science*, Physics D 60 1-4, pages 94–111. North-Holland, 1992.

[2] R. E. Bank. *PLTMG: A Software Package for Solving Elliptic Partial Differential Equations Users' Guide 7.0*, SIAM, Philadelphia,1995.

[3] M. Berzins, P. H. Gaskell, A. Sleigh, W. Speares, A. Tomlin, and J. M. Ware. An adaptive CFD solver for time dependent environmental flow problems. pages 311-318 in Numerical Methods for Fluid Dynamics V, Eds K. W. Morton and M. J. Baines, Clarendon Press, Oxford 1995.

[4] M. Berzins and J. M. Ware. Solving convection and convection reaction problems using the M.O.L. *Appl. Num. Math.* , 20:83–99, 1996.

[5] R. M. Furzeland, P. C. Rem, and R. F. Van der Wijngaart. General purpose software for multi-dimensional partial differential equations. Tech. Rep., Shell Research Amsterdam, 1989.

[6] E. Gallopoulos, R. Houstis, and J. R. Rice. Future directions in problem solving environments for computational science. Technical report, NSF Workshop on Research Directions in Integrating Numerical Analysis, Symbolic Computing, Computational Geometry and Artificial Intelligence for Computational Science, Washington, D.C., April 1991.

[7] B. Joe. GEOMPACK - A software package for the generation of meshes using geometric algorithms. *Adv. Eng. Soft.*, 13(5/6):325–331, 1991.

[8] J. Pan and C. G. W. Sheppard, A theoretical and experimental study of the modes of end gas autoignition leading to knock in an SI engine. *S.A.E. paper* 94-2060 (1994). S.A.E., Warrendale, PA 15096, USA

[9] S. V. Pennington and M. Berzins. New NAG library software for first-order partial differential equations. *ACM Transactions on Mathematical Software*, 20(1):63–99, March 1994.

[10] P. R. Pratt. Problem Solving Environments for the Numerical Solution of P.D.E.s Ph.D. Thesis, University of Leeds, 1995.

[11] P. R. Pratt and M. Berzins. Shock Preserving Quadratic Interpolation for Visualisation on Triangular Meshes. *Comput. and Graphics*, 20(5) 1996.

[12] L. E. Scales. NAESOL: User's guide. Internal report, Shell Research Ltd, Chester, 1993.

[13] H. J. Stetter. Tools for scientific computation. *Zeitschrift für Angewandte Mathematik und Mechanik ZAMM*, 73(12):335–348, 1993.

[14] C. M. Walshaw and M. Berzins. Dynamic load balancing for PDE solvers on adaptive unstructured meshes. *Concurrency Practice and Experience* 7:7-28, 1995.

[15] J. M. Ware. *The Adaptive Solution of Time-Dependent Partial Differential Equations in Two Space Dimensions.* PhD thesis, School of Computer Studies University of Leeds, 1993.

4

Electrochemical Modelling and Software Genericity

Gert Nelissen
Patrick F. Vankeirsbilck[1]

ABSTRACT
This contribution presents the major design abstractions identified in the course of an object-oriented C++ software project, named ElEMD, for solving a system of partial differential equations (PDEs) numerically. The main application area of this project is electrochemical modelling where the complex interaction between electrochemical mass and charge transport on one hand and the flow phenomena in the electrolytic solution on the other hand are to be modelled in an integrated manner. The most important abstractions in ElEMD deal with the analytical equations to be solved, the local discretization schemes, the material properties, iterative schemes, etc. They were often the result of evolving requirements imposed on the project.

4.1 Introduction

The physical phenomena occurring in a considerable number of industrial application domains, such as fluid dynamics, structural mechanics, electromagnetism, thermal analysis, etc. are described by a system of partial differential equations (PDEs). For most of these application domains, a number of well-established grid based numerical techniques exist to solve such a system of equations: the Finite Difference Method (FDM), the Finite Element Method (FEM), the Finite Volume Method (FVM) and the Boundary Element Method (BEM). The Finite Difference Method has not been considered in the work discussed in this text despite its high degree of accuracy and efficiency. The reason is that the Finite Difference Method lacks geometric flexibility unless it is used in a multiblock context. In the latter case, the complexity is moved towards the grid generation process.

[1]Vrije Universiteit Brussel, Pleinlaan 2, B-1050 Brussels, Belgium. Email: {gnelisse,pvkeirsb}@vnet3.vub.ac.be.

4.1.1 Complexity

Due to the ever increasing demand for accurate modelling of industrial processes, and pushed by the ever growing computing power of modern workstations and by progress in numerical algorithms, the software packages that solve systems of partial differential equations tend to become more and more complex. This complexity originated in the following areas:

1. The use of unstructured grids to increase the geometric flexibility of the solver, and ease grid adaptivity with multigrid techniques requires complicated data structures. Adaptivity techniques are optimal in the sense that a uniform level of accuracy is obtained using a minimal number of elements through grid adaptivity. Multigrid guarantees a convergence rate which is independent of the number of elements in terms of work units, ref. [18,21] and, hence, convergence acceleration is obtained.

2. Composite multigrid methods (ref. [15]) further require more complex data sets on the element level at the interface between refined and non-refined zones.

3. The implementation of higher order schemes (quadratic or cubic finite elements, higher order ENO Godunov schemes ref. [1,14,25], ...) are often a better trade-off between the number of elements and the desired accuracy. These schemes lead to a more complex element type because either more degrees of freedom exist per element or the stencil covers more layers of neighbouring elements.

4. The solution of different coupled problems (e.g. aeroelasticity, heat transfer in and around integrated circuits, electrochemical processes, etc.) requires the use of the most appropriate method for each subproblem and imposes that the coupling is established automatically by the system. The need for interior field coupling is easily demonstrated with the following example. The boundary layers in an electrochemical problem is between 10 to 100 times smaller than the hydrodynamic boundary layer occurring at the same location. This means that, in order to achieve optimal results and performance, the electrochemical and hydrodynamic problem are to be solved on grids with a different degree of resolution for the boundary layer, by using two different domains occupying the same location in space.

   Different domains may further be used to model the same physical problem while using more or less complex mathematical models. For instance, a simple and economic potential flow model is sufficient to model the far field in a fluid dynamics problem while, at the same time, a Navier-Stokes model is needed in the vicinity of solid boundaries. Thus, computing resources are allocated only where they are necessary.

5. The combination of zones where approximations in a different number of dimensions are needed involves polymorphic data sets. A typical example is the calculation of the magnetic field in an electrical machine (2D or 3D) fed by a complex external electrical network (composed of 1D segments). Again, resources are allocated where they are really required. One does not need a 3D model to simulate an electrical network.

4.1.2 Generic Frameworks – Object-Orientation

Evidently, a highly generic software framework is required in order to cope with the complexity mentioned above. In view of the difficulties observed in procedural programming paradigms to express genericity, the object-oriented paradigm is nowadays believed to be a promising alternative for creating generic software systems, also within the scientific computing community (ref. [2]). The present project started out in the late 80s as one of the pioneering attempts in this direction.

However, to achieve the promising benefits associated with object-orientation, a thorough design is essential. In view of the different approaches to use object-orientation in numerical modelling seen in the past (ref. [8, 17, 25]), the authors have tried to discover the commonalities between these approaches and incorporated them in their developments. Such commonalities often turned out to be good candidates for abstraction. Furthermore, many original abstractions were added to these candidates.

These efforts resulted in a framework called ELFMD, which stands for Elements in Multiple Dimensions using Multiple Discretizations combining Multiple Domains. Although the primary application area is electrochemical modelling, the scope of the project has grown much larger and now covers areas such as fluid mechanics and modelling of cathodic protection underground pipelines and off-shore structures. The range of available space discretization techniques in the ELFMD software system is restricted to FEM, BEM and FVM.

4.1.3 Overview

A very brief introduction to electrochemical modelling together with the requirements that arise from such modelling will be given in Section 4.2. The different discretization techniques involved are shortly addressed in Section 4.3. A number of the more important abstractions encountered during the development process of ELFMD will be presented in more detail in Section 4.4 followed by some critical remarks in a separate section.

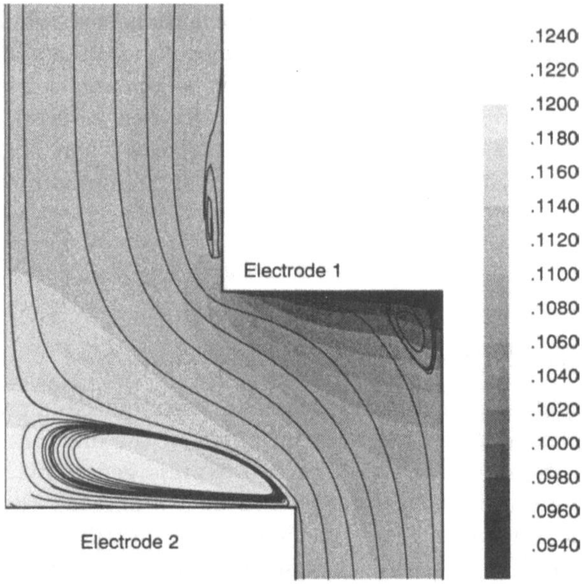

FIGURE 4.1. A typical electrochemical cell. The grey scale represents the electric potential distribution and the black lines are particle traces in the flow of the ionic solution.

4.2 Electrochemical Modelling

An almost unexplored example that closely follows the observations made in Section 4.1.1 is found in electrochemistry (ref. [20]), one of the most recent application areas in the field of numerical simulation. Determining the current density and potential distributions within electrochemical cells is important for both the analysis of experimental data obtained at individual electrodes, and the design and scale-up of useful industrial cell configurations. A typical example of an electrochemical cell is given in Figure 4.1.

Until a couple of years ago, only geometrically and physically simplified models were available, ref. [9, 12]. Recently however, some crucial progress has been achieved in this area, ref. [4, 5, 23] by using techniques that have been known for more than a decade in fluid mechanics. They have now been applied and extended to electrochemistry in close collaboration with fluid mechanics modelling experts. The requirements imposed on a state-of-the-art electrochemical solver for dilute solutions can be summarized as follows:

1. The incompressible electrolyte flow, governed by the Navier-Stokes equation (4.1) must be modelled in 2 and 3 space dimensions.

$$\begin{aligned} \frac{\partial \vec{u}}{\partial t} + \left(\vec{u} \cdot \vec{\nabla}\right) \vec{u} &= \vec{\nabla} P + \mu \vec{\nabla} \left(\vec{\nabla} \cdot \vec{u}\right) \\ \vec{\nabla} \cdot \vec{u} &= 0 \end{aligned} \tag{4.1}$$

with

$$
\begin{aligned}
t &= \text{the time,} \\
\mu &= \text{the viscosity of the electrolyte,} \\
P &= \text{the pressure potential field,} \\
\vec{u} &= \text{the velocity field.}
\end{aligned}
$$

In many cases, additional phenomena such as temperature effects and natural convection induced by temperature effects and concentration gradients have to be included in this model.

2. The mass and charge transport of N ions (industrially relevant cases contain up to 10 species) due to diffusion, convection and migration have to be modelled in conjunction with the above Navier-Stokes equations (4.1) and the conservation of charge. This is achieved by the following set of equations:

$$\frac{\partial c_i}{\partial t} + \vec{u} \cdot \vec{\nabla} c_i = z_i F . \vec{\nabla} \cdot \left(\mu_i c_i . \vec{\nabla} U \right) + \vec{\nabla} \cdot \left(D_i . \vec{\nabla} c_i \right) + R_i \,, \quad i = 1, \ldots , N \tag{4.2}$$

$$\sum_{i=1}^{N} z_i c_i = 0 \tag{4.3}$$

with

$$
\begin{aligned}
c_i &= \text{the concentration for ion } i, \\
D_i &= \text{the diffusion coefficient for ion } i, \\
F &= \text{Faraday's constant,} \\
R_i &= \text{a source term describing the reactions of ion } i, \\
& \text{in the electrolytic solution,} \\
\mu_i &= \text{the mechanical mobility of ion } i, \\
U &= \text{the electrical potential in the electrolyte,} \\
\vec{u} &= \text{the velocity field in the electrolyte,} \\
z_i &= \text{the charge number of ion } i.
\end{aligned}
$$

It is important to note that, in case the convection effects are so dominant that the concentration gradients can be omitted, eq. (4.2) reduces to the Laplace equation with the electric potential in the electrolytic solution as the only unknown. Charge transport is then only caused by migration effects due to electrical potential differences.

3. The models mentioned above are to be coupled with the modelling of the electric potential distribution V in the electrodes and the outside electrical network, governed by the Laplace equation (4.4):

$$\Delta V = 0. \tag{4.4}$$

4. The temperature distribution in the electrolyte and the electrodes is governed by the following equation:

$$c_p.\rho \cdot \left(\frac{\partial T}{\partial t} + \vec{u}.\vec{\nabla}T\right) = \vec{\nabla}.\left(\lambda\vec{\nabla}T\right) + Q_s \tag{4.5}$$

with

$c_p.\rho$	=	the heat capacity,
T	=	the temperature,
$\vec{u}$	=	the velocity field,
λ	=	the heat conduction,
Q_s	=	the local heat production.

Temperature effects are due to ohmic losses in electrodes and the chemical reactions that occur in the cell and at the electrodes. Note that the electrolyte flow interacts with these effects through the term in $\vec{\nabla}T$ in eq. (4.5).

5. The electrochemical reactions at the electrodes are modelled by highly non-linear relations (overpotential relations) between the potential of the electrodes and the associated values in the electrolyte: a typical example is the Butler-Volmer overpotential relation, see also ref. [20]:

$$J_n = J_{no} \cdot \left(e^{\alpha.k_1.(V-U)} - e^{-(1-\alpha).k_2.(V-U)}\right) \tag{4.6}$$

with

J_{no}	=	the exchange current density depending on the concentration of the ion i,
α	=	the transfer coefficient,
k_1, k_2	=	physical parameters of the reaction,
V	=	the electric potential in the electrode,
U	=	the electric potential in the electrolyte, adjacent to the electrode.

These effects give rise to a system (up to 20 equations) of coupled PDEs. The coupling between the equations (4.1) to (4.6) occurs through the material properties (for instance a temperature dependent conductivity), through temperature and concentration effects such as natural convection and through non-isothermal reactions in the bulk and at the electrode interfaces of the electrochemical solution.

4.3 A Generalized Approach to Numerical Modelling

4.3.1 Solution Strategy

Solving a general system of PDEs efficiently implies the use of the most appropriate numerical method for each of the sub-problems (see as an example Section 4.2) and even for each of the individual terms in the equations. The different numerical techniques need therefore to be combined in a flexible way. Such a combination of techniques transforms the system of analytical PDEs into a system of algebraic equations in the case of an implicit solution strategy:

$$\mathbf{A}(\mathbf{x}) \, . \, \mathbf{x} = \mathbf{b}(\mathbf{x}) \tag{4.7}$$

with

$\mathbf{A}(\mathbf{x})$ = the (non-linear) system matrix,
$\mathbf{x}$ = the vector of the unknowns or the degrees of freedom,
$\mathbf{b}(\mathbf{x})$ = the right hand side vector.

Using any explicit time marching method, one gets:

$$\mathbf{x} := \mathbf{f}\left[\mathbf{r}(\mathbf{x})\right] \tag{4.8}$$

with

$\mathbf{r}$ = the residual vector,
$\mathbf{f}$ = the function describing how the new solution vector is to be computed (e.g. a four step Runge-Kutta method).

Eqs. (4.7) and (4.8) remain valid regardless of the equation(s) to be solved, the dimension in space of the problem or the order of accuracy of the space discretization.

The complete ElEMD software system has been designed based on the observation that solving PDEs using any numerical method essentially reduces to two, more or less independent, steps:

- the system matrix assembly
- the iterative solution process

The first step deals with the evaluation of the entries in the matrix **A** and the vector **b** or the evaluation of the residuals in a set of solution nodes. Since ElEMD is restricted to grid based solvers, computing the matrix entries or residuals can be carried out by computing contributions to these items on an element by element basis. Different discretization methods correspond to different manners of computing the contribution for each

Term Type	**Discretization Technique**
Convection	MDUM, SUPG, upwind FVM
Diffusion	FVM, FEM, BEM
Migration	FVM, FEM
Source Terms	FVM, FEM, BEM

BEM	: Boundary Element Method
FEM	: Finite Element Method
FVM	: Finite Volume Method
MDUM	: Multi-Dimensional Upwind Method (ref. [22])
SUPG	: Streamwise Upwind Petrov-Galerkin Method (ref. [16])

TABLE 4.1. Available discretization techniques for different term types.

set of equations to be solved. The second step represents the solution of eqs. (4.7) and (4.8) using some iterative solution method (e.g. a Newton-Raphson scheme). Existing procedural libraries are used to perform the internal matrix inversions in ELFEMD. Object-oriented matrix libraries such as ISIS++ (ref. [7]) are currently under investigation.

4.3.2 Discretization Techniques

Furthermore, different sets of equations often include similar terms such as convection, diffusion, source terms, etc. An overview of the most common methods to discretize these different terms is given in Table 4.1.

4.3.3 Intergrid and Interdomain Operators

In addition, some numerical techniques require intergrid operators. Typical examples are multigrid methods and the coupling of different domains. The coupling of domains can be achieved in different ways: interior field coupling and boundary coupling.

- Interior field coupling involves the transfer of the solution variables in one domain to another domain that occupies (partly) the same location in space as the first one, see Figure 4.2. Note that a similar coupling is needed when performing multigrid computations where the nodes of the different refinement levels do not necessarily coincide. Non-coinciding nodes on different mesh levels offer the advantage that each mesh level can be generated independently of the previous one and, hence, can be optimized with respect to the solution known on another level. The cost is that more expensive interpolation schemes are required to perform the solution transfer, see e.g. ref. [21].

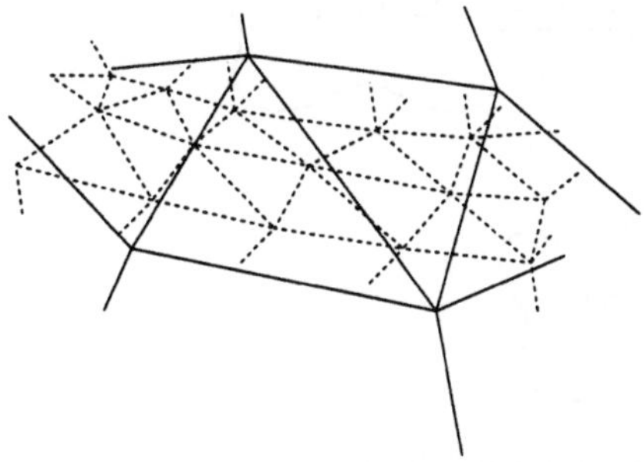

FIGURE 4.2. Interior field coupling in the case of two grids with different degree of resolution occupying the same position in space.

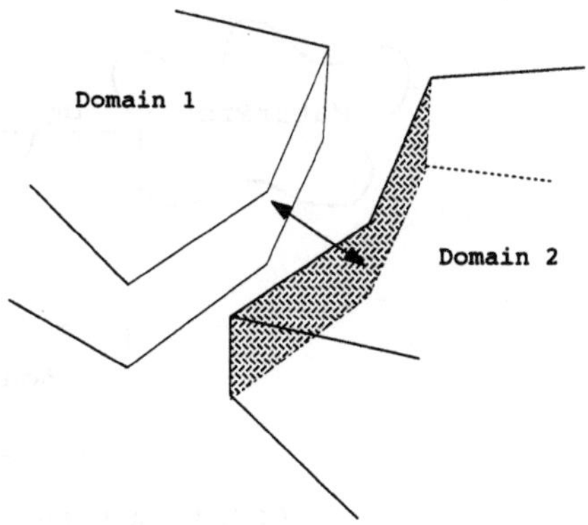

FIGURE 4.3. Coupling domains through their common boundary.

- To be able to simultaneously solve different equations in different neighbouring domains, information has to be exchanged between these domains via their boundaries. This means some coupling mechanism across the common boundary has to be devised, see Figure 4.3. This mechanism expresses the relation between the solution variables in the two neighbouring domains as a weighted residual problem, see ref. [19]. This allows the treatment of linear relations (such as a continuity relation) and non-linear relations (such as (4.6)).

4.4 Abstractions

The general observations made in the previous sections are at the basis of the design of the ELEMD software system and led to the key abstractions addressed in this section. Note that all the class diagrams are following the Booch notation, ref. [3].

4.4.1 Equation

4.4.1.1 Analytical Equation

A crucial abstraction in ELEMD and, to the belief of the authors, in any system with an enhanced range of applicability, is the `Equation` abstraction. It models a system of analytical PDEs.

The `Equation` abstraction represents the association between the material properties (`MaterProp` class) and the discretized formulation of the equation by means of a series of elementary contributions (`Contribs`, see Section 4.4.1.2) as shown in Figure 4.4. The `MaterProp` class is an abstract class. Concrete classes represent some physical quantity as a function of the local solution variables and the position in space.

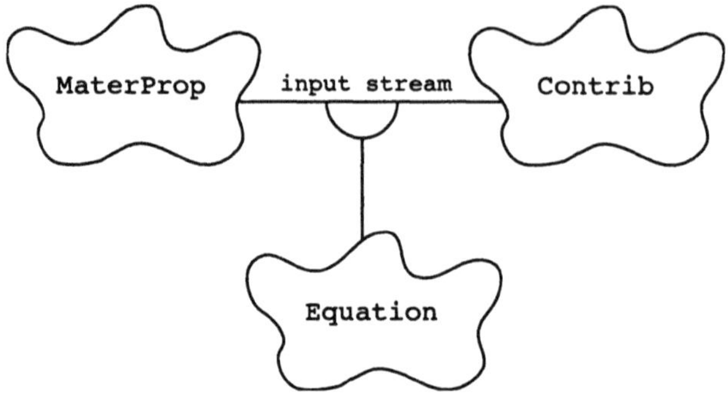

FIGURE 4.4. The Equation association.

The `Equation` reads all material properties including their function prescription from an input stream and assigns them to the appropriate `Contrib` which becomes aware of the material property.

It is important to notice that there can be more solution variables than genuine unknowns in a system of PDEs. For instance, the fluid velocity components are known solution variables to an electrochemistry problem (4.2) and are therefore not considered as unknowns in case the influence of the electrochemistry on the fluid flow is neglected.

4.4.1.2 Contribution Abstraction

The `Contrib` abstract class is a key abstraction in the realization of the system matrix assembly step discussed in Section 4.3. It is the base class of a number of primitive `Contrib` classes used to compute matrix or residual contributions related to a number of primitive terms in the analytical `Equation`:

$$\begin{array}{ccc} \frac{\partial N}{\partial x} & \frac{\partial^2 N}{\partial x^2} & \frac{\partial^2 N}{\partial x \partial y} \\ \frac{\partial N}{\partial y} & \frac{\partial^2 N}{\partial y^2} & \frac{\partial^2 N}{\partial x \partial z} \\ \frac{\partial N}{\partial z} & \frac{\partial^2 N}{\partial z^2} & \frac{\partial^2 N}{\partial y \partial z} \end{array}$$

Using combinations of primitive `Contrib` classes and adding `MaterProp` instances, application dedicated `Contrib` classes can be generated and used in some `Equation Discretizer`, see Figure 4.5.

The `Contrib` can, under the action of the `Local Discretizer` (see Section 4.4.4), present the local solution variables and the position in space to the `MaterProp` instance(s) it is associated with. The latter can then compute the local value of the property to be used in the ultimate matrix or residual contributions.

This demonstrates the open character of the ELEMD framework with respect to the equations to be solved.

As an example, the 3D isotropic diffusion term $\sigma(u).\Delta u$ leads in the case of a Galerkin FiniteElement discretization to a dedicated `Contrib` class

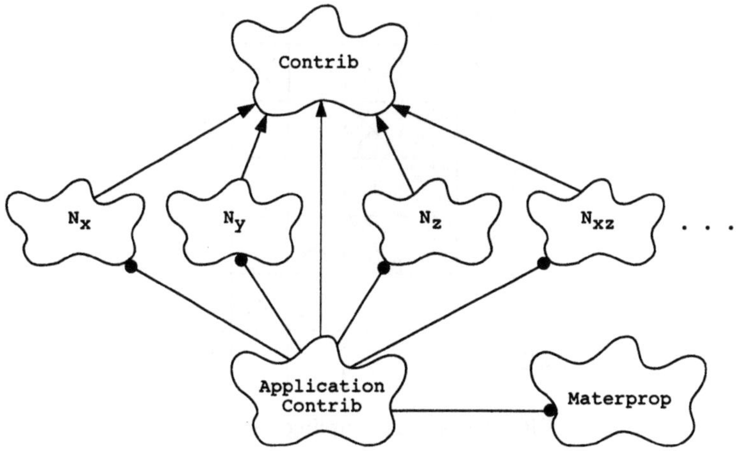

FIGURE 4.5. The `Contrib` hierarchy.

which computes:

$$\sigma(u)\left[\frac{\partial N_i}{\partial x}.\frac{\partial N_j}{\partial x}+\frac{\partial N_i}{\partial y}.\frac{\partial N_j}{\partial y}+\frac{\partial N_i}{\partial z}.\frac{\partial N_j}{\partial z}\right]$$

where the diffusion coefficient σ is implemented in a `MaterProp` class as, for instance, a linear function of the unknown u. The subscripts i and j are the local indices of a single element and N is the shape function used to interpolate u over the element.

4.4.1.3 Equation Discretizer

The `Equation Discretizer` abstraction prescribes how the analytical `Equation` in a given `Domain` is to be discretized.

The discretization of the analytical equations results, for each pair of coupled nodes belonging to the discretization, in an algebraic system of NEQ equations and NVAR unknowns, see also Figure 4.6. The systems associated with all possible pairs of coupled nodes constitute the complete system matrices in eq. (4.7) or the complete residual formulation of eq. (4.8). An `Equation Discretizer` instance is able to focus on a subset of these equations and a subset of the unknowns. It thereby prescribes the coupling pattern between the various variables in the different equations at hand.

For efficiency reasons, a specialized `Equation Discretizer` instance will work in decoupled mode when focusing on the square light shaded subset and will operate in a fully coupled mode when focusing on the rectangular dark shaded subset. The decoupled mode prevents the `Equation Discretizer` from computing contributions that are known to be zero on the off-diagonal positions. Furthermore, it is possible to focus on a non-square subset of variables and equations, as seen in Figure 4.6.

The `Equation Discretizer` stores a reference to a `Contrib` instance for each entry in the coupling pattern. The selection of the concrete `Contrib`

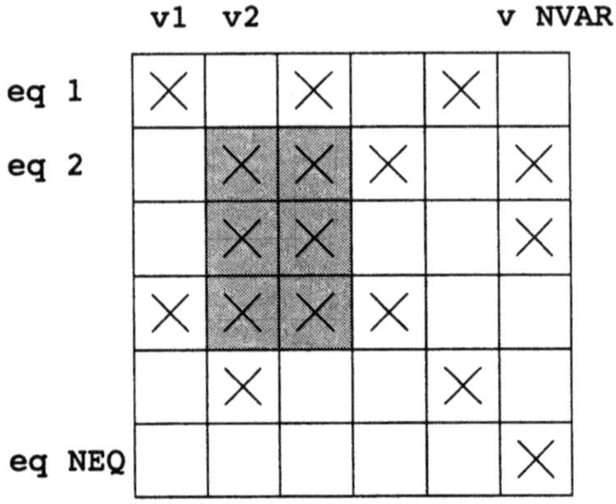

FIGURE 4.6. A coupling pattern.

instances is determined by the actual specialized `Equation` instance, see Section 4.4.1.1.

Through specialization, dedicated **`Equation Discretizers`** have been developed for the Finite Element, Boundary Element and MDUM (residual distribution) methods.

4.4.2 Data Representation – Space Discretization

As already stated, the scope of ELEMD is restricted to grid based solvers, i.e. solvers for which the physical space where a problem is to be solved is subdivided into a large number of elementary volumes, called elements. The classes used to model this kind of space discretization are outlined in this section.

4.4.2.1 Grid Representation

A hierarchical structure is used to represent all the grid data. The complete data set consists of a number of domains each containing a number of finer or coarser grid levels which, in turn, group a number of elements, points or solution nodes, as shown in Figure 4.7. A set of container classes is used to build such an aggregation hierarchy.

Tree The `Tree` class is used to represent the complete problem to be solved. It is composed of one or more `Domain` instances and points to a coupling scheme for the different domains.

Domain Each `Domain` instance contains enough data to be able to behave as a process of its own and, hence, could be placed on a separate CPU. This approach reflects the domain decomposition paradigm in parallel computing. The `Domain` knows what equation is to be solved, how this equation is to be solved (what discretization is used), what the space dimension of the problem is, etc. In its turn, each `Domain` instance is composed of one or more `Level` instances to be able to perform

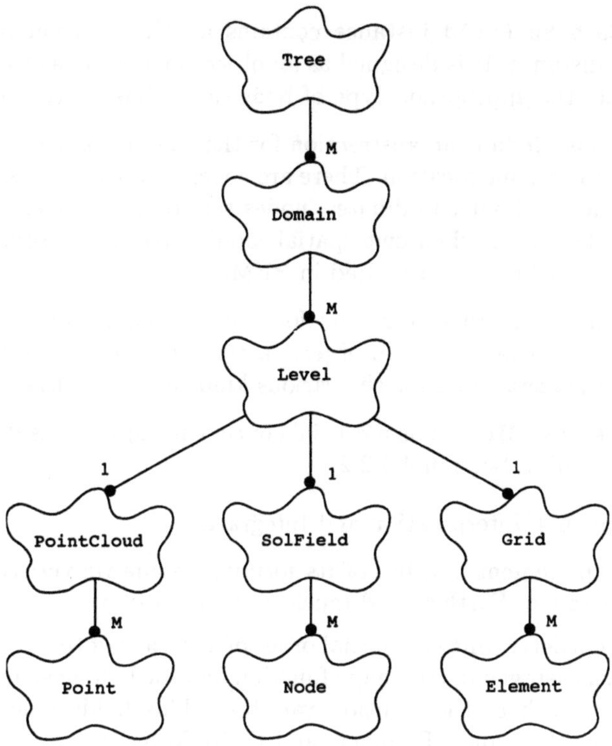

FIGURE 4.7. The data hierarchy.

multigrid calculations. `Domain` instances can represent overlapping and/or non-overlapping computational zones.

Level Each `Level` instance contains a `Grid` grouping the `Elements`, a `PointCloud` grouping the geometric `Points` and a `SolField` grouping the `Nodes` (solution points). In case all elements of a `Level` instance are conformal and isoparametric, the geometric and solution points coincide, allowing the `PointCloud` to serve simultaneously the geometric points and solution nodes. This leads to a reduction of the storage space requirements, see Section 4.4.2.2.

PointCloud The `PointCloud` class groups all the geometric `Points` in a particular `Level`. The class `PointCloud` can construct the appropriate point type depending on the type of problem that is solved. Classes with such capabilities are called object factories.

Point A `Point` groups the 3 spatial coordinates of a geometric point and holds a reference to the `PointCloud` instance it belongs to. Specializations have been made for points storing a set of solution variables, points storing a residual vector, points able to impose boundary conditions, etc.

SolField Each `SolField` instance contains all the solution nodes of a `Level` instance. It is designed as an object factory so as to be able to generate the appropriate type of `Node` depending on the context.

Node The class `Node` is an abstraction for the concept of a set of solution variables at some location. There are concrete node classes representing nodes without coordinates, nodes referring to a geometric point and nodes storing their own spatial coordinates. The association with spatial coordinates is required in BEM.

Grid Each `Grid` instance stores its forming elements. Furthermore, the `Grid` class behaves like an object factory so as to create the appropriate `Element` types for the various kinds of `Local Discretizers`.

Element Because `Element` is a crucial concept in E$_L$EMD it is discussed in more detail in Section 4.4.2.2.

4.4.2.2 Elements, Interpolation and Integration

Essentially, an element is a list of its forming geometric vertices (`Point`) and solution `Nodes`. Further, and much more important:

- Elements determine the spatial order of accuracy of the discretization through their geometric shape function and the kind of shape function used for each of the solution variables. This holds true for Finite Elements, Boundary Elements and Finite Volumes.
- Interpolation is modelled by specializations of the `FatherElem` class. A specialized `FatherElem` class represents a standard element in a local normalized coordinate system that is used to ease the evaluation of the interpolation functions and their integration over the element using some numerical integration scheme (see also Section 4.4.3). Specialized `FatherElem` classes represent the different interpolation functions over a specific element shape. Figure 4.8 depicts the classification of these different types.

  Several specialized `Integrator` classes exist to integrate a function over an element. However, not all `Integrator` classes can be used in combination with all `FatherElem` types. The `FEPair` concept models the restriction on the possible Integrator-FatherElem combinations. Both the number of Gauss Points (#GP) and the number of nodes of the `FatherElem` determine the possible combinations. It also contains the necessary data and functionality to actually compute the integral of some function over an element. The `FEPair` abstraction is a powerful and flexible method to assign different shape functions and associated integration methods to a single element. In the case of a non-conformal and/or non-isoparametric element, Figure 4.9, each unknown can have a different shape function and, hence, the element is referencing a series of `FEPairs`. In Figure 4.9, the markers indicate the position of the geometric or solution nodes.

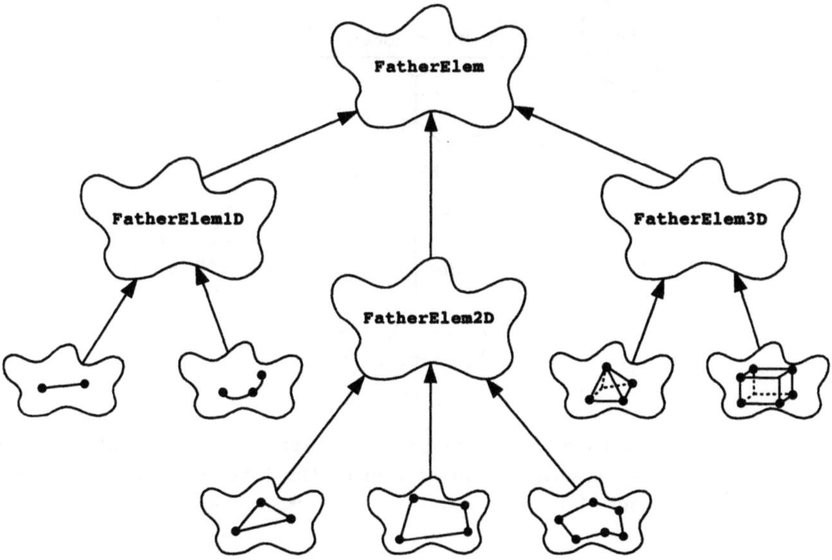

FIGURE 4.8. The `FatherElem` hierarchy.

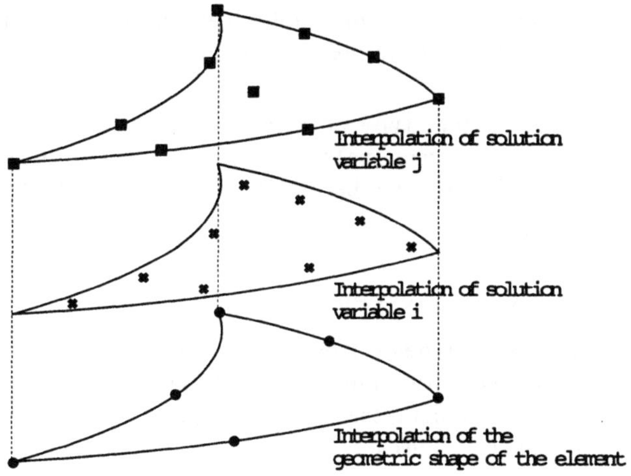

FIGURE 4.9. A non-conformal, non-isoparametric element. The marks denote the nodes used to construct the different shape functions.

Concrete `FEPair` classes are parameterized with the actual type of `FatherElem` and the actual type of `Integrator`. Parameterization avoids numerous function invocations based on dynamic binding leading to enhanced run-time efficiency.

Note also that all `FatherElem` and `Integrator` concrete classes are Singletons, see ref. [11], which ensures that each of these classes can allocate only one instance in memory.

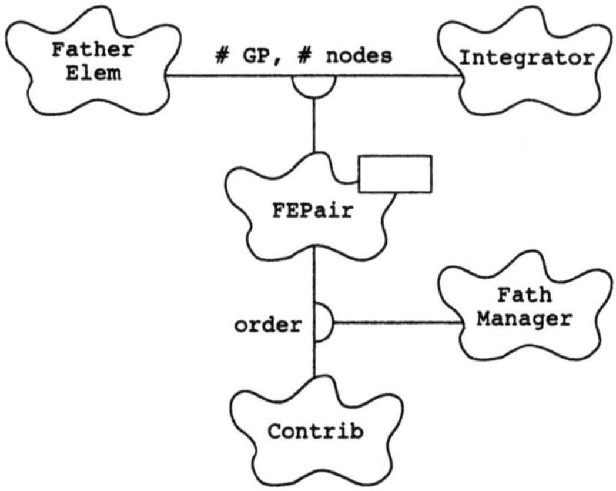

FIGURE 4.10. The associations modelled by the **FatherManager**. **# GP** stands for the number of Gauss quadrature points used by the **Integrator** class. **# nodes** represents the number of nodes used by the **FatherElem** class.

It is very important to notice that the number of unknowns can vary between the different nodes of a non-conformal, non-isoparametric element. This means that a more intelligent mapping scheme between the solution nodes and the position in the system matrix is needed.

- Finally, each concrete type of element uses a **Local Discretizer** to compute the different matrix or residual contributions over the element, depending on the discretization technique at hand, see also Section 4.4.4.

4.4.2.3 Separation Geometry – Physics

A statement which is often heard from people dealing with Finite Element codes is that a strict separation must be maintained between the geometry and the physics of the problem, see e.g. ref. [17]. The authors do agree with this statement in general. However, they believe that this leads to a considerable memory space overhead in the *frequently* occurring situation where all elements in the same **Level** instance are conformal and isoparametric. In this very common case, the separate **SolField** instance (storing the physics of the problem) is not needed since its tasks can as well be executed by the already available **PointCloud** instance (storing the geometry of the problem). This then implies that the **Points** become aware of the solution variables and therefore the same objects are able to hold the geometric and solution variables, thus eliminating one of the list objects. Further memory savings in the conformal and isoparametric case are achieved through the fact that additional bookkeeping associated with the different shape functions on an element by element basis can be avoided.

4.4.3 Father Manager

The association between the different `Contrib` instances in a given `Domain` instance and the appropriate `Integrators`, and therefore also the `FEPairs`, is established by the `FatherManager` class, see Figure 4.10.

This `FatherManager` class stores for each specific type of `FatherElem` instance occurring in the current `Domain` instance an array of concrete `FE-Pair` instances. This array contains as many `FEPairs` as there are `Contrib` entries in the `Equation Discretizer` of the current `Domain` instance. Furthermore, it takes the initiative to select the appropriate `FEPairs` in the arrays depending on the complexity of the `Contribs` it observes in the `Equation Discretizer`, see Figure 4.10. This means that the optimal integration method can be used for each kind of contribution in all situations.

The `FatherManager` class acts then as a well-organized pool of `FEPairs` towards the `Local Discretizers`. The `FatherManager` class is in this respect a typical example of the Flyweight pattern as described in ref. [11]. The Flyweight pattern uses a central data base that is shared by a huge number of objects (`Element` instances in this case) instead of storing that data repeatedly in each individual object. The abstract class representing an array of `FEPairs` acts as a Flyweight while concrete `FEPair` array classes act as a ConcreteFlyweight. The `FatherManager` class plays the role of the FlyweightFactory and simultaneously holds the Flyweight pool (i.e. the central data base). All `Elements` of a `Domain` instance (and they can be numerous) will share that single Flyweight pool. In this manner, dramatic savings can be obtained with respect to memory resources.

4.4.4 Local Discretizer

`Local Discretizers` deal with the discretization scheme to be used in an `Element` of some `Grid`. They actually indicate the way of computing local matrix or residual contributions in the matrix assembly corresponding to the residual update phase of some solution process. The class inheritance depicted in Figure 4.11 shows the different local discretization schemes used in ELEMD.

Note that from a given level in the hierarchy, the classes become parameterized with either `Point` or `Node`, as indicated by the `NODE` labels in Figure 4.11. This parameterization allows one to use all classes both in a conformal, isoparametric context and a fully non-conformal non-isoparametric context. This context is defined by each individual `Domain` instance and cannot be changed at run-time.

The `GenDiscret` class is the actual abstraction of a local discretizer. It is initialized by the `Element` and is used by an `Equation Discretizer` instance, see Section 4.4.1.3. It also holds a reference to a `Functor` which passes the contribution to the appropriate destination.

The class `GenBEM` is used for general Boundary Element Methods. The class `BEM` is a concrete class able to compute contributions for finite computational domains. The class `InfiniteBEM` is a concrete class able to com-

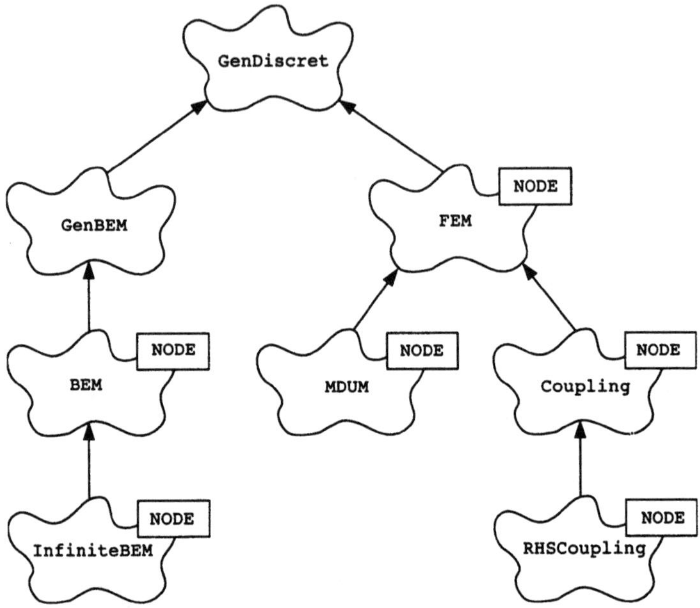

FIGURE 4.11. The local discretizer hierarchy.

pute contributions for infinite computational domains which is one of the most powerful features of the Boundary Element Method.

Any standard Finite Element matrix contributions can be computed using the `FEM` class.

The complexity of MultiDimensional Upwind Methods, also called residual distribution schemes for discretizing convection terms, is encapsulated in the class `MDUM` which is derived from the class `FEM`. The class `MDUM` interacts intensively with the `Equation` abstraction in order to obtain information concerning the local eigenvalues and eigenvectors of the convection term at hand.

The class `Coupling` implements the operators discussed in Section 4.3.3 and is used by `Elements` positioned on a dedicated coupling domain. This is a domain with no thickness that lies in between the boundaries of the two real domains and that deals with the coupling relationship between them, see also ref. [10, 19] and Figure 4.3. The class `RHSCoupling` is derived from the `Coupling` class and is designed to compute contribution to the right hand side (e.g. natural boundary conditions and source terms) of a Finite Element system coupling a heterogeneous set of domains.

4.4.5 Solvers – Time Discretization

`Solvers` model the second step mentioned in the boxed text of Section 4.3. The `Solver` abstraction is at the basis of a hierarchy of concrete classes

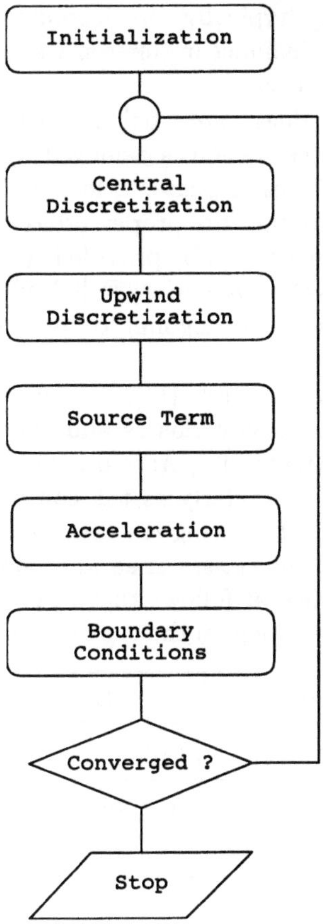

FIGURE 4.12. A typical `Solver` sequence.

that manage the solution of the discretized equations on either a `Tree` instance or a `Level`. The choice between the two is performed through parameterization (C++) of the `Solver` classes.

`Solver` classes exist for explicit time marching, Newton-Raphson iterative solving, etc. All of these basically perform some kind of iteration loop in which very similar sequences take place, see Figure 4.12. Each phase in the sequence is a reference to some abstract functor class and can vary depending on the kind of application. This means that a user can easily plug in new phases while reusing the overall solution scheme.

In the initialization phase, matrices can be allocated or some overall variables can be computed.

In the following step, matrix and/or residual contributions are being computed originating from different discretizations. If a discretization is not present then nothing is done.

The source term phase is responsible for the calculation of the right-hand-side matrix or residual contributions corresponding to some source term (e.g. heat due to chemical reactions, external magnetic force, etc.).

In the acceleration step, typically, the Jacobian of the system matrix is computed as is needed for instance in Newton-Raphson methods. In explicit methods, nothing is done here.

Multigrid acceleration does *not* belong to the acceleration phase of Figure 4.12. Multigrid acceleration is achieved by letting the `Solver` act upon a complete `Tree`. In each phase, a loop is then performed over all `Level` instances while invoking the proper prolongation and interpolation operators. Different sequences can be provided representing different types of multigrid cycles (V-cycles, W-cycles, Full Multigrid cycles, ...). In a sense multigrid acceleration is orthogonal to the acceleration phase depicted in Figure 4.12.

The Composite pattern, see ref. [11], has been used for the design of the relation between the `Solver` classes and the `Tree`, `Domain` or `Level` classes (Section 4.4.2.1) they can operate on. This means that parts of the complete structure behave similarly as the complete structure itself with respect to a `Solver` client. Following the terminology in ref. [11], the class `Level` acts as a Leaf and the classes `Tree` and `Domain` play the role of the Composite. From the above, it follows that `Solver` instances can work in three different modes depending on the type of Composite they operate on:

Composite	**Mode**
`Tree`	multidomain / multigrid
`Domain`	single domain / multigrid
`Level`	single domain / single grid

4.4.6 Boundary Conditions

No sound abstraction for boundary conditions has been established yet in ELEMD. Nevertheless, it can be stated that boundary conditions reduce to modifying matrix or residual contributions such that the boundary conditions are satisfied.

For instance, in case of a Dirichlet boundary condition, a complete row is set to zero except for the position corresponding to the imposed variable of a specific node while the imposed value is placed at the appropriate place in the right-hand-side matrix when using the Finite Element method.

4.5 Critical Remarks

The authors have no formal education in software design and engineering. They have tried to apply the object-oriented paradigm such as they found it in the literature for the specific application domain they are dealing with. A posteriori, they have looked back on their work as a whole and have tried to increase reusability and maintainability by introducing more abstraction.

Object-orientation has not always been glitter and bliss to the authors. The enhanced maintainability has been obtained at the cost of loss of run-time efficiency. In an earlier version of ELEMD a slow-down factor of two was observed compared to a procedural FORTRAN code in the case of a standard TVD upwind Finite Volume computation. Comparing a procedural C++ electrochemical solver (i.e. no classes, no virtual functions) to an object-oriented solver showed about 30 % overhead in ELEMD. These observations seem to confirm the results obtained by Haney in ref. [13].

The current design of ELEMD does not use long (global) arrays to store space coordinates and solution fields that are shared by almost all other components of the system such as it is done in ref. [17]. Avoiding long arrays has the advantage that grid adaptivity becomes much simpler. No global shifts in long arrays are needed when introducing new nodes for instance. It is to be noted on the other hand that long arrays offer much more opportunities for vectorization and for optimal memory access. Storing the solution variables on a node by node basis as in ELEMD causes much indirect memory access.

4.6 Conclusions

A number of abstractions such as they occur in ELEMD have been presented. These are believed to be essential to build any generic and open framework for solving partial differential equations (PDEs) based on grid discretization. In designing those abstractions, the authors found the design patterns described in ref. [6, 11] to be of relevance to the domain of scientific computing. However, the story concerning abstractions and design patterns for generic PDE solvers is far from closed. A consensus concerning a good PDE solver architecture is still lacking in the object-oriented scientific computing community.

Furthermore, the question still remains whether scientific computing is willing to pay in run-time efficiency for shorter development cycles in object-orientation. Fortunately, this price has a tendency to decrease in view of the progress being made in the use of class templates and the fact that with the oncoming ANSI/ISO C++ standard, compiler vendors will be more eager in improving their compiler optimization tools, see for instance ref. [24].

Acknowledgments: This work is partly financed by the European Commission (BRITE/EURAM contracts BRE2-CT92-00179 and BRPR-CT95-0008) and by a research grant from the Flemish Institute for the Promotion of Scientific and Technological Research in the Industry (IWT-SB-941146-DRIESENS).

The authors wish to thank Prof J. Deconinck and all the other members of the Computational Electrochemistry Group at the Vrije Universiteit Brussel for their very good cooperation over the last years. The input of all these different people with their different engineering backgrounds helped them to render ELEMD as generally as possible.

4.7 REFERENCES

[1] René Abgrall. Design of an Essentially Non-Oscillatory reconstruction procedure on Finite Element type meshes. Technical Report 91-84, ICASE, December 1991.

[2] John J. Barton and Lee R. Nackman. Scientific & engineering programming in C++. *C++ Report*, 5(7):26–31, September 1993.

[3] Grady Booch. *Object-Oriented Analysis and Design with Applications*. ISBN 0-8503-5340-2. Benjamin/Cummings, Redwood City, CA, 1994. Second Edition.

[4] Leslie Bortels, Johan Deconinck, and Bart Van Den Bossche. The multi-dimensional upwinding method as a new simulation tool for the analysis of multi-ion electrolytes controlled by diffusion, convection and migration. Part I. Steady state analysis of a parallel plane flow channel. *J. Electroanal. Chemistry*, 404:15–26, 1996.

[5] Bart Van Den Bossche, Leslie Bortels, Johan Deconinck, Sven Vandeputte, and Annick Hubin. Numerical steady state analysis of current density distributions in axisymmetrical systems for multi-ion electrolytes: Application to the rotating disc electrode. *J. Electroanal. Chemistry*, 411:129–143, 1996.

[6] Frank Buschmann, Regine Menier, Hans Rohnert, Peter Sommerlad, and Michael Stal. *A System of Patterns*. ISBN 0-471-95869-7. John Wiley & Sons, 1996.

[7] Robert L. Clay and Kenneth J. Perano. ISIS++: Using object-oriented design for improved component integration and solution adaptibility. In *SciTools'96 International Workshop on Modern Software Tools for Scientific Computing, see* *http://www.oslo.sintef.no/SciTools96/*, Oslo, Norway, 16-18 Sept. 1996.

[8] T.S. Collins and J.C. Browne. MaTRiX++: An object-oriented approach to the hierarchical matrix algebra. In Allan Vermeulen, editor, *Second Annual Object-Oriented Numerics Conf.*, Sunriver, Oregon, April 24-27 1994. Rogue Wave Software.

[9] Johan Deconinck. *Current Distribution and Electrode Shape Changes in Electrochemical Systems.* Springer-Verlag, Berlin, 1992.

[10] Thierry Driesens, Gert Nelissen, and Johan Deconinck. A generalised approach of Boundary Element and Finite Element Methods using object orientation. In *18th World Conference on the Boundary Element Method*, Braga, Portugal, 1996.

[11] Erich Gamma, Richard Helm, Ralph Johnson, and John Vlissides. *Design Patterns : Elements of Reusable Object-Oriented Software.* ISBN 0-201-63361-2. Addison-Wesley, 1995. Second Edition.

[12] Maria Georgiadou and R. C. Alkire. Anisotropic chemical pattern etching of copper foil. III. Mathematical model of reaction and transport phenomena. *J. Electrochem. Soc.*, 141:679, 1994.

[13] Scott W. Haney. Is C++ fast enough for scientific computing? *Computers in Physics*, 8(6):690–694, November 1994.

[14] Ami Harten. High resolution schemes for hyperbolic conservation laws. *J. of Comp. Phys.*, 49:357–393, 1983.

[15] M. Heroux, S. F. McCormick, S. McKay, and J. W. Thomas. Applications of the Fast Adaptive Composite grid method. In S.F. McCormick, editor, *MultiGrid Methods - Theory, Applications and Supercomputing*, volume 110. M. Dekker, 1988. Lect. Notes in Pure and Appl. Math.

[16] Claes Johnson. *Numerical Solution of Partial Differential Equations by the Finite Element Method.* ISBN 91-44-25241-2. Studentlitteratur, Lund, Sweden, 1987.

[17] Hans Petter Langtangen. Diffpack: Software for partial differential equations. In Allan Vermeulen, editor, *Second Annual Object-Oriented Numerics Conf.*, Sunriver, Oregon, April 24-27 1994. Rogue Wave Software.

[18] D. J. Mavriplis and A. Jameson. Multigrid solution of the Euler equations on unstructured and adaptive meshes. Technical Report 87-53, ICASE, Hampton,Virginia, July 1987.

[19] Gert Nelissen, Thierry Driesens, and Johan Deconinck. A generalised approach for the non-linear coupling of Finite and Boundary Element Methods. *J. of Comp. Phys.*, 1996. submitted.

[20] John S. Newman. *Electrochemical Systems.* Prentice Hall, Englewood Cliffs, N.J., 1991. ISBN 0-13-248758-6.

[21] Carl F. Ollivier-Gooch. Multigrid acceleration of an upwind Euler solver on unstructured meshes. *AIAA Journal*, 33(10):1822–1827, October 1995.

[22] Henri Paillère. *Multidimensional Upwind Residual Distribution Schemes for the Euler and Navier-Stokes Equations on Unstructured Grids.* PhD thesis, Université Libre de Bruxelles, Brussels, Belgium, June 1995.

[23] Zhong H. Qiu. *Boundary Element Method for Predicting Current Density Distributions in Electrochemical Systems.* PhD thesis, Wessex Institute of Technology, UK, 1995.

[24] Arch D. Robison. C++ gets faster for scientific computing. *Computers in Physics*, 10(5):458–462, September 1996.

[25] Patrick Vankeirsbilck. *Algorithmic Developments for the Solution of Hyperbolic Conservation Laws on Adaptive Unstructured Grids.* PhD thesis, Katholieke Univ. Leuven, 1993.

5

An Object-Oriented Adaptive Finite Element Code: Design Issues and Applications in Hyperthermia Treatment Planning

Rudolf Beck
Bodo Erdmann
Rainer Roitzsch[1]

ABSTRACT
The finite element code Kaskade has been developed for the solution of partial differential equations in one, two, and three space dimensions. Its object-oriented implementation concept is based on the programming language C++. Adaptive finite element techniques are employed to provide solution procedures of optimal computational complexity. This implies a posteriori error estimation, local mesh refinement and multilevel preconditioning.

One major concept of the implementation is the separation of 'geometric' and 'algebraic' entities. The former ones mainly comprise typical problem-dependent classes like the mesh, the finite element, and the materials. The mesh structure and node distribution may be rather complex due to the nested refinement levels created in the adaptive solution process. By using a global node-numbering strategy we transfer the geometry-based features into algebraic structures; the latter include special sparse matrix classes and vector templates.

Beyond the formal description of the code we present an application arising in a current research project. Here a real-life problem has to be solved, involving the calculation of electromagnetic fields and temperature distributions in human bodies in order to support hyperthermia treatment planning in a clinical environment.

[1]Konrad-Zuse-Zentrum für Informationstechnik (ZIB), D-14195 Berlin, Germany. Email: {Beck,Erdmann,Roitzsch}@zib.de.

5.1 Introduction

In this chapter we are going to explain the philosophy governing the implementation of the finite element code Kaskade. This will be done against the background of a real-life problem: the numerical simulation of electromagnetic fields and the resulting temperature distributions in human bodies. Such calculations are presently carried out in a joint research project for hyperthermia treatment, which is part of a cancer therapy applied at the Rudolf-Virchow-Klinikum in Berlin [8,33].

Like many other problems in science and engineering, these field calculations require the solution of partial differential equations in a region filled with different media of rather complicated shapes. Whereas on certain simply structured domains analytical solutions can be formulated, in our case numerical methods are indispensable. To this end, the domain has to be covered with a grid that restricts the available degrees of freedom to a finite number. The most popular ones of such discretization techniques are finite difference, finite element, and boundary element methods. All of them have certain merits and drawbacks, but we are not going to discuss them here. With regard to our applications, we have decided in favor of the finite element method; its definite advantages are the capabilities of a good geometrical approximation of external and internal boundaries and adaptive mesh refinement, which, quite naturally, may also be the key to powerful multilevel techniques for the solution of the arising equation systems.

In general, adaptive refinement algorithms are driven by error estimators, which have to render an approximation of the local discretization error in the calculated field. Then the finite element grid is improved in regions where the solution does not have the required precision. Such a device may be of crucial importance, as in practice the user of numerical codes does not have sufficient information or knowledge to construct problem-specific discretizations or to check the reliability of a numerical solution.

During the past decades, an immense progress in finite element methods has been reported for many problems in science and engineering, e.g. in mechanics, hydrodynamics, diffusion phenomena and electrodynamics [1, 13,22,24]. As the numerical algorithms and data structures for all of them have a great deal in common, a finite element program should exploit this fact and thus restrict application specific features to a minimum.

An early adaptive finite element code was PLTMG [2], which is written in Fortran and is restricted to linear triangular elements in 2D. Its linear solvers are presently based on the hierarchical basis multigrid algorithm described in [3].

A hierarchical basis preconditioner [34] was introduced in a first version of Kaskade [15]. It was designed for elliptic problems and written in Pascal, whereas the following extensions [17] switched to C, a more widespread and flexible programming language. Further developments, e.g. for parabolic problems [10], lead to separate code versions and thus a comprehensive revision and re-integration became desirable. The decision was, for reasons which are explained in Section 5.2, to create an object-oriented version based on C++.

Meanwhile quite a number of adaptive finite element codes working on unstructured grids has come up; it is not possible to give a comprehensive overview or a qualified comparison here. However, we want to mention the package ug [6] (implemented in C), a prototype in tackling parallelization, and an object-oriented approach based on the C++-library Diffpack, with its focus on the rapid implementation of problem solvers for complex applications [14].

In the following section we are going to give an overview of the basic capabilities and the implementation concept of Kaskade. Section 5.3 will be devoted to hyperthermia treatment planning.

5.2 Code Structure

5.2.1 General Considerations

As mentioned in the introduction, our first aim was to re-integrate separate C-based versions of the 'Kaskade family'. The new version should serve as a generic toolbox both for the development of new algorithms within numerical research projects and for the creation of application-specific codes. Many researchers developing numerical algorithms spend a vast amount of time on writing codes and debugging them, so our main goal was to facilitate 'rapid prototyping' in order to save human labour. Gaining a maximum of computational efficiency was not of primary importance.

We planned to use this toolbox for extensions to certain nonlinear applications, including obstacle problems and phase transitions, and to electromagnetic field calculations. The latter ones generally require complex numbers, which are not directly available in the 'native' languages Pascal and C of Kaskade.

In the former versions of Kaskade geometric and algebraic entities were closely intertwined. Nearly all numerical data were located at the relevant 'sites' within the mesh (points, edges, etc.) and accessed via pointers. At a first glance this seems to be quite a natural approach in a programming language like C. However, we think that it has some drawbacks: numerical algorithms like equation solvers and preconditioners had to be implemented with respect to the geometric data structures they were linked to. This made their design clumsy and unflexible. Furthermore, structures accessed by pointers are often hard to debug and error-prone. So we decided to attempt a thorough decoupling of mesh, finite elements, solvers and preconditioners in an object-oriented environment.

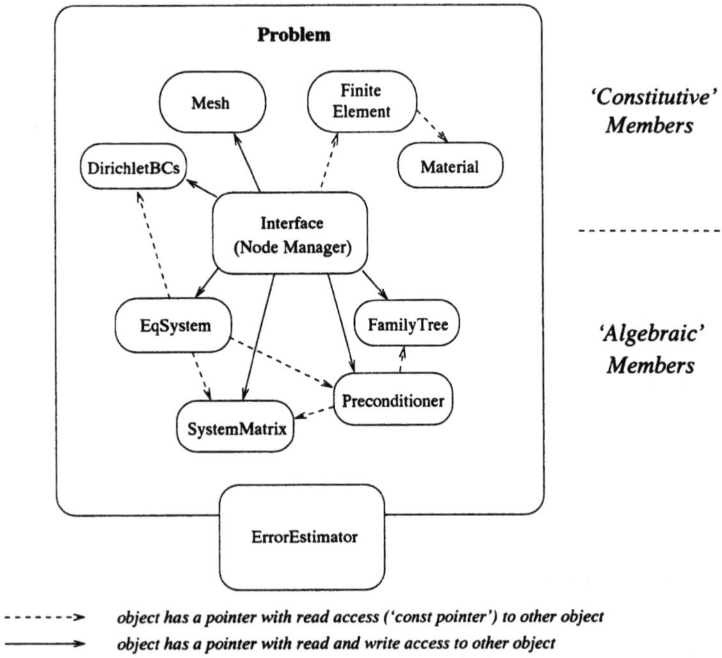

FIGURE 5.1. The basic problem class with its members.

5.2.2 *The Problem Classes*

We wanted to start with class structures that represent a finite element context in a natural way. The main element in the code hierarchy is the abstract base class `Problem`. Its data members (see Figure 5.1) are the 'ingredients' necessary to define and solve specific problem types by a finite element method. More precisely, these members are pointers to abstract base classes; it is the task of a specific problem type to construct appropriate objects taken from the relevant class hierarchies.

The 'constitutive' members for the problem definition are the triangulation (class `Mesh`), the finite element (class `FiniteElement`), the material properties, and the essential boundary conditions (classes `Material` and `DirichletBCs`). The other members of `Problem`, i.e. the algebraic classes, participate in the numerical solution procedures. It is worth noting that the class `ErrorEstimator` must have access to the complete `Problem` itself, as in the most general case an error estimator has to be regarded as a 'secondary' solution step.

We have introduced the class `Interface` to achieve the desired separation of mesh, finite elements, and algebraic entities as mentioned above. The `Interface` performs the 'node management'. Its main tasks are the distribution of global node numbers in the mesh, mapping them onto the local degrees of freedom of the finite elements, and the initiation of data manipulations in other classes.

```
class Problem
{
  public:
    Problem(char* fileName, symmetryType symmetry,
                              int spaceDimension);
           // fileName: input file for the geometry of the domain
    virtual Bool solve(Real reqPrecision);
  protected:
    virtual void assembleGlobalMatrix();
    virtual void assembleElementMatrix(const PATCH* p,
       const Jacobian& Jac, Matrix<Num>& A, Vector<Num>& b) = 0;
    ...
};
```

TABLE 5.1. Basic member functions of the problem classes. `Matrix` and `Vector` are template classes; the term `Num` denotes a real or complex number.

With regard to Figure 5.1 it should be noted that access operations are implemented via virtual function calls and thus do not explicitly depend on the underlying data structures within the classes. In this sense write access means that non-constant member functions may be called to request internal data manipulations. Such calls are restricted to the `Interface` in order to concentrate them within a distinct module. So class hierarchies, like the mesh, finite element, or system matrix may widely be regarded as stand-alone code and can easily be extended, replaced, or extracted for the use within other programs.

All application-specific problem classes are derived from the abstract base class `Problem`. Up to now we have implemented several static problem types (heat conduction, Helmholtz equation, time-harmonic Maxwell's equation, obstacle problems) and parabolic types (transient heat conduction, two-phase Stefan problems and a porous media equation).

The classes for static problems comprise only rather short pieces of code, as they mainly use inherited functions. Transient problems are collected in a common branch, sharing routines which organize the time steps. The nonlinear types (Stefan-, porous media, and obstacle problems) are typical free boundary problems which are solved by monotone multigrid algorithms [21,25] (see the section about preconditioners). They are equipped with an additional member defining their characteristic nonlinear behavior.

To shed some more light on the problem class, we sketch the declaration of its basic member functions in Table 5.1. The only task of the '`main`' program section is to construct a specific problem class, which creates the members depicted in Figure 5.1. Then the virtual function `solve(reqPrecision)` is called, where the parameter `reqPrecision` determines the allowed discretization error of the solution.

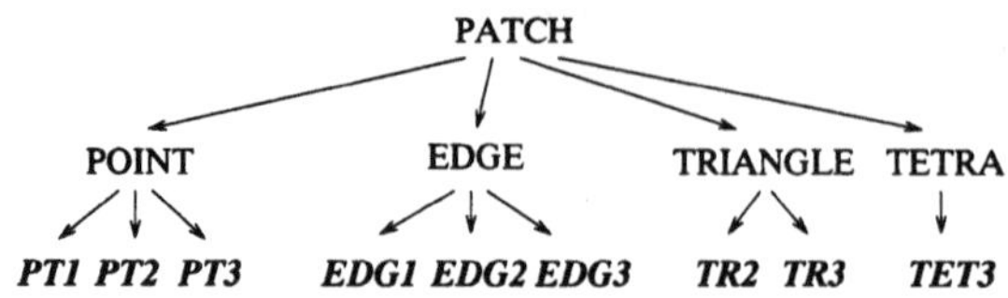

FIGURE 5.2. Classes within the finite element mesh. Numbers added to class names denote the space dimension.

For an elliptic problem like static heat conduction, the procedure `solve` organizes the typical steps of an adaptive solution process (a loop containing global assembling, system solution, error estimation and mesh refinement) by calling the appropriate virtual functions of the members. Other problem classes may have to modify this procedure. Transient problems, for example, have an additional outer loop for the time steps.

The assembling functions are a central point where the different components of the problem have to 'work together'. How this interplay is organized will be shown in the section about the interfaces.

5.2.3 Mesh, Finite Elements, and Interfaces

The Mesh

All our finite element meshes exclusively comprise simplexes, which are derived from the base class `PATCH` (see Figure 5.2). The *'geometric' elements* of the mesh (tetrahedrons in 3D, triangles in 2D and edges in 1D space, as depicted in Figure 5.2) should not be mixed up with the *finite elements*, which define local degrees of freedom and interpolation functions.

All the patches of a mesh are stored in doubly-linked lists to allow the removal or replacement of any entry. This is an essential requirement in the case of adaptive refinement, if irregular elements are created for 'green closures' (see Section 5.2.5). There are separate lists for points, edges, etc.; stepping through the lists is done by abstract iterators.

As the patches are accessed via virtual functions, all operations involving the mesh can be implemented without taking into consideration the specific mesh class and space dimension. For example, a loop through all points of the mesh looks as follows:

```
PATCH* patch;
Mesh()->resetPointIter();
while ((patch=Mesh()->pointIterAll()))  { ... }
```

The only data member in the base class `PATCH` is an integer, the global node number. It is set and read by the interface classes to map the global degrees of freedom to the local ones of the finite elements (see below).

Derived patch classes include information about the patch hierarchies in the mesh: each patch has pointers to all lower-order simplexes (patches) on its boundaries (a 2D-triangle for example has pointers to its edges and points, an edge to its points). Space coordinates are stored in the points.

Thus the elements in the mesh are able to supply sufficient information about their location in space, boundary properties, and the complete set of their global node numbers.

To capture the multilevel structure within the mesh, additional information about the refinement hierarchy is needed. So each element has a pointer to its first 'son' element if it is refined, the other sons follow immediately in the respective lists where they are stored. Mesh refinement schemes are discussed in Section 5.2.5.

Of course, more compact structures can be chosen to save space. On the other hand, if information that is not explicitly stored has to be retrieved algorithmically, this may slow down operations involving the mesh and lead to more complicated codes.

The Finite Elements

The finite element classes define the local shape functions and provide assembling routines for the variational terms of the functional. They do not contain any geometric information with regard to the mesh, so they have to be defined only for a reference element, e.g. on the unit triangle or unit tetrahedron.

We have implemented 'standard' elements with linear, quadratic and cubic nodal basis functions in one, two, and three space dimensions: line, triangle, and tetrahedron (all of Lagrange type with C^0-continuity). For the calculation of electromagnetic fields we have added edge elements of the lowest order, sometimes also called Whitney-forms [11, 26].

The Interfaces

As mentioned above, the interface is responsible for the node management in the mesh. For each finite element type a specific interface class has to be created which 'knows' how to distribute and to read the global node numbers in the patches.

Furthermore, the interface has to call initializing functions in algebraic classes for adjusting their data structures after every refinement step.

Global node numbers are set by simple loops through the relevant patches of the mesh. They have to be retrieved element-wise when the system matrix is assembled; the first example in Table 5.2 shows this member function of the class `Problem`. Here the virtual function `getGlobalNodes` is called before the element data are stored in the global equation system. The procedure for reading global node numbers is usually rather simple. To give a typical example, we have added the code of the interface class `StdElemInt`, which is related to linear Lagrange-type elements. In `getGlobalNodes` first the points of the element are requested and then their node numbers are stored in a vector.

For higher-order elements, also the node numbers of the edges or the inner ones of the element would have to be retrieved in an analogous way. Another situation can arise if there are several degrees of freedom located at a point, edge, etc. Also in this case it is sufficient to store one single

```
void Problem:: assembleGlobal()
{
    PATCH* patch;
    const dim = element->NoOfNodes();
    Matrix<Num> AElem(dim,dim);     // element stiffness matrix
    Vector<Num> bElem(dim);         // element rhs vector
    Vector<int> globalNodes(dim);   // global node numbers
    Jacobian Jac;                   // Jacobian for transform.
                                    //  from reference-element
                                    //    to element in mesh
    Mesh()->resetElemIter();
    while ((patch=Mesh()->elemIterAll()))  // loop for ele-
    {                                      //   ments in mesh
        patch->compJ(Jac);                 // compute Jacobian
        assembleElementMatrix(patch, Jac, AElem, bElem);
        interface->getGlobalNodes(patch, globalNodes);
        eqSystem->store(AElem, bElem, globalNodes);
    }
}


void StdElemInt:: getGlobalNodes(const PATCH* patch,
                                 Vector<int>& globalNodes) const
{
    int i;
    PATCH points[MaxNoOfNodes];
    patch->getPoints(points);
    for (i=1; i<=patch->noOfPoints(); ++i)
      globalNodes[i] = points[i]->getNode();
}


void StaticHeatConduction:: assembleElementMatrix(
                        const PATCH* p,         // element in the mesh
                        const Jacobian& Jac,
                        Matrix<Num>& AElem,     // element matrix
                        Vector<Num>& bElem)     // element rhs
{
     element->assembleEllipticTerm(A, Jac);
     element->assembleSource     (b, Jac);
     element->assembleCauchyBCs (p, A, b);
     element->assembleNeumannBCs(p, b);
};
```

TABLE 5.2. Examples for global matrix assembly, retrieval of global node numbers, and element matrix assembly.

number, as those of further degrees of freedom can be defined uniquely by adding appropriate offsets. Thus is it possible to implement any finite element type within this scheme.

The function `assembleElementMatrix(...)` in Table 5.2 is the member of a specific problem class (static heat conduction in this case) and defines the weak form of the differential equation to be solved. Its task is to call the relevant assembling routines of the finite element. The first function call requests the assembling of the Laplacian; to get contributions from Cauchy- and Neumann-type boundary conditions the `PATCH p` is required to detect the location and the type of an outer boundary.

5.2.4 The Algebraic Classes

System Matrices

Sparse matrix structures play an important role in finite element codes. They are a natural consequence of locally defined basis functions, allowing the storage space for system matrices to be of order N, where N is the total number of degrees of freedom in the mesh. Combined with an 'order-1 solver' for the equation system this yields a solution strategy of optimal complexity with respect to memory space and execution time.

Our sparse matrix classes are derived from an abstract base class `SystemMatrix`. Their most important member functions carry out the following procedures: multiplication with a vector, smoothing operations (like Jacobi- and SOR-relaxation), LU-decomposition, and forward-backward substitution. For the latter operations we have included the Harwell-MA28 direct sparse matrix solver [16] and produced a variant for complex numbers.

A specific feature of finite element matrices is that they are symmetric or symmetrically populated. So we use implementations which try to minimize the memory space for these cases.

Our first type, the class `SparseMatrix`, stores all data in vectors (see Figure 5.3). For a symmetric matrix the upper triangle vector U is identified with L and does not need extra space. L is stored row-wise, whereas U is stored column-wise. Thus for symmetrically populated matrices the column vector contains the column indices for the entries of L and the row indices for those of U. A disadvantage is that the location of all non-zero entries must be known when the matrix is constructed. So an additional routine is necessary to determine the fill-pattern from the mesh structure.

To overcome this drawback, we also have created a slightly different type, the class `MLSparseMatrix`. It maintains lower and upper triangle via linear lists. Each list entry contains the column index of the entry and one or two off-diagonal data terms (one in the symmetric case). This scheme needs extra storage for the pointers in the list. On scalar processors its performance is about 30 % below that of the `SparseMatrix`; as modern workstations are sometimes supplied with vectorization capabilities, we intend to drop this

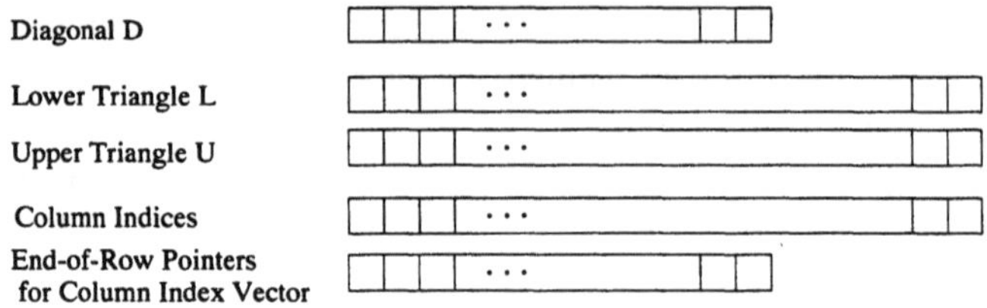

FIGURE 5.3. Data arrangement in the class `SparseMatrix`.

matrix type in the near future. The prefixed letters `ML` indicate that we use it in multilevel-preconditioning. It is supplied with a Galerkin procedure to compute the matrices of coarser grid levels from the fine grid matrix. Entries that do not take part in local smoothing operations of a multigrid cycle may easily be removed (see the section about preconditioners).

Additionally, we have implemented the derived class `BlockSparseMatrix`. Here a nested dissection algorithm is employed to create an algebraic block decomposition; this procedure is completely self-contained and does not require any information about the underlying geometry. The blocks may have a user-specified overlap, which improves convergence considerably [31]. Of course, alternative ordering algorithms exist; for some basic types see [18]. However, nested dissection is rather easy to implement and tends to minimize the size of the boundaries between the blocks; a desirable property for keeping overlapping blocks small. During the node ordering phase, a vector containing its node numbers is created for each block. Then the entries of each block are extracted and collected in vectors forming the input for the MA28 sparse matrix solver. Here an incomplete factorization is carried out with respect to a drop tolerance; i.e. fill-in is suppressed for entries below a specified value. Taking one per cent of the maximum diagonal entry of the matrix is often a good choice for the drop tolerance.

Solvers

Linear equation systems may be handled by a variety of solvers based on Krylov-subspace iterations (see, e.g., [5,20]): conjugate gradients, conjugate residuals, symmetric QMR, CGS, BiCGStab, GMRES, QMR, TFQMR, and CGNR.

For the nonlinear problems a special variant of the Gauss-Seidel iteration has been added (see the remarks about the nonlinear multilevel preconditioner below).

The solvers are functions of the class `EqSystem`, which is equipped with a pointer to the preconditioner. Each solver requests the action of the preconditioner by a simple call to its member function `invert`:

```
preconditioner->invert(e,A,r);
```

where `A` is a pointer to the system matrix, `r` the residual, and `e` the result.

For a multilevel preconditioner this operation usually is one V-cycle. By this construction the codes for solvers are nearly completely independent of the preconditioner classes; any solver can be combined with any preconditioner. Of course, the user has to decide whether a combination makes sense in a specific problem context.

Multilevel Preconditioning and Grid Transfer

Typical tasks in multilevel preconditioning are prolongation and restriction, i.e. the transfer of vectors within the grid hierarchy determined by the refinement levels of the mesh. From an algebraic point of view, these operations are defined by transfer matrices: a vector u_l of level l is transferred to level $l+1$ by the matrix-vector product $u_{l+1} = P_l^{l+1} u_l$. In most cases it is a good choice to use the transpose ${P_l^{l+1}}^t$ for restriction [19].

As transfer matrices are not symmetrically populated, we could not use the sparse matrix classes described above, but created new types (also based on linked lists), which are members of the `FamilyTree`. There is one transfer matrix for each refinement level in the `FamilyTree`; the relevant data are supplied by the interface after every step of mesh refinement.

Beyond a pointer to the `FamilyTree`, each multilevel preconditioner has a stack of sparse matrices (called the smoothers), one for each refinement level. These smoothers are obtained from the system matrix (called the fine-grid matrix in this context) by a Galerkin procedure, which is implemented in the sparse matrix classes and requires the transfer matrices of the `FamilyTree`. The 'canonical' way of computing the smoother A_l of level l is given by $A_l = {P_l^{l+1}}^t A_{l+1} P_l^{l+1}$. Another possibility is to assemble the smoothers of all levels directly, but this procedure is very costly.

The codes for the preconditioners are rather compact. There are two main branches in the class hierarchy: one for multiplicative, the other one for additive versions. The classes on top of these branches organize the multigrid cycles, including calls to the grid transfer functions of the `FamilyTree`. Classes derived from these additionally define smoothing operations. In most cases these are calls to the appropriate relaxation routines of the sparse matrices (e.g. a Jacobi- or SOR-relaxation step).

Another important aspect when working with adaptively refined grids is the 'smoothing pattern'. In general it is not necessary to smooth each node of a certain level. If the operation is restricted to those which are 'new' on this level (i.e. not present on coarser ones), this results in the hierarchical basis multigrid method if a multiplicative V-cycle is employed [3]. In the additive case one obtains the hierarchical basis preconditioner of Yserentant [34]. These types are efficient in 2D space, but deteriorate in 3D. Here the smoothing pattern has to be extended to the immediate neighbors of the new nodes, giving the local multigrid method [29] and the BPX-preconditioner [12] (the latter is related to the additive V-cycle). In Kaskade the user is free to combine different patterns, smoothing operations, and additive or multiplicative cycles.

Beyond preconditioners with standard Jacobi- or SOR-smoothing, we have implemented a 'transforming' version based on the Kaczmarz iteration (see, e.g. [32]), which may be used for indefinite systems.

If the smoothers are constructed as `BlockSparseMatrix`-types (see the section about system matrices), they can be decomposed algebraically, resulting in block-type versions of the respective preconditioners.

A special solution technique, based on monotone multigrid methods, was implemented for the nonlinear free boundary problems [21, 25]. It carries out a nonlinear SOR-iteration on the fine grid with a subsequent linearized multigrid V-cycle. An active set strategy can be used to take into account inactive nodes in contact with an obstacle. Such nodes are identified by a flag, which is set or reset during each SOR-iteration step on the fine grid. As the active set and the data entries of the fine grid matrix may change due to nonlinear corrections, the smoothers for the following V-cycle have to be computed anew immediately afterwards. Here the Galerkin procedure is slightly modified to take into account the active set, which defines 'truncated' basis functions for the coarser grids. Furthermore, prolongation and restriction are 'switched off' for inactive nodes. Loosely speaking, parts of the mesh are dynamically cut out and restored in each iteration. This solver could be implemented without great effort, as only algebraic classes are affected and all extra information about nodal configurations can be handled by simple vectors. Thus the basic technique can be used in various problem contexts dealing with active set strategies.

5.2.5 Error Estimation and Adaptive Mesh Refinement

The estimation of discretization errors serves for two purposes: to check the quality of a numerical solution and to yield local information for mesh refinement. In general, error estimators are problem dependent and so various types have been proposed in the past years (see, e.g., [1]). However, we think that two basic procedures may cover a large field of applications: the stress-projection method of Zienkiewicz and Zhu [37] and hierarchical defect correction schemes [15, 36].

The first method is based on the recovery of stress fields given by derivatives of the solution at superconvergent locations. For polynomials these are the points of Gauss-Legendre quadrature rules. Due to the derivation procedure, the approximation order of this stress field is lower than that of the computed discrete solution and may lack continuity requirements. By an L_2-projection into a suitable basis continuity can be restored and the approximation quality of the stress is improved. A simple comparison of original and improved stress yields the local error estimates. This approach is very attractive if solution and derived stress can be expressed in the same finite element basis, whereas in other cases its implementation may be cumbersome.

On the other hand, hierarchical defect correction schemes require elements of higher order than those employed in the prior solution procedure. The problem is solved once again in this extended finite element basis and the result compared to the previous one. In general this procedure is very costly, but in most cases it is quite sufficient to carry out an approximate solution only for the hierarchical degrees of freedom in the extended basis.

We have implemented basic algorithms for both types, collected in the class hierarchy of the `ErrorEstimator`. The codes are quite compact, they involve standard algebraic classes and procedures. For the second estimator, however, the implementation of hierarchical elements of higher order is necessary; in Kaskade this was done for Lagrange-elements up to third order.

After local error estimation, all elements i of the mesh with errors ϵ_i above a certain threshold η are refined. Usually $\eta = 0.25 \max_i \epsilon_i$ is a satisfactory choice.

Meanwhile, mesh refinement algorithms have become fairly standard. For two space dimensions we have implemented the well-known 'red-green refinement' [4] and the 'bisection scheme' [29]. The latter avoids the transition between non-nested grids as no irregular green closures are created. Both algorithms guarantee stability with respect to the degradation of the interior angles of the triangles.

Our practical experiences did not exhibit a significant advantage for one of these types. So we restricted ourselves to red-green refinement in three space dimensions, where, in our opinion, the implementation is easier. Here the subdivision technique introduced in [9] guarantees stability. However, the slightly simpler (but not necessarily stable) 'shortest edge' strategy [35] did not perform worse in our applications.

5.3 Applications in Hyperthermia Treatment Planning

Hyperthermia treatment as a cancer therapy requires an appropriate temperature distribution in the body of an individual patient: the temperature in the tumor should be above 42°C, whereas the healthy tissue has to remain below this level. In our research project the heating is induced by radio waves which are generated by an antenna array. The goal of numerical simulations is to provide a-priori information for an optimal adjustment of wave amplitude and phase of each antenna.

To achieve the desired temperature distribution, i.e. to focus the heating onto the tumor, field and temperature calculations have to be carried out for all independent excitation states of the antennas. The results yield the input for a subsequent optimization procedure, where a quality functional is minimized by an adequate superposition of the fields.

So the numerical simulation involves two steps: calculation of electromagnetic fields and – based on these – the resulting temperature distribution. In both cases we use a tetrahedral mesh for the discretization of the solution area to get a good resolution of the complicated tissue boundaries. Figure 5.4 shows the torso of a patient and the surrounding antenna array. The model was generated on the base of medical image data (CT-scans), where only a partial representation of the body was available.

In radio-frequency hyperthermia the governing physical laws are Maxwell's equations describing the electromagnetic fields radiated by the antennas (serving as heat sources in the patient's body), and a heat transfer equation modelling the essential transport mechanisms in human tissue.

5.3.1 The Steady-State Solution of Maxwell's Equations

The antennas of our hyperthermia applicator are driven with a fixed angular frequency ω. We neglect initial transients of the fields and are left to solve Maxwell's equations in the frequency domain. For the finite element method we use a variational principle based on the double-curl equation, where an electrical field $\mathbf{E}$ is sought which causes the functional

$$F(\mathbf{E}) = \frac{1}{2} \int \left\{ \frac{1}{\mu} (\nabla \times \mathbf{E})(\nabla \times \mathbf{E}^*) - \omega^2 \epsilon \mathbf{E}\mathbf{E}^* \right\} dV$$

to be stationary. Here ϵ denotes the dielectric constant, μ the permeability. In lossy media, currents can be taken into account by introducing ϵ as a complex quantity. A similar approach, but with a different finite element type, was also used in [27].

Care has to be taken to enforce the physical continuity relations of the fields correctly in numerical calculations. Besides the tangential component of $\mathbf{E}$, also the normal component of the electric flux density $\mathbf{D}$ has to be continuous, as in dielectric media no free charges are present. To fulfill these requirements, we employ edge elements, sometimes also called Whitney forms [11,26] (in mathematical terms, these elements are $H(curl)$-conforming). In conjunction with the variational principle above, such basis functions enforce the required continuity relations.

For the electric field calculations we had to extend the finite element mesh to an area covering the complete arrangement of patient and antenna array. The 'infinite' space surrounding this spherical region is approximately taken into account by second order absorbing boundary conditions [23], which are based on a radial field expansion [7]. Furthermore, the grid spacing has to be small enough to resolve the 'local wave length' λ of the field in the different media, which is given by $\lambda = \frac{2\pi}{\omega\sqrt{\epsilon\mu}}$. To this end, it was necessary to perform a-priori mesh refinement steps preceding the numerical calculations.

The application of the code to the solution of Maxwell's equations was quite straightforward. The main work was to extend the finite element classes by a new one for edge elements and to add the related interface. Furthermore, an new problem type had to be derived and the mesh classes were supplied with a procedure for the a-priori refinement steps described

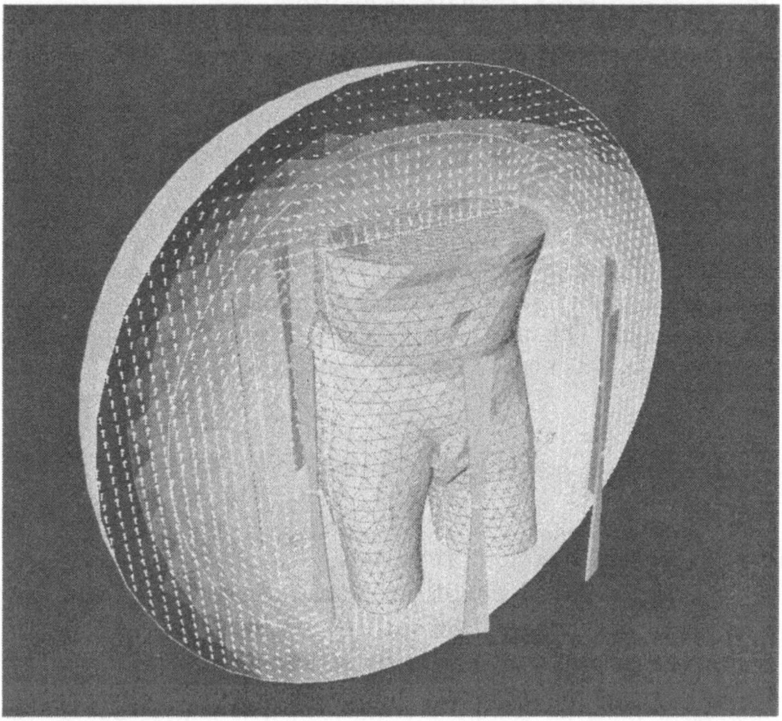

FIGURE 5.4. Electric field generated by the antenna array (initial mesh). Elements covering torso and antennas have been removed.

above. Here, of course, the basic mechanisms of the usual refinement algorithm are used.

For error estimation we chose the stress-projection method ([37], see Section 5.2.5). In our application, by virtue of $\mathbf{H} = -\nabla \times \mathbf{E}/(i\mu\omega)$, the magnetic field $\mathbf{H}$ plays the role of the stress, which may be projected into the same basis as used for $\mathbf{E}$.

As the linear equation systems are indefinite, we employed a conjugate residual solver. Due to the complex geometry and the a-priori refinement steps, already the initial mesh comprises a large number of nodes (some tens of thousands), which increases heavily during the adaptive refinement. As a factorization of the coarse grid matrix proved to be inefficient, we resorted to block-SOR preconditioning with ILU-factorization for the blocks (algebraic mesh coarsening spoils the convergence by aliasing effects). Presently we are still working on more refined preconditioning techniques.

5.3.2 The Bio-Heat-Transfer Equation

The average power Q deposited by electromagnetic waves in a medium with electric conductivity σ is given by

$$Q(x) = \frac{1}{2}\,\sigma(x)\,\mathbf{E}(x)\mathbf{E}^*(x)\ .$$

It is a rather difficult task to model the heat transport in a human body; the main problem is to establish an approximate physical model with adequate parameters. The most widely used approximate description of heat transport is given by Pennes's bio-heat-transfer equation [28]

$$\rho_t\, c_t\, \frac{\partial T}{\partial t} = \nabla(k\nabla T)\ -\ \rho_b\, c_b\, \rho_t\, W\,(T - T_b) + Q\ ,$$

where	ρ_t, ρ_b	:	density of tissue and blood
	c_t, c_b	:	specific heat capacitance of tissue and blood
	T, T_b	:	temperature of tissue and blood
	k	:	thermal conductivity of tissue.

The perfusion W is an empirical coefficient; it is related but not equal to the arterial perfusion based on mass flow rates [30]. We are planning to carry out further investigations with temperature-dependent nonlinear models for W.

For our optimization purposes we are only interested in stationary temperature distributions. So we may set the time derivative on the left hand side in the heat transfer equation to zero and are left to solve a boundary-value problem. The source density Q in the above equations is determined by the results of the electric field calculations.

For the temperature calculations we could use a standard problem type of Kaskade. An extension had to be made with respect to data handling; we created a new class for reading the output data of the field calculations and for interpolating them onto a different mesh, as the refined grids do not match for both solution steps.

The initial mesh being quite large, we employed linear Lagrange-type elements; the equation systems were solved with a conjugate gradient routine accelerated by a multilevel preconditioner with SOR-smoothing. Due to the (dominating) perfusion term, which has the structure of a mass matrix, the systems are quite well-conditioned and an exact coarse grid solver was not needed. For error estimation we used a hierarchical defect correction scheme with quadratic Lagrange-type elements (see Section 5.2.5). Solving the extended quadratic problem by one single Jacobi relaxation sweep was quite satisfactory.

5.4 Concluding Remarks

We think that our approach of separating mesh and algebraic classes not only yields a flexible finite element code, but also naturally reflects abstract viewpoints, which consider multilevel splittings not directly related to the underlying geometry, but rather in a general discrete function space. The language C++ proved to be helpful in realizing our concept; we were able to implement our applications rather quickly within a basic framework by some appropriate extensions of class hierarchies.

In our opinion such hierarchies provide a good basis for code organization. They form kind of a library both in an 'internal' and in an 'external' sense: derived classes can use non-private member functions of their ancestors, which reduces code duplication. On the other hand, the public member functions of the classes form a well-defined interface for the user. So we could extend or even replace class hierarchies rather efficiently.

We also appreciated the concept of virtual functions and used it frequently; typical examples are assembling routines for element stiffness matrices or the smoothing operations of multilevel preconditioners. In modular languages such 'generalized' function calls are often implemented via function pointers; a technique, which we regard as less clear and more liable to errors.

Clearly, we did not consider numerical performance to be the ultimate goal; to improve the execution speed, low level routines should be coded without using object-oriented features. Due to our experience, portability still remains a sore point of C++.

Beyond activities mentioned in the previous sections, we are planning to work on timestepping algorithms for wave equations and chemical reaction-diffusion systems in the near future.

A prototype of the code with restricted capabilities (no solver for Maxwell's equations and no block decomposition) is available via Internet:

`ftp://ftp.zib.de/pub/kaskade/3.x`

Acknowledgments: Part of this work was supported by the Deutsche Forschungsgmeinschaft, Sonderforschungsbereich 273 ("Hyperthermie: Methodik und Klinik").

5.5 References

[1] I. Babušhka, O. C. Zienkiewicz, J. Gago and E. R. de A. Oliveira, editors. *Accuracy Estimates and Adaptive Refinements in Finite Element Computations.* Wiley, 1986.

[2] R. E. Bank. PLTMG 7.0 – A Software Package for Solving Elliptic Partial Differential Equations. Users' Guide 7.0, SIAM, 1994.

[3] R. E. Bank, T. Dupont and H. Yserentant. The hierarchical basis multigrid method. *Numer. Math.*, 52, pp. 427-458, 1988.

[4] R. E. Bank, A. Sherman and A. Weiser. Refinement algorithms and data structures for local mesh refinement. In R. Stepleman et al., editor, *Scientific Computing.* IMACS / North Holland, Amsterdam, 1983.

[5] R. Barret, M. Berry, T. Chan, J. Demmel, J. Donato, J. Dongarra, V. Eijkhout, R. Pozo, C. Romine and H. van der Vorst. *Templates for the Solution of Linear Systems: Building Blocks for Iterative Methods.* SIAM, 1993.

[6] P. Bastian, K. Birken, K. Johannsen, S. Lang, N. Neuss, H. Rentz-Reichert and C. Wieners. UG – a flexible software toolbox for solving partial differential equations. Technical report, Institut für Computeranwendungen III, Universität Stuttgart, 1996.

[7] A. Bayliss, M. Gunzburger and E. Turkel. Boundary conditions for the numerical solution of elliptic equations in exterior regions. *SIAM J. Appl. Math.*, 42, pp. 430-451, 1982.

[8] R. Beck, H. C. Hege, M. Seebass, P. Wust, P. Deuflhard and R. Felix. Adaptive finite element codes for numerical calculations in hyperthermia treatment planning. In C. Franconi, G. Arcangeli and R. Cavaliere, editors, *Proc. of the 7th Int. Congress on Hyperthermic Oncology, Rome, Italy*, 1996.

[9] J. Bey. Tetrahedral grid refinement. *Computing*, 55, pp. 355-378, 1995.

[10] F. Bornemann. An adaptive multilevel approach to parabolic equations I. *Impact of Comput. Sci. Engrg.*, 4, pp. 1-45, 1992.

[11] A. Bossavit. Whitney forms: a class of finite elements for three-dimensional computation in electromagnetism. *Inst. Elec. Eng. Proc., Part A*, 135, pp. 493-500(8), 1988.

[12] J. H. Bramble, J. E. Pasciak and J. Xhu. Parallel multilevel preconditioners. *Math. Comp.*, 55, pp. 1-22, 1990.

[13] S. C. Brenner and L. R. Scott. *The Mathematical Theory of Finite Element Methods.* Texts in Applied Mathematics, Vol. 15, Springer, 1994.

[14] A. M. Bruaset, E. Holm and H. P. Langtangen. Increasing the efficiency and reliability of software development for systems of PDEs. In E. Arge, A. M. Bruaset and H. P. Langtangen, editors, *Modern Software Tools for Scientific Computing*, pages 247–268. Birkhäuser, 1997.

[15] P. Deuflhard, P. Leinen and H. Yserentant. Concepts of an adaptive hierarchical finite element code. *Impact of Comput. Sci. Engrg.*, 1, pp. 3-35, 1989.

[16] I. S. Duff. MA28 – A Set of FORTRAN Subroutines for Sparse Unsymmetric Linear Equations. Technical report, AERE–R.8730, Harwell, 1980.

[17] B. Erdmann, J. Lang and R. Roitzsch. KASKADE Manual Version 2.0: FEM for 2 and 3 Space Dimensions. Technical report, Konrad-Zuse-Zentrum für Informationstechnik Berlin, 1993.

[18] A. George and J. W. -H. Liu. *Computer Solution of Large Sparse Positive Definite Systems.* Prentice Hall, Englewood Cliffs, New Jersey, 1981.

[19] W. Hackbusch. *Multi-Grid Methods and Applications.* Springer, Berlin, 1985.

[20] W. Hackbusch. *Iterative Solution of Large Sparse Linear Systems of Equations.* Springer, Berlin, 1993.

[21] R. H. W. Hoppe and R. Kornhuber. Adaptive multilevel methods for obstacle problems. *SIAM J. Numer. Anal*, 31(2), pp. 301-323, 1994.

[22] J. Jin. *The Finite Element Method in Electrodynamics.* John Wiley & Sons, New York, 1993.

[23] V. N. Kannelopoulos and J. P. Webb. A numerical study of vector absorbing boundary conditions for the finite element solution of Maxwell's equations. *IEEE Microwave Guided Wave Lett.*, 1, pp. 325-327, 1991.

[24] H. Kardestunger and D. H. Norrie, editors. *Finite Element Handbook.* McGraw-Hill, New York, 1987.

[25] R. Kornhuber. *Adaptive Monotone Multigrid Methods for Nonlinear Variational Problems.* Wiley-Teubner, 1997.

[26] J. C. Nedelec. Mixed finite elements in $\mathcal{R}^3$. *Numer. Math.*, 35, pp. 315-341, 1980.

[27] K. D. Paulsen, X. Jia and J. M. Sullivan. Finite element computations of specific absorption rates in anatomically conforming full-body models for hyperthermia treatment analysis. *IEEE Trans. Biomed. Engrg.*, 40(9, pp. 933-945), 1993.

[28] H. H. Pennes. Analysis of tissue and arterial blood temperatures in the resting human forearm. *Journal of Applied Physiology*, 1, pp. 93-122, 1948.

[29] M. C. Rivara. Design and data structure of a fully adaptive multigrid finite element software. *ACM Trans. Math. Software*, 10, pp. 242-264, 1984.

[30] R. B. Roemer and A. W. Dutton. A new tissue convective energy balance equation for predicting tissue temperature distributions. Presented at: ICHO VII, Rome, Italy, April 9-13, 1996; University of Utah preprint, Salt Lake City, 1996.

[31] B. Smith, P. Bjørstadt and W. Gropp. *Domain Decomposition. Parallel Multilevel methods for Elliptic Partial Differential Equations.* Cambridge University Press, Cambridge, 1996.

[32] G. Wittum. On the convergence of multigrid methods with transforming smoothers. *Numer. Math*, 57, pp. 15-38, 1990.

[33] P. Wust, M. Seebass, J. Nadobny, P. Deuflhard, G. Mönich and R. Felix. Simulation studies promote technological development of radiofrequency phased array hyperthermia. *Int. J. Hyperthermia 12*, pp. 477-494, 1996.

[34] H. Yserentant. Hierarchical bases give conjugate gradient methods a multigrid speed of convergence. *Appl. Math. and Comp. 19*, pp. 347-358, 1986.

[35] S. Zhang. Multilevel Iterative Techniques. Ph.D. Thesis, Pennsylvania State University, 1988.

[36] O. C. Zienkiewicz, J. P. de S. R. Gago and D. W. Kelly. The hierarchical concept in finite element analysis. *Computers and Structures*, 16, pp. 53-65, 1983.

[37] O. C. Zienkiewicz and J. Z. Zhu. The superconvergent patch recovery (SPR) and adptive finite element refinement. *Comp. Meth. Appl. Mech. Eng.*, 101, pp. 207-224, 1992.

6

On the Efficient Implementation of Multilevel Adaptive Methods

U. Rüde[1]

ABSTRACT
Adaptive multilevel methods are difficult to implement, because their basic software components are strongly interrelated and thus the modularization is nontrivial. Program design is further complicated by efficiency considerations. These issues are addressed in the abstract mesh concept and its implementation in a patch-adaptive multigrid program.

6.1 Introduction

The efficient solution of partial differential equations (PDE) requires a combination of different components:

- an accurate discretization
- adaptivity
- a fast solver.

The research efforts of the past decades have brought tremendous progress in these areas.

6.1.1 Mathematical Issues in Efficient PDE Solvers

Discretization techniques, like the finite element method (FEM), are now well-established mathematically and in practice. Efficient and reliable error estimators based on an FEM discretization can be exploited to construct systematic mesh refinement strategies.

Many practical problems require adaptivity. Structural singularities caused by re-entrant corners or jumps in the material coefficients or problems with sharp transitions require locally increased resolution. In all these

[1]Department of Mathematics, University of Augsburg, Universitätsstrasse 14, D-86159 Augsburg, Germany. Email: ruede@math.uni-augsburg.de.

cases (unless explicit knowledge of the solution is available and can be used appropriately) the meshes must be *locally* adapted. Uniform meshes permit only a uneconomic approximation of the problem.

Even with adaptive meshes, the resulting discrete systems are so large that the fastest of the available solution techniques must be used. Among the most attractive solvers are *multigrid algorithms.* With multigrid it is possible to solve discretized elliptic partial differential equations (PDE) in typically below hundred floating point operations per unknown. Multigrid can also be implemented such the program needs just little more memory than what is required to store the data and the solution. In many examples such a solver is several orders of magnitude cheaper than, say, a direct solver based on the banded matrix structure.

The operation count of hundred operations per unknown is conservative for simple problems, say like Laplace's equation in a domain of simple shape, where special problem features can be exploited to construct even more efficient methods. The hundred operations per unknown are estimated under the conditions that the typical multigrid efficiency with convergence rates of around 0.1 per V-cycle can be obtained and each V-cycle applies the equivalent of two or three elementary relaxation steps on each level. Each such relaxation is equivalent to a matrix-vector multiplication with the stiffness matrix. In many cases, including systems of PDE and problems with variable coefficients, a relaxation can be implemented with 10 – 20 operations, provided that the problem is discretized carefully. With a nonlinear multigrid method one may even solve (not too badly) nonlinear problems within that complexity estimate.

Of course, general problems may be more expensive to solve. Difficult problems for which the 100 operations per unknown are (much) too optimistic include those with strong nonlinearities and many convection dominated systems of PDE. Note however, that the complexity estimate may also be spoiled by certain design decisions. For example, a discretization of the Laplacian in 3D with trilinear finite elements will lead to a stiffness matrix with 27 entries per row, whereas a seven-point difference stencil can provide the same accuracy at lower cost. Clearly it is also essential to use a multigrid solver or one with similar computational efficiency.

6.1.2 Computer Related Issues in Efficient PDE Solvers

The practical solution of PDE has also greatly profited from the continuing evolution of computer technology and architecture. Even low cost personal computers can easily compete in speed with what was a high end super-computer just a decade ago. As it seems, the average annual speed improvement for microprocessors of about 80% [6] will continue for some more years and further progress may be obtained by parallel computers.

Presently, modern workstations reach a peak-performance of up to 10^9 Operations per second (1 GFlop). Thus, from the operation count alone, systems with a million unknowns, say arising from the discretization on a grid equivalent to a cube with 100 × 100 × 100 nodes should be solvable in approximately a 0.1 seconds on a high end workstation. Practical experience, however, shows that this is *much* too optimistic. Even well-written programs are usually several orders of magnitude slower. We will discuss some of the aspects of machine-efficient implementations of numerical software in Section 6.4 below.

It is also interesting to reflect how little the basis of numerical software development has changed. Much of numerical programming is traditionally done in a plain procedural style and the programming methodology is influenced by the shortcomings of the older FORTRAN versions. More recently, however, software aspects of numerical applications have gained wider scientific interest.

Numerical programmers are faced with an increasing complexity of the algorithms. Traditional software development techniques are beginning to pose severe restrictions on the implementation of these advanced algorithms. Of course, any problem can in principle be coded by traditional techniques, given enough man-power to do so. However, since man-power is probably the most limiting resource, new approaches are sought to handle the software complexity, and a part of the numerical analysis community has begun to explore the suitability of advanced techniques from computer science for numerical applications. Projects like Diffpack [5], KASKADE [3] or PETSc [1] that are discussed in this proceedings volume are typical examples where these considerations have been one of the driving forces.

This paper will discuss some of the issues arising in the construction of efficient and flexible PDE solver software. Section 6.2 will outline the general structure of multilevel adaptive methods and discuss why a good software design is non-trivial. Section 6.3 will then present concepts for data abstraction that may useful when implementing multilevel programs, and Section 6.4 will focus on efficiency aspects. In Section 6.5 some programming techniques for the abstract mesh structures will be discussed, and Section 6.6 will briefly introduce the patch-adaptive multigrid method.

6.2 Multilevel Implementations

Figure 6.1 shows the general logical structure of many linear or nonlinear adaptive multilevel codes. A multilevel *solver* is employed to provide an approximate solution which is in turn used in an *error estimator* whose results will then drive the *mesh generation* module. With the new grid, another solver step may be started, new errors estimated, and new meshes generated, etc., until a satisfactory overall accuracy is obtained.

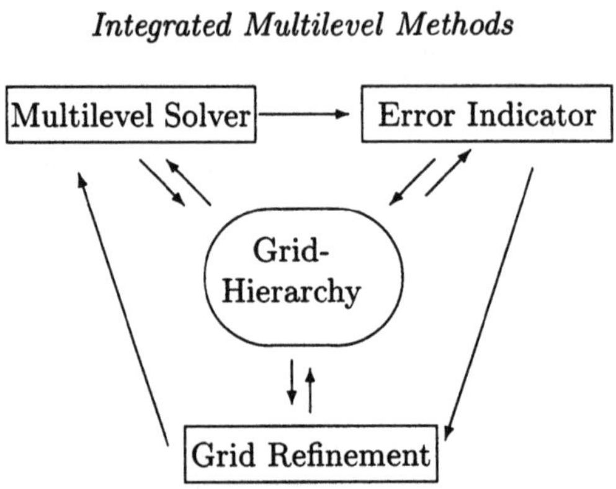

FIGURE 6.1. General structure of multilevel methods.

Since the solution is performed by an iterative multilevel method, the three modules cannot be kept independent. The power of the solver depends essentially on

- a sequence of progressively coarser approximation spaces (often nested) to build the coarse grid correction (this is the key feature of any multilevel algorithm).
- a good *initial guess*, that is the solution from the previous refinement step: Only a *nested iteration* is capable of providing errors with accuracy on the level of truncation errors within a fixed number of floating point operations per unknown.

These requirements clearly impose restrictions on each of the basic modules and make their interface more complex than in a non-multilevel setup. For example, the solver must know about all levels in the discretization, and must usually be adapted to the discretization technique. The mesh generator cannot simply work from a geometric description of the domain, but it must interface to the error estimator and it must conform to the meshes of previous refinement steps such that the overall structure is suitable for a multilevel method. This makes the design of software considerably more difficult and prohibits the use of numerical PDE solvers developed along the classical vertical modularization:

Grid generation → Discretization → Solver → Post-processing

Unfortunately, it is the very interaction between the mesh levels and the modules that make multilevel methods efficient.

Multilevel codes are often developed for specialized situations only. For example, the classical multigrid codes developed in the early 1980s, like MG00 [7], treat scalar equations on logically rectangular grids only. Here the basic mesh structure is coded into the program. Other multilevel codes like PLTMG [2] are restricted to the treatment of scalar equations.

In this context and in this paper we focus on software aspects and how data structures should be designed so that there are no constraints on what type of equations can be *represented* within the software system. Note that this question addresses only the software and data structures, and not the algorithms. Of course (but not in this paper) also the algorithms must be studied since eventually they will determine what kinds of equations can be *solved*.

More advanced multilevel algorithms, like KASKADE, or PLTMG are often based on unstructured mesh finite element discretizations. As we will discuss below (see Section 6.4), these techniques may result in severe efficiency problems.

In summary, the general problem seems to be that traditional software is not designed flexibly enough to handle complex situations without programming everything anew. On the other hand, we will show that the programming techniques that are currently being used to increase the flexibility almost inevitably lead to a severe performance penalty.

Of course, the flexibility and quality of a software design is not a well-defined mathematical term and thus it may be necessary to first reflect on what we wish to accomplish. For the purposes of this paper, we will focus on just a few exemplary criteria that we believe are important to understand why multilevel PDE programs are limited in their applicability.

One issue in software design is the *locality* of code. Programs are easier to write, read, and maintain, when

> *related data structures and algorithms are textually represented close together.*

Subroutines partly serve this purpose since they liberate the programmer from the order in which the program statements are executed. Thus code may be placed where it belongs logically, and not where it is executed. Object oriented languages additionally permit the programmer to put data structures and algorithms together in a class. (Standard) Pascal in contrast, forces the programmer to describe data structures and the corresponding subroutines at places that are potentially far apart. In this sense, Pascal is not supporting good program designs.

In our view, another main objective of good program design is to

> *avoid multiply replicated code fragments.*

A piece of code that is repeated several times would probably be better represented on a more abstract level. Multiple occurrences of a code fragment disperse a central idea all over the program rather than making it explicit by stating it at a single place. With code replications, changes are difficult, and — worse, since the abstract concept is not made explicit — it may not even be realized that the program could be generalized.

Generalizability of software is often the central issue to the efficient solution of real life problems. In just too many cases, software must be developed from scratch because none of the other programs is directly applicable or extensible to the problem at hand. The typical case may be that though programs are available that can solve the right kind of equations, they are unable to cope with the shape of the domain and therefore thousands of lines of code must be redeveloped.

To improve flexibility and thus the chances for code-re-usability, it is essential to avoid code replications. To deal with this issue, the most common technique is again to use subroutines to encapsulate the replicated piece of code and to make it callable from as many places as required. Using parameterized functions, the flexibility of this concept can be improved. Technically, subroutines can be implemented by copying the code to the calling point and substituting parameters, when necessary. This is called *inlining* and is similar to textual expansion as it can be achieved by macro preprocessors. In this case, subroutines do not create any overhead for the calling mechanism, but they also do not have the effect of reducing the machine code length. Regular subroutine calls create overhead, and care should be taken not to use them at too low granularity when inlining would be more advisable.

Subroutines address the modularity in the *algorithmic structure*, but modularity may also be needed in the *data structures*. When the same pattern of data declarations (e.g. declaration of variables or subroutine parameters) is replicated, the typical modularization device is to collect these variables in a *structure* (or *record* in Pascal). Structures are not available in all programming languages, and many FORTRAN77 programs show how this affects the software design. Larger FORTRAN77 programs tend to have long subroutine parameter lists, repeating over and over the same pattern of data types and declarations, making changes awkward and error prone.

Object oriented programming can additionally address the case, when replicated code patterns encompass both data and algorithms, since *classes* contain both the data and the algorithms. Instances of a combined data/algorithm pattern are then termed *objects* and can be invoked from a given class at as many points as needed. Clearly, other elements of object oriented programming, like inheritance and polymorphism, can be used for the same purpose: try to avoid having the same program pattern occur at too many places.

6.3 Data Abstraction Concepts for Multilevel Adaptive Methods

In light of the general issues of the previous section, we state and discuss three special criteria that we will use to discuss the software design for multilevel PDE solvers.

Question 1: Is the program flexible in the type of equations to be solved, and if yes, what portions of the program need to be changed when the number of state variables changes, in particular when one wants to extend the program from a scalar PDE to a system of PDEs?

This issue is relatively simple to deal with. Instead of fixing the data type `double precision` or `float` for the unknown (and right hand side, etc.) we must introduce abstractions. The solution value associated with a node of the mesh becomes a data type `Value` that can be defined as `double precision` in the case of scalar problems. Additionally, we need a data type `RhsValue` for the right hand side and `Connector` for the describing the entries in the stiffness matrix. In the case of a scalar PDE all these are defined as `double precision`, however, in the case of systems of PDE they become vectors and matrices of corresponding dimension. Of course it is especially nice to have a language like C++ that permits the overloading of operator symbols (`+ - * /`) for operations between, say, `Values` and `Connectors`. Thus, in the case of a scalar PDE, a Connector and a Value are multiplied as two floating point numbers, but when the `Connectors` are matrices and `Values` are vectors, the operation symbol `*` represents a matrix-vector multiplication. Thus, except the changed data declarations, the code looks the same as in the scalar case.

Of course this concept does not automatically result in an efficient algorithm for all systems of PDE, when the algorithms used in the scalar case are mathematically not suitable for systems of PDE. Then, of course, also the algorithms must be generalized. However, we have succeeded in *representing* the problem with the same software, making changes only where natural (in the description of `Value` and `Connector` classes), not throughout the program in each traversal of the grid. This simplifies a programmer's task and frees him or her to focus on the necessary mathematical changes and not to be overwhelmed by technical changes.

The next two questions are interrelated.

Question 2: Is it possible, and if yes, what has to be changed when the program is extended from two dimensions to three?

Question 3: Is it possible, and if yes, what must be changed when the same program shall once be employed on a simple uniform rectangular grid, and when it is applied to fully unstructured meshes?

Questions 2 and 3 are motivated by an interesting observation. Despite their complexity, the core of a multigrid algorithm is essentially independent of the special mesh structure. Most of the processing (relaxation, residual computation, interpolation, and restriction) is done in loops of the *abstract form*

```
For all nodes in the mesh
    For all neighbors of the node
        Compute some weighted average from neighboring values
```

which in a *concrete multigrid algorithm* may look like

```
Real g[n,n];
    ⋮
// Here is a Gauss-Seidel-Method for the Poisson-Eq:
For i= 1 to n
   For j= 1 to n
      g[i,j]= 0.25*(g[i-1,j]+g[i-1,j]+g[i,j-1]+g[i,j+1])
```

In any conventional multigrid implementation, this loop construction or an equivalent code fragment is repeated over and over again, so that a change requires almost a complete redesign of the program. Several approaches have been taken to address this problem.

Many advanced implementations of multilevel methods have discarded (logically) uniform data structures because they seem to be inflexible. These codes are based on unstructured grids, and, — at the expense of considerable extra software complexity — these codes can deal with all kinds of mesh structures that are special cases. Clearly, uniform meshes can be dealt with in such programs as special cases.

From the point of view of functionality the problem seems to be solved. In terms of program efficiency, however, unstructured grids present a significant performance penalty. On the basis of nodes processed per second, an unstructured code may be a factor of 10 or more slower than a structured grid code. We therefore need to discuss the factors that determine the run-time efficiency on current computer architectures.

6.4 Efficiency

Let us first return to the problem of representing systems of PDE addressed in Question 1 in Section 6.3. An alternative approach to deal with a system of PDEs would be not to use structures (records), but a global problem parameter `Dimension` (of type integer) and arrays of length `Dimension`. The parameter `Dimension` can now be changed from 1 (for a scalar PDE) to whatever the number of PDEs in the system is. This approach would be used typically in a FORTRAN code and results in a program where the operations for a node would be implemented by loops from 1 to `Dimension`.

Let us additionally assume that we wish to use an algorithm, where the loops of length `Dimension` are necessarily the innermost ones. This is the case in many problems and in particular for systems of reaction-diffusion equations (as they occur in the multi-group neutron diffusion simulations) where the coupling of the unknowns in each node is strong compared to the diffusion to neighboring nodes. Here an approach to solve the diffusion equation for the components independently and then to iterate between the different components is much slower than a genuine direct multigrid approach to the system that would better reflect the physical nature of the

Innermost loop	32	16	8	4	2	1
Execution time	1.01	1.01	1.34	1.80	3.26	5.40

TABLE 6.1. Execution times (for an Indigo2, R4400, 200 MHz) with constant number of passes, varying passes in innermost loop.

problem. Here the unknowns associated with a node form a block and are relaxed *simultaneously* by inverting the block. For the software structure this means that the innermost loops must be those that run from 1 to `Dimension`.

Unfortunately, this results in an unpleasant (and not well-known) efficiency problem. As can be easily tested, the efficiency of a loop decreases drastically for *short* loops. This is shown in Table 6.1, where a simple nested loop is executed such that the total number of passes is kept fixed, while the upper limit of the innermost loop varies from 1 to 32. Of course, a performance deterioration is only noticeable when the loop is executed many times, as in our case, and when the compiler finds it impossible to optimize it appropriately, as is typically also the case. Table 6.1 shows that while loops of length 32 and 16 are still equally fast, the performance starts to deteriorate quickly when the length gets to be 4 or lower. When the innermost loop has an upper limit of one, corresponding to the representation of a scalar PDE, then the overhead is a factor of more than 5. Thus an attempt to use this technique to implement a general code must be expected to result in a quite significant overhead spoiling the performance.

Even more overhead results when the treatment of each node is encapsulated in subroutines. A straightforward multigrid implementation would for example use a subroutine for relaxing a single unknown by the block-relaxation step described above. Function calls on a too low granularity level that are called millions of times will spoil the performance, and this effect is more dramatic the more advanced the CPU architecture is and the more it depends on internal parallelism, pipelining and super-scalar instruction issue. This would be even more severe when virtual functions were used. Virtual functions cannot be inlined, so there is little chance to optimize when they occur in a tight loop.

It is also interesting to analyze the situation from a more fundamental perspective. What we must accomplish is to make the dimension of the PDE an exchangeable parameter of the software. When we use structures of fixed size to represent a node with its variables, then the instantiation of the parameter is performed at compile-time. In other words, the compiler knows everything about the data structure of a node and can exploit this information to generate efficient code.

In the other case when arrays and loops are used to represent the unknowns at a node and process them, this information is more hidden. With present optimizer technology it is likely that the compiler cannot deduce enough information about the loops to optimize them appropriately. The best optimization would involve complete inlining and unrolling of the innermost loops, thus resulting in a straight sequence of machine language operations for the node-operation. Presently available compilers, however, will not usually perform this optimization, but will treat `Dimension` as a run-time variable and will thus generate general purpose loop code.

In some sense, the difference between these two scenarios is similar to that of a compiler versus an interpreter. When structures are used, efficient machine code can be generated, but in the case of loops, the same information (namely that the loop ranges from 1 to, say, 1000) must be re-analyzed and re-interpreted millions of times within a single program run.

Next, let us return to the issues involved with unstructured meshes and discuss why they lead to such large overhead on many advanced CPUs. First, note that unstructured meshes result in inefficient short inner loops since, e.g., the number of entries in each row of the stiffness matrix is variable, and thus a (short) loop must be used with variable limits. For a uniform mesh, the same operation can be implemented by just a fixed sequence of elementary machine instructions without any loop.

Additionally, unstructured meshes necessarily require indirect addressing and dynamic memory allocation/deallocation and thus may lead to memory fragmentation. On a uniform grid, simple arrays can be used to represent the data structure. Nested loops and straightforward indexing can be used to access the data. All together, we have observed performance penalties of approximately one order of magnitude, and in some cases up to two orders of magnitude, when running an unstructured mesh code to simulate a (logically) uniform grid.

Thus the penalty associated with unstructured meshes is in a range that cannot be simply neglected. The penalty may be so large that it compensates all advantages of the more flexible mesh structure. In many practical examples, a significant number of nodes can be saved by adaptive meshing. However, we are not aware of many cases where adaptivity results in a savings of 99% of the unknowns compared to a uniform mesh — but exactly that would be necessary to compensate for an efficiency loss of a factor 100 as it may be caused by a poor implementation of unstructured meshes.

While it is of course debatable how much unstructured codes can be improved by more careful and efficiency-aware implementation, there is one fundamental criticism that remains. Unstructured meshes that simulate uniform meshes, result in the run-time interpretation of information that could in principle (and should) be pre-evaluated at compile time. Thus, if uniform meshes can be used for the problem (and it is quite likely that this is the case at least for parts of the domain) then we expect that they are the superior alternative in terms of run time.

How much more efficient uniform grids are, depends on the hardware and on the system software. The general trend in computer hardware to more internal parallelism of CPUs will increase the advantage of uniform data structures even further.

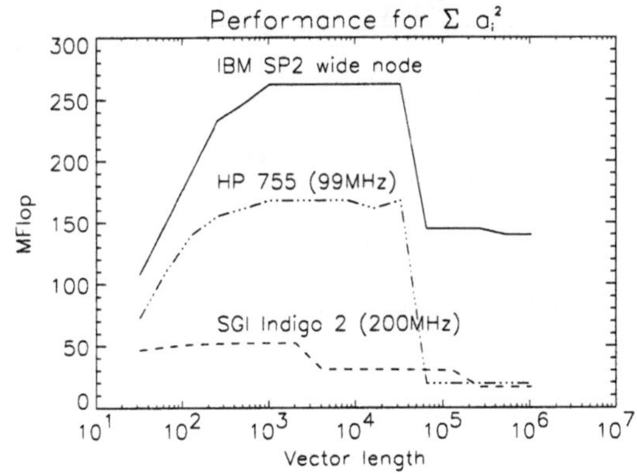

FIGURE 6.2. $\sum_1^n a_i^2$ for $n = 32, 64, \ldots, 1048576$.

We next demonstrate the relevance of these issues by presenting a simple performance benchmark. To clearly demonstrate the basic problem, we do not consider complex adaptive PDE algorithms, but instead focus on a simple computational linear algebra kernel, as it could be used as part of an adaptive PDE code. Our example is the computation of a scalar product (actually just a norm)

$$s = \sum_{i=1}^{n} a_i a_i \tag{6.1}$$

that we modify to (a genuine scalar product now)

$$s = \sum_{i=1}^{n} a_{r_i} b_i, \tag{6.2}$$

where r_i is a randomly chosen permutation vector to simulate the effects of indirect addressing and non-unit-stride memory access.

Fig. 6.2 shows the performance in MFlop ($= 10^{-6}$ times the number of floating point operations performed per second) for three different, RISC-based workstations. In all cases we vary the data set size and execute the code multiple times to keep the overall execution time close to constant. The loop has been unrolled to improve the usage of super-scalar and pipelined execution.

Note that all modern cache-based architectures result in a sharp drop of performance, when the cache size is exceeded. For small data sets, all machines come reasonably close to their peak-performance of 266, 198, and 50 MFlop, respectively, but this drops to a mere fraction when the cache size is exceeded.

In the case of (6.2), we simulate the effect of unstructured grids and indirect addressing by introducing an index array r_i holding a random permutation of the numbers $1, 2, \ldots, n$ and use this to address the vector a

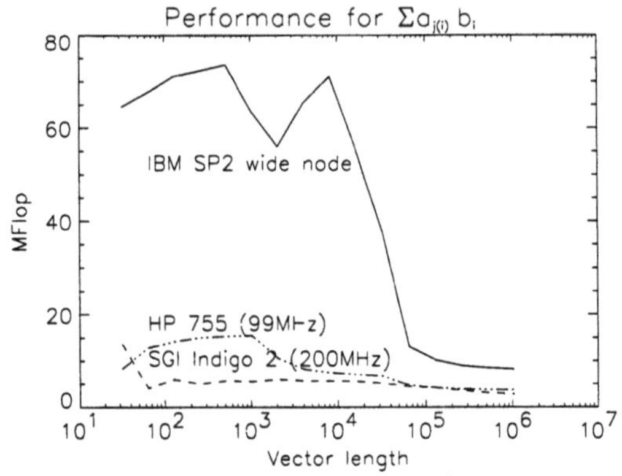

FIGURE 6.3. $\sum_1^n a_{ir_i} b_i$ for $n = 32, 64, \ldots, 1048576$.

as a_{r_i}. The vector b is still accessed with unit stride, as before. The results are shown in Fig. 6.3. Here, the performance for long arrays drops down to below 3 MFlop for the SGI, 4 for the HP, and 8 for the IBM and thus all machines are below 5% of their peak performance. When the data fits in the cache, the performance is somewhat better but still dramatically lower than before. This example shows that besides the general need to be memory- and cache-aware, uniform structures have performance advantages that should not be given up lightly.

Of course the benchmark is only a very simplified model for real PDE computations where usually more work is done at each node, and where the data is not usually allocated at random. Still it seems that the need for indirect addressing, combined with global sweeps through the data is potentially a very serious performance killer. Furthermore, it seems that the continuing development of computer architecture may make this problem even more pronounced.

On the other hand, the mathematical nature of PDE requires adaptivity. Thus we believe that it is important to find a strategy to implement adaptivity but to avoid the price of unstructured meshes wherever possible. Thus, we try to design a software system that can use uniform meshes (and their inherent efficiency) wherever they are appropriate, but we provide means for adaptivity and even unstructured meshes, when these are required. Additionally, we must attempt to limit the size of the data sets in order to improve the cache usage.

6.5 Abstract Mesh Data Structures

The discussion in the previous section has motivated us to look at meshes in a more abstract form. Both, simple (but efficient) uniform meshes, and unstructured meshes do have severe disadvantages. Therefore it is interesting to see whether it is possible to use an abstraction that permits the exchange of the mesh data type in a multilevel program. In other words, we plan to make the mesh data type an exchangeable parameter in a multilevel program.

Note again, that this is motivated by efficiency considerations. No techniques must be used that are inherently unable to support the highest possible efficiency for the implementation. Technically this means that as much as possible of the parameterization must be evaluated at compile time. Ideally, the compiled result should be as efficient as a code on a simple rectangular grid using a 5-point difference discretization, when this is the grid module used. However, the code should also be able to operate on an unstructured mesh that then provides the full flexibility.

Furthermore, since uniform meshes may be suitable for parts of the domain, while unstructured ones are necessary for other parts, we wish to have the software structure such that the different mesh types can coexist and cooperate.

Iterator Macros. We first discuss an approach that is independent of the programming language and at the same time as efficient as possible. The idea is simply to replace the loop constructs by macros that expand to the loop headers and loop-ends, respectively. With such macros, a loop in the program may look like

```
FOR_ALL_NODES(i)
   FOR_ALL_NEIGHBORS(k)
      ...
   END_ALL_NEIGHBORS
END_ALL_NODES
```

where `FOR_ALL_NODES(i)` is being expanded to suitable loop construct, and `END_ALL_NODES` contains all the statements syntactically required to terminate the loop. More generally, this approach leads into more or less involved pre-compilation techniques and the design of special purpose pre-compilers. Taken to an extreme, we could define our own special purpose language.

A prototype software based on this approach has been developed by the author around 1990 using C as a base language and the standard Unix macro-expander *M4* to construct the preprocessor. The concept has been applied successfully so that the same algorithmic description of multigrid could be used both on uniform rectangular meshes and on unstructured triangular finite element meshes. In both cases, the efficiency was essentially the same as for directly coded programs. Despite the efficiency advantage of the technique, the approach has not been developed further, partly because the involved code expansion made program maintenance and debugging increasingly difficult. Some of the techniques involved are described in Jänsch et al. [9].

Iterator Classes. Using an object oriented design (in C++), it is natural to make the meshes a class, and providing access to it using member functions. *Iterator* classes provide an abstraction of what an index is to an array. In the simple case, when meshes are just multidimensional arrays, the corresponding iterator is essentially just a tuple of integers for indexing the array.

An iterator data type must provide the natural operations, like incrementation and decrementation. Using C++ we have the additional benefit that overloading the indexing brackets `[ · ]` and operators like `++` and `--` results in a code that is very close to the usual syntax.

Making this approach efficient requires the intensive use of function inlining and a careful implementation of all classes. In principle, the compiler can obtain all information to generate optimal efficient code, however, our experience is that in this respect currently available compilers differ much and do not always provide the desired (and possible) efficiency. Of course, the hope is that the compilers will continue to improve.

Other mechanisms to implement the desired functionality efficiently can exploit *templates.* Templates constitute the C++ implementation of generic programming; that is, to parameterize a program with *compile-time* instantiation. In principle, this is exactly what is required in our case, and programming techniques like Veldhuizen's *expression templates* [15] show how powerful this may be. Expression templates rely on the template mechanism having a full computational engine lying behind it. Thus it is possible to help the compiler by doing all kinds of compile time evaluations. One particular use is to pre-evaluate complex expressions and convert them to a form not requiring temporary objects and the corresponding copy operations that often spoil C++ performance. However, this technique is still new, and it is unclear how useful it can be in solving the performance problems discussed in this paper.

6.6 Patch-Adaptive Multigrid

Built on the above considerations, a new prototype adaptive multigrid code is currently being developed by H. Lötzbeyer [10], see also [11]. The basic design decision in this project is to use uniform meshes as much as possible while still providing reasonable flexibility in arranging the meshes.

In classical multigrid literature it has always been emphasized that multigrid is capable to obtain adaptivity using locally uniform structures, see e.g. [4, 12] and [8, Chapter 15]. The patch-adaptive multigrid method (PAM) described here is partly motivated by these ideas, and partly by the *virtual global grid concept* developed by Rüde [13].

To provide adaptivity, the PAM follows the concept of virtual global grids. Here adaptivity is seen as a means for reducing the complexity by *leaving out* nodes from a fully refined mesh, when it is found that individual nodes do not contribute efficiently to the overall accuracy. These nodes are then called *ghosts*. Together with the *live* nodes they form a *virtual* global grid. The ghosts are not represented in memory and are (usually) not accessed by the algorithms so that they do not contribute to the computational complexity of handling a virtual global grid.

It is important to see ghost nodes as *virtual* data structures. In a so-called variational framework it is natural to assign each ghost a solution value defined by interpolation from coarser grids. If the interpolation is consistent with the discretization and the multilevel algorithm, it can be shown that ghost nodes conform with the multilevel algorithm in the following sense: A multigrid algorithm executed only on the live nodes is identical to one executed on the full grid (that is including the ghosts) when the smoothing of the ghosts is skipped. Thus the adaptive algorithm can also be interpreted as a perturbed full grid algorithm algorithm whose only difference is that all operations on ghost nodes are skipped (see Rüde [14]).

Note that the ghost nodes are important to define the operations performed at neighboring live nodes correctly. A live node may need to access the values from neighboring ghosts for computing residuals or a step of relaxation. The concept of ghost nodes is particularly useful to define the processing at the interfaces between refined and unrefined regions consistently. Without the ghost node concept, so called *slave nodes* must be introduced at the interfaces and our experience is that often either the approximation quality suffers or the iterative solver performance deteriorates when subtle errors are made in the treatment of slave nodes. In the final implementation the virtual global grid concept will be equivalent to using slave nodes in a special form that guarantees both the approximation quality of the resulting discretization and the efficiency of the multigrid solution method.

As described so far, the virtual global grid concept would require to distinguish between ghost and live nodes on an individual node-by-node basis. Clearly this would result in performance problems similar to those of an unstructured mesh. When we want to do better, we must modify the concept. Therefore, the PAM strictly groups unknowns in *patches*, where a patch is typically a 32×32 (in 2D) uniform sub-grid of nodes. Each mesh level is now composed from one (the coarsest grid) or several of these patches. The nodes in a patch are always treated collectively, thus defining the minimal granularity of all processing. This is essential for maintaining high efficiency on modern computer architectures. On the other hand, since a patch is a fairly heavy weight object, we can now use all other techniques from object oriented programming without sacrificing efficiency. Thus patches are implemented as classes and objects, and their member functions may be virtual, where this is appropriate. Here the overhead for dynamic binding is negligible, since the work performed by such a member function is relatively large.

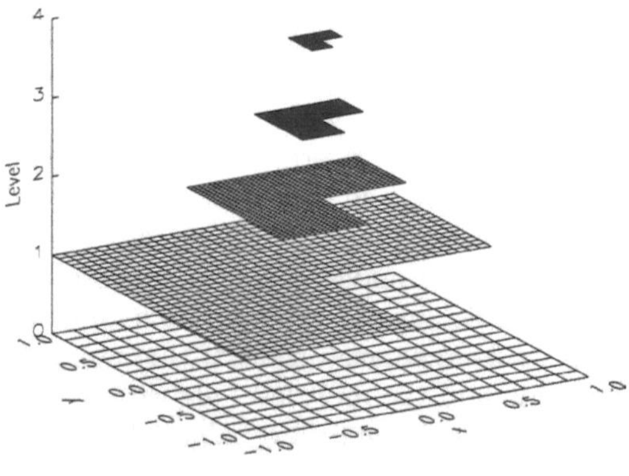

FIGURE 6.4. Patch structure for adaptive discretization of L-domain.

A typical adaptive patch structure is shown in Fig. 6.4. If the flexibility to arrange patches does not suffice to resolve complex problem features, the technique may be extended further to provide special unstructured patches to resolve these details. However, the goal is always to process the bulk of information with uniform data structures.

As a solver for the patch-adaptive virtual global grid structure, we propose a variant of the multilevel adaptive relaxation described in [13]. For this method the overhead is again reduced by strictly grouping the nodes in patches and performing the operations uniformly within these patches.

The patches can also be used to increase the data locality with the goal of making better use of cache-based systems. This is accomplished by performing several operations for each patch whenever possible. Typically, a multilevel algorithm requires several smoothing steps of each grid level before either proceeding to coarser or finer scales. While a conventional implementation would typically use several sweeps of the entire global mesh, we can now use a single sweep over all patches but perform several smoothing steps for each patch. Though the algorithms are not equivalent (the patch oriented approach would lead to a deterioration of performance when many relaxation steps are employed), our experience is that for a few sweeps the patch-oriented technique provides almost identical convergence at significantly reduced CPU time. The time required to perform a relaxation step on a single patch is dominated by the time required to transfer the data from the main memory to the cache. The processing itself is much cheaper. Therefore subsequent relaxation steps on the same patch are much faster than the first one. Detailed timing experiments are reported in Lötzbeyer [10].

6.7 Conclusions

Some aspects of the design of flexible yet efficient implementations of multilevel methods have been discussed. The core idea is to use abstract data types to implement various mesh structures and use implementation techniques that do not create run-time overhead. We devise techniques that result in compile-time instantiation mechanisms for low-granularity structures and light-weight objects. Expensive programming techniques, like virtual functions with late binding, must be restricted to medium- and heavy-weight objects. In the case of PDE solvers, this leads to the technique of patch-adaptive grids, where all processing is performed in terms of uniform patches of nodes.

Acknowledgments: I thank an anonymous referee for many helpful remarks.

6.8 References

[1] S. Balay, W. Gropp, L. C. McInnes, and B. F. Smith. Efficient management of parallelism in object-oriented numerical software libraries. In E. Arge, A. M. Bruaset, and H. P. Langtangen, editors, *Modern Software Tools for Scientific Computing*, pages 163–202. Birkhäuser, 1997.

[2] R. E. Bank. *PLTMG: A Software Package for Solving Elliptic Partial Differential Equations – Users' Guide 6.0*, volume 7 of *Frontiers in Applied Mathematics.* SIAM Books, Philadelphia, 1990.

[3] R. Beck, B. Erdman, and R. Roitzsch. An object-oriented adaptive finite element code: Design issues and applications in hyperthermia treatment planning. In E. Arge, A. M. Bruaset, and H. P. Langtangen, editors, *Modern Software Tools for Scientific Computing*, pages 105–124. Birkhäuser, 1997.

[4] A. Brandt. Multi-level adaptive solutions to boundary value problems. *Math. Comp.*, 31:333–390, 1977.

[5] A. M. Bruaset, E. Holm, and H. P. Langtangen. Increasing the efficiency and reliability of software development for systems of PDEs. In E. Arge, A. M. Bruaset, and H. P. Langtangen, editors, *Modern Software Tools for Scientific Computing*, pages 247–268. Birkhäuser, 1997.

[6] D. C. Burger, J. R. Goodman, and A. Kägi. On the declining effectiveness of dynamic caching for general-purpose microprocessors. Technical Report CS-TR-95-1261, Computer Science Department, Univ. of Wisconsion-Madison, 1995.

[7] H. Foerster and K . Witsch. Multigrid software for the solution of elliptic problems on rectangular domains: MG00 (Release 1). In W. Hackbusch and U. Trottenberg, editors, *Multigrid Methods*, volume 960 of *Lecture Notes in Mathematics*, pages 427–460, Berlin, 1982. Springer-Verlag.

[8] W. Hackbusch. *Multigrid Methods and Applications*. Springer Verlag, Berlin, 1985.

[9] C. R. Jänsch, U. Rüde and K. Schnepper. Macro expansion, a tool for the systematic development of scientific software. Bericht I-8814, Institut für Informatik, TU München, November 1988.

[10] H. Lötzbeyer. Objektorientierte parallele adaptive Mehrgitterverfahren auf semistrukturierten Gittern. Dimplomarbeit, Insitut für Informatik, TU München, 1996.

[11] H. Lötzbeyer and U. Rüde. Patch-adaptive multilevel iteration, 1996. Submitted to BIT, also available from mgnet: http://casper.cs.yale.edu/mgnet/www/mgnet.html.

[12] S. F. McCormick. *Multilevel Adaptive Methods for Partial Differential Equations*, volume 6 of *Frontiers in Applied Mathematics*. SIAM, Philadelphia, 1989.

[13] U. Rüde. *Mathematical and computational techniques for multilevel adaptive methods*, volume 13 of *Frontiers in Applied Mathematics*. SIAM, Philadelphia, PA, USA, 1993.

[14] U. Rüde. On the V-cycle of the fully adaptive multigrid method. In W. Hackbusch and G. Wittum, editors, *Adaptive methods — algorithms, theory and applications, Proceedings of the Ninth GAMM Seminar Kiel, Jan. 22–24, 1993*, volume 46 of *Notes on Numerical Fluid Dynamics*, pages 251–260, Braunschweig, Wiesbaden, 1994. Vieweg.

[15] T. Veldhuizen. Expression templates. *C++ report*, 7(5):26–31, June 1995.

7

Finite Element Kernel with Metaobject Protocol

Rostislav Chudoba[1]

ABSTRACT
This chapter presents an alternative approach to the design of numerical software exemplified by a finite element system. The design is introduced in the context of current approaches (procedural and object-oriented) to the numerical software development. Its salient features are (1) the ability to express commonalities between algorithmic structures and (2) the explicit treatment of the dependency between metadata and data involved in the computation. By capturing this dependency within a general metaobject protocol, the effort to instantiate general algorithms in a variety of settings is minimized. Furthermore, the implementation of an abstract algorithm is more straightforward than in the usually applied object-oriented design. Better correspondence between the mathematical formulation and the computer code is demonstrated by two interesting issues: the amount of dynamic processing during the computation is reduced and a user interface for the produced application is constructed automatically.

7.1 Introduction

The primary goal in the design of a complex numerical simulator is flexibility of the system. In order to increase it in a systematic way, general design concepts are applied. These concepts suggest a particular structuring of the implementation. In the algorithm-based area it is presently common to use the procedural and object-oriented approaches.

The fact that one design approach is better than another is justified by generally higher flexibility. It is a commonly accepted idea that the object-oriented approach results in a more flexible design than the procedural approach. The notion of flexibility, however, is inherently informal and therefore its evaluation is difficult. This is why we use the notion of *implementation locality* to characterize the flexibility of the system. This characteristic has been introduced by Kiczales [9] as follows:

[1]Institute for Statics and Dynamics, Ruhr-University Bochum, Germany. Email: `rch@mail.sd.bi.ruhr-uni-bochum.de`.

> *Extension of an implementation ought to take code proportional to the size of the change being contemplated. A simple customization ought to be incremental change.*

The presented design is based on the assumption that *if the implementation locality is preserved in all design situations, the highest possible flexibility is achieved.* It has been found productive to use the implementation locality as the criteria for evaluation of the design approach. It will be also shown that the currently applied approaches (i.e. procedural and object-oriented) violate implementation locality in important design situations. This allows us to illuminate the fundamental problems of these approaches.

In order to improve the implementation locality of an algorithm-based system, the presented design treats an algorithm as a primary abstraction. The approach is therefore referred to as *algorithm-oriented.* This may be seen in contrast with the object-oriented approach where object and class are the basic terms and data abstraction drives the design. Promoting an algorithm to a primary abstraction may be seen as a step back to the procedural approach where a procedural abstraction (e.g. function in C language or subroutine in FORTRAN) represents a concrete algorithm. The difference is that the algorithm-oriented approach allows us to preserve the abstract nature of a mathematical formulation. It is possible to represent an algorithm by a single piece of code containing only the necessary and sufficient specification, and to abstract from the issues that are related to its concrete application. In particular, the code of an algorithm may be *parameterized* so that it could later be *instantiated* in various usage contexts (i.e. it may generate several concrete procedures).

The notions of *parameterization* and *instantiation* are of fundamental importance: parameterization refers to the technique of how the parameters are propagated to a piece of code representing an algorithm, and instantiation is a process producing the special version of the algorithm from the parameters and preparing it for computation.

The present chapter attempts to introduce the algorithm-oriented approach in the broader context of numerical software design. Therefore, the explanation starts by introducing two simple examples of abstract algorithm formulation (Section 7.2). It is shown that the abstract mathematical formulation of an algorithm is inherently parameterized and that this feature ought to be preserved in the code.

In the subsequent two sections the introduced abstract algorithms are implemented using the procedural and object-oriented approaches. It is demonstrated that parameterization of an algorithm using these two approaches necessarily leads to the violation of the implementation locality.

In Section 7.5 the algorithm-oriented approach is presented in the form applied during the design of the finite element kernel. In the design, both the compile-time and run-time parameterizations are applied. The compile-time parameterization is realized by using the C++ templates. The templates were used in a similar way as in the design of the Standard Template Library (STL) [10] or in the Blitz++ Project [12,13].

Besides the compile-time parameterization, systematic support for run-time parameterization and instantiation has been recognized as necessary. For this purpose, the run-time *metaobject protocol* has been designed. In terms of the protocol the run-time parameters and instantiation process of an algorithm may be specified within the algorithm definition. The code for instantiation is used by the kernel to set up the data model before the computation starts. The introduction of the metaobject-protocol was inspired by the concepts of the meta programming [9].

The combination of these two types of parameterization allows us to tune the run-time flexibility and performance in the algorithm implementation as required.

The C/C++ language allows us to show examples of implementation in all the three approaches mentioned above and it is therefore used throughout the chapter to demonstrate the discussed concepts.

7.2 Example Problems

Throughout the chapter a few examples from the finite element algorithm will be used to compare the procedural, object-oriented and algorithm-oriented approach. First, we will try to implement a simple algorithm for a mapped quadrature, i.e. the Gauss numerical integration over a space domain in parametric coordinates with Jacobi transformation to a reference domain:

$$I = \int_{-1}^{1} f(\xi)\,|\boldsymbol{J}(\xi)|\;\mathrm{d}\xi = \sum_{1}^{m} w_i f(\xi_i)\,|\boldsymbol{J}(\xi_i)| \tag{7.1}$$

This algorithm formulation is *abstract* in the sense that the polynomial function and the expression for computing the Jacobi determinant remain unspecified. The formulation contains only the necessary and sufficient information, i.e. it only says that the result is produced by summation over the terms with the weighting factor w_i, the function value $f(\xi_i)$ and the Jacobi determinant $|\boldsymbol{J}(\xi_i)|$ at the particular point ξ_i.

From this formulation we produce a concrete algorithm by specifying the two abstract terms $f(\xi_i)$ and $|\boldsymbol{J}(\xi_i)|$. In the finite element algorithm applying the numerical integration we set [14]:

$$\boldsymbol{K} = \sum_{1}^{m} w_i \boldsymbol{B}(\xi_i)^T \boldsymbol{D} \boldsymbol{B}(\xi_i)\,|\boldsymbol{J}(\xi_i)| \tag{7.2}$$

where $\boldsymbol{K}$ is the element matrix (e.g. stiffness), and $\boldsymbol{B}$ and $\boldsymbol{D}$ are the geometric and material stiffness matrices respectively. In the mathematical notation the correspondence between (7.1) and (7.2) is straightforward. The abstract formulation is parameterized to produce concrete instances of an algorithm. When writing a general code, it is desirable to preserve this flexible nature of mathematical formulation in the implementation.

For this purpose we need a mechanism to produce the concrete implementations from (7.1) with a minimum of efforts. In other words, we wish that the algorithm (7.1) is available in a *parameterized* form with $f(\xi)$ and $|\boldsymbol{J}(\xi)|$ as formal parameters. Then, we require that a construction of a concrete algorithm (7.2) from (7.1) may be done only by supplying the parameters and that no additional adjustments are necessary. This process will be called *instantiation* in the sequel.

Although in (7.2) we have instantiated the mapped numerical integration for the simple finite element problem, the obtained algorithm is still not concrete. The terms in expression (7.2) need further specification, or more precisely: a next instantiation step. This reveals the recursive nature of instantiation. In the examples discussed later, we will use the isoparametric bilinear plane stress element as an example of a concrete setting of the algorithm (7.2).

7.3 Procedural Approach

Parameterized implementation of (7.2) using the traditional procedural programming (in the C language) is shown in the Figure 1. It is explained in two steps: first the facilities employed to parameterize an algorithm are specified, next the instantiation process is discussed.

7.3.1 Direct Parameterization

The prototype of the function `mapped_quad` contains

- instantiation parameters: offset of integration scheme within a statically initialized table (`integ_scheme`), pointer to the function representing the parameter $f(\xi)$ (`f_xi`) and pointer to the function returning $|\boldsymbol{J}(\xi)|$ (`jac_det`).
- input data: reference domain for which to compute the integral.

While the parameters control the execution of the algorithm, the input data affects the value of its result. In the context of the procedural approach, this distinction is purely formal since both the parameters and the data are received in form of run-time arguments. We should note, however, that data is usually altered for every call to the function `mapped_quad` and that the same parameters are used for several calls.

7.3.2 Direct Instantiation

With this implementation of algorithm (7.1) at hand, let us produce a concrete algorithm integrating the function $\frac{1}{2}(1-\xi^2)(1-\eta)$ over a quadrilateral

```
#define TYPE double                  /* type of the integrated expression */
#define NDIM 3                       /* number of spatial dimensions */

typedef struct IntegPoint{
  double weight;
  double coord[3];                   /* integration point coordinates */
}IntegPoint;

typedef struct IntegScheme{
  double      n_integ_points;        /* number of integration points */
  IntegPoint* integ_points;
}IntegScheme;

/* Static initialization of integration schemes */

IntegPoint line_1[] = {{2.0,              { 0.0,              0.0,0.0}}};
IntegPoint line_2[] = {{1.0,              {-0.577350269189626,0.0,0.0}},
                       {1.0,              {-0.577350269189626,0.0,0.0}}};
IntegPoint line_3[] = {{0.555555555555555,{-0.774596669241483,0.0,0.0}},
                       {0.888888888888888,{ 0.0              ,0.0,0.0}},
                       {0.555555555555555,{ 0.774596669241483,0.0,0.0}}};
IntegPoint quad_1[] = {{4.0,              { 0.0,              0.0,0.0}}};

IntegScheme static_integ_table[] = {
  {1,line_1}, {2,line_2}, {3,line_3}, {1,quad_1}  /* and so on */
};

void mapped_quad(int integ_scheme, double nodes[][NDIM],
                 double (*jac_det)( double* x, double nodes[][NDIM] ),
                 void (*f_xi)( double scale, double* coord, TYPE* sum ) )
{
  IntegScheme* is = &static_integ_table[integ_scheme];
  int i;
  for( i = 0; i < is->n_integ_points; i++ ){
    IntegPoint* ip = &is -> integ_points[i];
    double j_det = jac_det( ip->coord, node_list );
    f_xi( ip -> weight * j_det, ip->coord, sum );
  }
}
```

FIGURE 7.1. Procedural approach.

reference domain. For this purpose, we write the function with the required prototype:

```
void the_above_expr( double scale, double* coord, TYPE* value )
{
  *value += scale * ( 1.0 - pow(coord[0],2.0) ) * ( 1.0 - coord[1] ) / 2.0 ;
}
```

and another function (`bilin_jac_det`) to compute the Jacobi determinant for bilinear reference domain. Finally, the boundary nodes of the reference domain need to be specified. The complete instantiation may be written like this:

```
int main()
{
  double rdomain[][NDIM] = {
    { -1.0, -1.0 }, { -1.0,  1.0 }, {  1.0, -1.0 }, {  1.0,  1.0 } };
  double integ_val;
```

```
  mapped_quad( 3, &integ_val, rdomain, bilin_jac_det, the_above_expr );
  printf("Integration result: %g\n",integ_val);
}
```

At this point, we may observe the following features of the implementation:

- There is no distinction between the parameters and arguments in the implementation. The prototype of the function `mapped_quad` contains both parameters for instantiation and arguments for evaluation. As a result, the instantiation must be performed simultaneously with the evaluation.
- The representation of the argument `rdomain` depends on the parameter `jac_det`. In order to instantiate `mapped_quad` we need to know the details (number and selected ordering of nodes) of `bilin_jac_det`. It is not an easy task to extract this information from a non-transparent implementation of `bilin_jac_det`.
- The client of `mapped_quad` is always responsible for associating the instantiation parameters with the correct evaluation arguments, i.e. he or she has to associate the selected version of reference domain representation with the appropriate version of the function for computing the Jacobi determinant.

Because the described dependency between parameters and arguments plays an important role in our design, we examine the problem with a more complex example. In particular, we try to instantiate the `mapped_quad` to obtain the algorithm (7.2). We do this with a concrete function `BtDB_mtx_bilin_plain_stress` computing the term for isoparametric linear plain stress element.

The situation after instantiation is shown in Figure 7.2 in the form of a directed acyclic graph. A node in this graph represents a function and/or its result. The edges leading down to other nodes specify prerequisite operations to produce the result. Some operation nodes have attached additional information characterizing their execution. For example, the `forall` quantifier indicates an iteration (loop) over a set of variables. Further, parentheses indicate access to data (e.g. `E,G,nu` or `coord`) necessary for evaluation.

We observe that there are arguments required in the internal operation computing the material matrix (`compute_D_mtx`). However, there is no possibility to pass these arguments within the interface of the instantiated function:

```
mapped_quad(3, stiff_mtx, rdomain, bilin_jac_det,
            BtDB_mtx_bilin_plain_stress );
```

The reason is that the interface cannot be changed during the instantiation.

Of course, there are possibilities to supply the data in a different way. However, these approaches either bypass the static type checking or the interface of the function. Both these solutions violate the principle of implementation locality and thus are error prone and make the implementation less transparent.

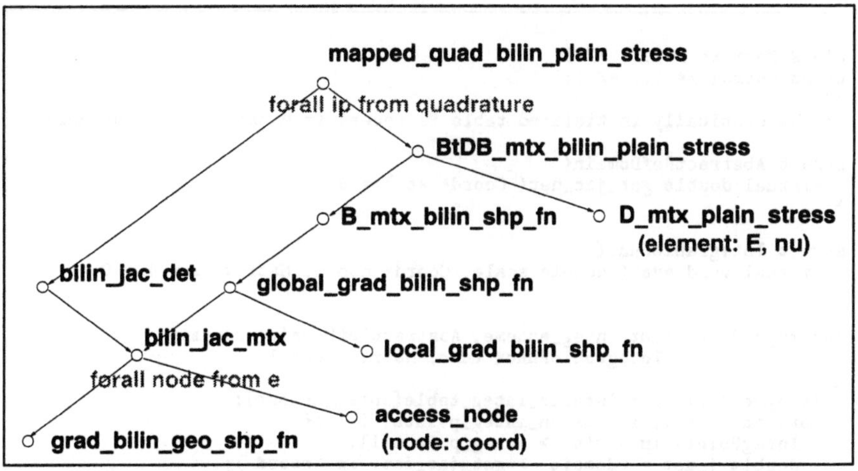

FIGURE 7.2. Mapped quadrature instantiated for bilinear plain stress element.

We may conclude that the supplied parameter may affect both the quality of the arguments (e.g. representation of the reference domain) and their quantity (e.g. `E, nu` for material matrix).

The described difficulties are the reason why we would hardly find an abstract implementation of `mapped_quad` in a procedural code for finite element systems. It is usually easier to code it manually every time it is applied. As a result, its implementation is replicated many times in the code.

This fact demonstrates that the procedural approach fails to serve our purposes. It allows us to perform only the first step of instantiation, i.e. to receive the algorithmic parameters. However, it cannot handle the dependencies arising in the underlying data structure. The instantiation must be completed manually by the programmer in a rather "creative" manner and this is obviously undesirable.

7.4 Object-Oriented Approach

The fundamental concept of the object-oriented paradigm is the explicit coupling of a structure and a behavior. This concept may effectively be used to improve the design from the previous section, because it allows us to capture the dependency between the parameters and arguments involved in the evaluation. Figure 7.3 shows the new implementation in C++. We explain first how to parameterize an algorithm using the object-oriented technique. Next, the outcome of the instantiation is discussed.

7.4.1 Indirect Parameterization

The first instantiation parameter specifying the integration order remains as in the preceding implementation because it is local to the algorithm.

```
class Matrix{ /* ... */ };
class Coord{ /* ... */ };
//
// The statically initialized table is reused from the previous example
//
struct AbstractRefDomain{
  virtual double get_jac_det( Coord* at ) = 0;
};

struct IntegrableFunc{
  virtual void eval( double scale, Coord* coord, Matrix* sum ) = 0;
};

int mapped_quad(int integ_scheme, AbstractRefDomain* rdomain,
                IntegrableFunc* func, Matrix* sum )
{
  IntegScheme* is = &static_integ_table[integ_scheme];
  for( int i = 0; i < is->n_integ_points; i++ ){
    IntegPoint* ip = &is -> integ_points[i];
    double j_det = rdomain -> get_jac_det( ip->coord );
    func -> eval( ip -> weight * j_det, ip->coord, sum );
  }
}
// Quadrilateral domain
class QuadLatDomain: public AbstractRefDomain{
  int         n_nodes;
  Coord*      nodes;
public:
  QuadLatDomain(int n_n_, Coord* nodes_ ): n_nodes(n_n_), nodes(nodes_){}
  double get_jac_det( Coord* at ); // concrete version
};

class BtDB_mtx: public IntegrableFunc{
  // Context data for BtDB - design parameters, nodes, etc.
public:
  void void eval( double scale, Coord* coord, Matrix* sum );
};
```

FIGURE 7.3. Parameterized implementation using abstract classes.

The other parameters are introduced using abstract classes. This reveals the methodology applied in the object-oriented design: the objective to build an abstract algorithm transforms to the objective to build a "flexible" class hierarchy. In the general case, the object-oriented analysis may be applied [6]. However, in the algorithm-based area we may identify a few simple rules how the algorithm parameters are transformed to the abstract classes and then merged in a class hierarchy:

Case 1: *Abstract class is motivated by an algorithm parameter.*

The second instantiation parameter of `mapped_quad` for computing the Jacobi determinant is obtained through the variable `rdomain`, see Figure 7.3. The function may be invoked by issuing `rdomain->get_jac_det`. Further, the `rdomain` hides the details of reference domain representation required by that function. As a result, the reference domain representation does not appear as a separate argument in the prototype of `mapped_quad`.

The appropriate version of data is passed together with the specified algorithm parameter. Although the data is invisible for `mapped_quad` it represents a context of its execution. Therefore, data that is accessed during the computation without being explicitly passed as an argument will further be referred to as *context data*[2].

By specializing the class `AbstractRefDomain`, various representations of the reference domain may be used for the same implementation of `mapped_quad`[3]. This is the desired situation: the abstract function remains untouched.

This example reveals the clear correspondence between the algorithm parameterization and abstract class definition: in order to preserve the abstract nature of an algorithm, its parameters should be represented as virtual functions of some abstract class. The structural dependency between the parameter and the particular version of context data may then be handled separately in the concrete subclass.

> Case 2: *Abstract class is motivated by type dependency between the algorithm parameter and evaluation argument.*

The interface of `mapped_quad` in Figure 7.3 contains an argument of type `Matrix`. This limits the usage context of the implementation too much. We also need to integrate expressions of other types, e.g. `Vector`. The simple solution is again to hide the representation behind an abstract class: `IntegrableVar`. Then, the definition of `IntegrableFunc` changes as follows:

```
struct IntegrableFunc{
  virtual eval(double scale, Coord* coord, IntegrableVar* sum) = 0;
};
```

Further, matrices, vectors (and their various representations) have to inherit from `IntegrableVar`. Although such an implementation is theoretically possible, it violates the aforementioned principle of *implementation locality*: the aim to write an abstract algorithm should not result in changes of unrelated portions of code or reorganization of class hierarchies. In the present example, the primitive (`double`) and abstract (`Matrix`) mathematical variables should be implicitly integrable.

> Case 3: *Abstract class motivated by unification of context data associated with other abstract classes.*

---

[2]In the object-oriented terms, the context data coincides with the structure of an object having the `mapped_quad` algorithm attached as a method. The notion of context data is introduced to emphasize the fact that in the algorithm-based system the attributes of an object are introduced "on demand" of a concrete algorithm. The structuring of context data may therefore be automated.

[3]Declaring the function `get_jac_det` as virtual means that its concrete version will be supplied later in a subclass. For the purpose of the `mapped_quad` it is enough to define `AbstractRefDomain`. Concrete reference domain representation may be defined later.

There may be more instantiations of **mapped_quad** sharing the same context data. In the presented example, the context data of **BtDB_mtx** contains the list of nodes, and the design parameters **E,nu**. The same context data is required for computing the body force vector:

```
class Ntb_vec{
  // The same context data as for BtDB_mtx
public:
  virtual eval(double scale, Coord* coord, Vector* sum) = 0;
};
```

Thus, we need to define both functions within a single class:

```
class Elem{
// shared context data
public:
  virtual BtDB_mtx(double scale, Coord* coord, Matrix* sum) = 0;
  virtual Ntb_vec( double scale, Coord* coord, Vector* sum) = 0;
};
```

However, the implementation of **mapped_quad** must now be changed as follows:

```
void mapped_quad(int integ_scheme, AbstractRefDomain* rdomain, Elem* elem,
                 void(Elem::*f_xi)(double,Coord*,IntegrableVar*),
                 IntegrableVar* sum);
```

Here, the context data (**elem**) and the algorithm parameter (**f_xi**) are passed as separate arguments. Nevertheless, any function from within the **Elem** scope having the required prototype may be used as an instantiation parameter of **mapped_quad**. Although the context data **elem** and parameter **func** are passed separately, the implementation of **mapped_quad** remains untouched for particular instantiations.

The real implementation is certainly much more complex and finer rules might probably be identified in the design process. However, their detailed discussion is not our primary interest. The above specified rules should demonstrate how the design of a class hierarchy is driven by the aim to preserve the abstract formulation of an algorithm. In other words, the class hierarchy is motivated by indirect parameterization of the incorporated algorithms.

7.4.2 Indirect Instantiation

The dependency between the parameter and the associated context data is resolved at a well defined place: in the object constructor. For example, let the abstract algorithm **mapped_quad** be instantiated with another abstract algorithm **BtDB_mtx**. The situation is displayed in the Figure 7.4. The instantiation with the expression $\boldsymbol{B}^T\boldsymbol{DB}$ does not produce a concrete algorithm but another abstract algorithm **mapped_quad_BtDB_mtx** with new parameters specified in square brackets. The new set of parameters is constructed as a union of all parameters appearing in the displayed algorithm graph.

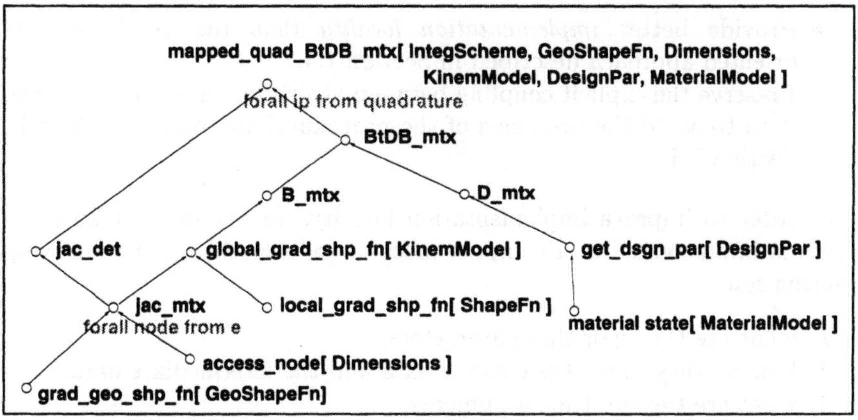

FIGURE 7.4. Instantiating with parameterized algorithm.

To produce a concrete algorithm, the new set of parameters must be specified. These parameters provide sufficient information to set up the structure of the context data. For example, the array of nodes is allocated depending on the selected set of shape functions and entries for material characteristics are set up according to the selected material model. This dependency is resolved in the constructor of the class `Elem`. We observe that the instantiation of an algorithm is performed indirectly by instantiating an object. As a result, the information "what algorithm motivates a particular context data structuring" is not explicitly available in the code. This may be seen as a violation of implementation locality: the dependency arising from the algorithm is specified at a different place in the code.

While the structure of the context data is set up separately from the algorithm instantiation, this is not the case for the intermediate data. The particular representation of intermediate matrices (e.g. the dimensions of $\boldsymbol{B}$, $\boldsymbol{D}$ and $\boldsymbol{J}$ from the discussed example) is known only during the evaluation. As a result, the evaluation code must be polluted by the management of representation dependencies between intermediate objects (e.g. propagation of matrix dimensions and construction of sparsity maps).

Summarizing, we may say that the object-oriented approach is able to reflect the abstract formulation of the algorithm (7.1) better than the procedural approach. However, the algorithm implementation is transformed to the design of a class hierarchy. Addition of new algorithms into the kernel may result in contradicting demands and initiate non-local changes of the implementation.

7.5 Algorithm-Oriented Approach

The goal behind the design of the finite element kernel with a metaobject protocol may be summarized as:

- Provide better *implementation locality* than the purely object-oriented approach described in Section 7.4.
- Preserve the explicit coupling between algorithm parameters and the data to avoid the problems of the procedural approach described in Section 7.3.

In order to improve implementation locality, we require that an algorithm is implemented in a compact piece of code specifying the following information:

1. what are the algorithm parameters,
2. how do they affect the context data and the intermediate data,
3. what are the run-time arguments,
4. the code for evaluation

The additional job of data organization should be taken either by the compiler (at compile-time) or by the run-time system provided by the kernel. The decision whether to apply compile-time or run-time parameterization comes from the requirements of run-time flexibility of the algorithm: if there is no need to alter the execution of the algorithm at run-time, we may use only compile-time parameterization. The next two paragraphs show how both versions of parameterization are realized. Subsequently, the new abstract implementation of the mapped quadrature is shown and its features are discussed.

7.5.1 Compile-Time Parameterization and Instantiation

The implementation of an algorithm is performed within a C++ class definition. The above specified information is attached to the class at compile-time via the embedded type definition/aliasing. This technique has been extensively used in [3, 10, 13].

The layout of a class defining an algorithm looks as follows:

```
class Function{
public:
  typedef  TypePar1   ap_tp;     // type of an algorithm parameter
  typedef  TypePar2   ctx_tp;    // type of the context data
  typedef  TypePar3   ret_tp;    // type of the return value
  typedef  TypePar4   arg_tp;    // type of arguments
  void inst( ap_tp&,    ctx_tp::mo&, ret_tp::mo&, arg_tp::mo& );
  void eval( ctx_tp&,   ret_tp&,      arg_tp&      );
};
```

This type of structure will further be referred to as an *extended function prototype.* All the concrete functions in the system should be wrapped with such a definition. Based on this conventional prototype, the clients (callers) of the above defined `Function` may reuse the type attributes from its extended prototype. In other words, they may reflect the features of the called function in their own extended prototype. A simple example is demonstrated by a propagation of a return type during the compile-time instantiation:

```
template < class AlgParFunc >
class ClientFunction{
public:
  typedef  AlgParFunc::ret_tp   ret_tp;  // reused return type
  // following structure as above
};
```

This approach clearly avoids the problem encountered in connection with direct instantiation used in the procedural approach (Section 7.3.2), since the extended prototype may change during the compile-time instantiation.

7.5.2 Run-Time Parameterization and Instantiation

In the extended function prototype, the algorithm parameters `::ap_tp` are separated from the arguments `::arg_tp`. This allows us to separate the run-time instantiation from the evaluation. The former is performed in the `inst` method and the latter in the `eval` method. The task of the `inst` method is to transform the algorithm parameter `::ap_tp` to *metaobjects.* These are the run-time objects specifying the representation of data accessed during the evaluation of the algorithm. A typical example of a metaobject is the sparsity map of a sparse matrix. It uniquely defines the representation and interpretation of the data for the algorithms defined on the matrix. Another example is a metaobject specifying a size of a homogeneous array with all entries having the same structure. For inhomogeneous arrays, the metaobject consists of a set of other metaobjects. This kind of metaobjects may be used to describe the structure of the context data.

As the prototype of the `inst` method indicates, the type of an object is associated with the type of a metaobject through the type attribute `::mo`. Therefore, any class representing a fundamental mathematical abstraction used in an abstract algorithm (i.e. matrix, vector, tensor and their various representations) must have a type attribute `::mo` specifying the appropriate metaobject type. For example, the full matrix is implemented as follows:

```
class FullMtx{
  class mo: public MetaObject{
    int n_col, n_row;
  public:
    void set_dim( int n_c, int n_r );  // construct the metaobject
    void fix();                        // fix the structure
    size_t get_size();                 // required by run-time system
  };
public:
  void setup( mo& );  // set up the object layout based on the metaobject
  // the usual matrix operations follow
};
```

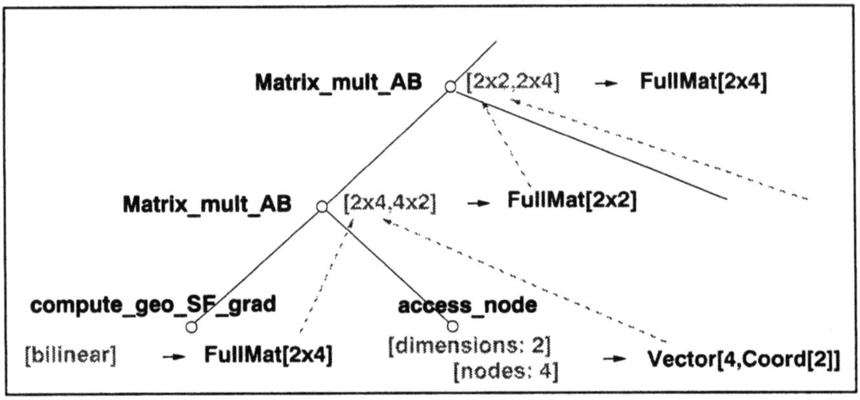

FIGURE 7.5. Metaobject propagation for intermediate objects.

With such a definition of a matrix we may write the extended prototype of a matrix multiplication as follows:

```
class FullMtx_mult_AB{
public:
  typedef MOPVoid       ap_tp;      // no algorithm parameter
  typedef FullMtx       ret_tp;     // return type is matrix
  typedef MOPVoid       ctx_tp;     // no context type
  struct  arg_tp{
    struct mo{ FullMtx::mo A; FullMtx::mo B; };
    FullMtx* A; FullMtx* B;
  };
  void inst( ap_tp&,       ctx_tp::mo&, ret_tp::mo&, arg_tp::mo& );
  void eval( ctx_tp&,      ret_tp&,     arg_tp&      );
};
```

The instantiation method receives a pair of matrix metaobjects in the local structure `arg_tp::mo` and sets the dimensions of the resulting matrix in `ret_tp::mo`. This demonstrates the propagation of metaobjects during the instantiation. Figure 7.5 exemplifies this on a portion of a finite element algorithm from the Figure 7.4 shown earlier. The matrix dimensions are propagated throughout the graph from the initial algorithm parameters specifying the shape functions and dimensionality.

We observe, that the `inst` method allows us to resolve the dependency between metaobjects locally, within the algorithm implementation. In contrast to the object-oriented approach, the instantiation is completed before the computation starts (e.g. the propagation of matrix dimensions is resolved separately from the evaluation).

As already specified, the run-time instantiation is performed in terms of the metaobject protocol. This protocol specifies a set of services required from a metaobject for the automatic set up of the data model. As a result, the metaobject type may be an arbitrary C++ class providing the required functionality: (1) construction methods (`set_dim`) (2) `fix` method and (3) `get_size()` method. The specified services are interpreted by the run-time system of the kernel during the run-time instantiation to set up the structure of the context data and the stack-like management of intermediate objects.

7.5.3 Implementation Example

In order to give an idea how the introduced notions fit together, we show the algorithm-oriented implementation of the mapped quadrature in Figure 7.6. The main features of the implementation are summarized as follows:

- The implementation combines the compile-time and run-time parameterizations. The integrated expression `Fxi` and the expression for Jacobi determinant `JacDet` are specified as compile-time parameters. The integration scheme and the run-time parameters of `Fxi` and `JacDet` constitute the set of run-time parameters.

- The composition of run-time parameters is a recursive operation. It has been pointed out previously in Figure 7.4 that if an instantiation parameter (e.g. `Fxi`) requires run-time parameters specified in `Fxi::ap_tp`, then the run-time parameter of the instantiated algorithm `mapped_quad` should unify its local parameters with the parameters of the integrated function (`Fxi::ap_tp`). In C++, such a unification can be achieved by making `Fxi::ap_tp` a virtual base class of `mapped_quad::ap_tp`. Thus, the parameter composition recursively extends the class hierarchy of algorithm parameters.

- The parameter composition is further assisted by the template classes `AP` and `APRef`. They allow us to inspect the structural relationships between the parameters. This information is then used to generate an input data structure (a simple object-oriented database) and the associated user interface.

- The return type `mapped_quad::ret_tp` coincides with the return type of `Fxi::ret_tp`. The argument type `mapped_quad::arg_tp` is constructed as a unification of all argument types needed to evaluate the supplied expressions for `Fxi` and `JacDet`. As for algorithm parameters, the unification is achieved through virtual base classes.

- The context data type `ctx_tp` is constructed as a set of contexts required by a parameter `Fxi`. In case that `Fxi` requires state parameters in integration points, it may simply specify their type as its `Fxi::ctx_tp` type attribute. The wrapper class `MOPHomogCtx` instructs the run-time system of the kernel to treat the context data of `mapped_quad` as a homogeneous array with a single metaobject for all entries. This possibility may effectively be used in time-dependent and non-linear problems to specify the requirement on material state representation at integration points. Again, the requirements on the context data are specified locally in the definition of the specific algorithm and the `mapped_quad` implementation remains untouched.

- Using this concept, the context data is structured depending on the structure of the algorithm, more precisely on its iteration (loop) levels. In the general multilevel computation the iteration includes the levels of domain, subdomain, element, quadrature, integration point

```
template < class Fxi, class JacDet >
class mapped_quad
{
public:
  // Algorithm parameter is exported to the interface via the
  // following entries
  class ap_tp: public AP<ap_tp>,
               public virtual APRef<Fxi::ap_tp>,
               public virtual APRef<JacDet::ap_tp>
  {
    // coincident with IntegScheme
  public:
    ap_tp(){}  // export to user interface
  };
  class   arg_tp: public virtual Fxi::arg_tp,
                  public virtual JacDet::arg_tp{}
  typedef Fxi::ret_tp             ret_tp; // return type
  typedef MOPHomogCtx<Fxi::ctx_tp> ctx_tp; // for state representation
  // Run-time instantiation
  void inst( const ap_tp& ap, ctx_tp::mo& ctx_mo,
             ret_tp::mo& mo_sum_, arg_tp::mo& arg_mo )
  {
    ap_tp* integ_scheme = ap;
    // number of integration points is derived from the integ. scheme
    ctx_mo. set_dim( integ_scheme -> get_n_ip(); );
    jac_det. inst( ap, ctx_mo, arg_mo, mo_sum_ );
    f_xi.    inst( ap, ctx_mo, arg_mo, mo_sum_ );
  }
  // Evaluation
  void eval( ctx_tp& ctx, ret_tp& sum, arg_tp& args )
  {
    for( int i = 0; i < integ_scheme -> get_n_ip(); i++ ){
      //
      const MANumIntegPoint* ip = integ_scheme -> get_ip(i);
      double J_det;

      args. x = ip->get_coords();
      get_J_det. eval( J_det, NULL, args );

      args. scale = ip->get_weight() * J_det;
      f_xi. eval( sum, ctx[i], args );
    }
  }
private:
  JacDet                              jac_det; // function object
  Fxi                                 f_xi; // function object
  ap_tp*                              integ_scheme; // private metaobject
};
```

FIGURE 7.6. Parameterized implementation algorithm-oriented approach.

and layer. Algorithms at each level may specify requirements on the state representation. Their request is caught at the corresponding iteration level by the run-time system during the instantiation. The "on demand" set up of the context data allows us to insert an additional iteration level with a minimum effort. For example, an algorithm updating a state of a material point may be applied either at the integration point level or at the layer level of a layered shell finite element. The change does not require any explicit management of state parameters since they are structured automatically during the run-time instantiation.

Figure 7.7 demonstrates the overall instantiation process in the designed system. Starting from the abstract algorithms and fundamental mathematical classes (i.e. matrices, vectors) the compile-time instantiation is performed. During this step, the auxiliary classes are used to generate the user interface and the underlying input structure. The generated interface is connected to an external mesh generator. The elements of discretization are connected to the active algorithm parameter. Then, the run-time instantiation is performed to set up the state representation and communication with global mathematical objects. After that, the computation may start.

7.6 Conclusions

The present chapter examines the approaches to the design in algorithm-based problem domains. A comparison of procedural, object-oriented and algorithm-oriented approaches is provided using the simple algorithm for computing the mapped numerical quadrature. It is demonstrated that the algorithm-oriented approach results in the best support for implementation locality of the abstract (i.e. parameterized) implementation of an algorithm.

The described algorithm-oriented approach is realized using the facilities of C++. In particular, the type parameterization and embedded type definition/aliasing is employed at compile-time. Further, the metaobject protocol complements the concept with the uniform run-time data management. Based on this protocol, the kernel of the system provides services for algorithmic parameter and metaobject management. These services are hidden behind a set of types that may be declaratively associated with the type attributes of an algorithm. As a result, there are no imperative commands and the communication between the algorithm and the kernel is almost factored out from the code implementing the algorithm.

The data model produced by the kernel during the run-time instantiation is organized in homogeneous arrays reflecting the layout of the loops in the overall algorithm. Also the intermediate data is neatly allocated in a stack-like manner. This allows us to reduce the amount of dynamic memory management during the computation.

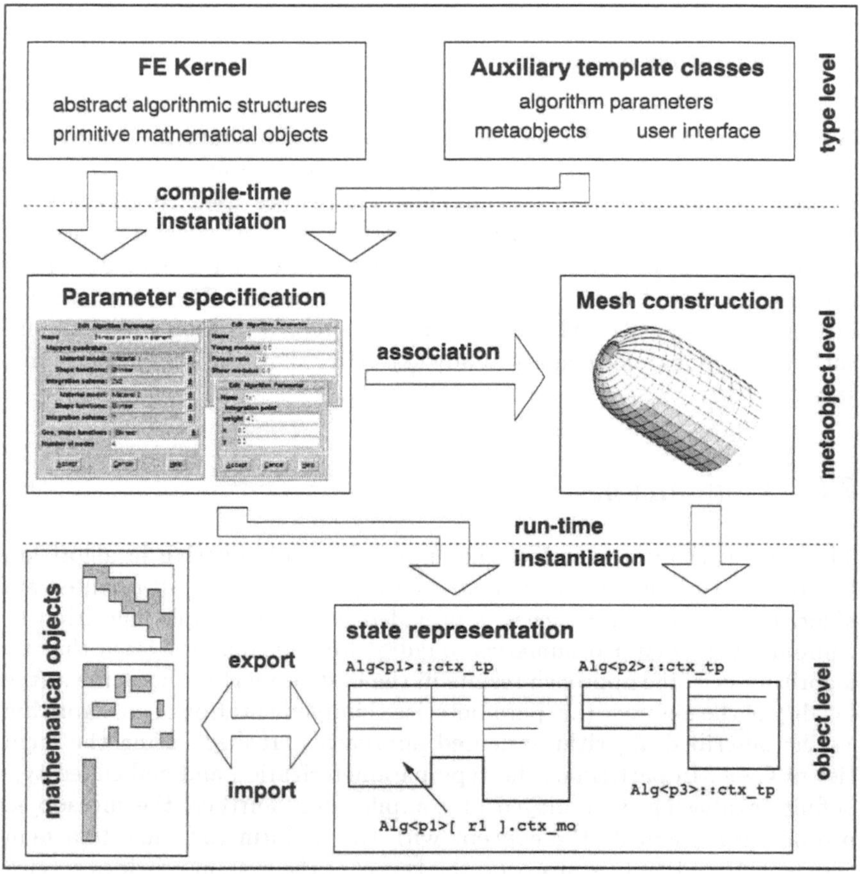

FIGURE 7.7. Overall instantiation process.

Although the examples used throughout the chapter are quite simple, it should be obvious that the discussed concepts are generally applicable to abstract algorithmic structures.

The complete support for compile-time and run-time parameterization allows us to tune the run-time flexibility and performance as required without affecting the layout of the implementation. One can argue that the distinction between compile-time and run-time parameters in the algorithm implementation should not be necessary. Indeed, it would be convenient to let the client of a parameterized algorithm decide what parameters should be fixed at compile-time and what should be exported to the run-time interface. Unfortunately, such a layout cannot be realized in C++; its realization would require much more effort.

Acknowledgments: This work has been carried out in the framework of the project "Object-Oriented Finite Element System" supported by the German Science Foundation and it was also partially supported by the Czech Grant Agency (Project No. 103/96/1271).

7.7 References

[1] E. Arge, A. M. Bruaset, P. B. Calvin, J. F. Kanney, H. P. Langtangen, and C. T. Miller. On the efficiency of C++ for scientific computing. In M. Dæhlen and A. Tveito, editors, *Mathematical Models and Software Tools in Industrial Mathematics*, pages 93–119. Birkhäuser, 1997.

[2] R. Barrett, M. Berry, T. F. Chan, J. Demmel, J. Donato, J. Dongarra, V. Eijkhout, R. Pozo, C. Romine, and H. Van der Vorst. *Templates for the Solution of Linear Systems: Building Blocks for Iterative Methods.* SIAM, Philadelphia, PA, 1994.

[3] J. Barton and L. R. Nackman. *Scientific and Engineering C++: an Introduction with Advanced Techniques and Examples.* Addison-Wesley, 1995.

[4] R. Chudoba. *Sensitivity Analysis of RC Framed Structures: Object-Oriented Approach.* PhD thesis, Czech Technical University Prague, November 1996.

[5] R. Chudoba, P. Krysl, and Z. Bittnar. Explicit finite element computations: Object-oriented approach. In *Proceedings of the VI. Int. Conf. on Comp. in Civil and Building Engng.*, pages 139–145, Berlin, 1995. Balkema Publishers.

[6] P. Coad and E. Yourdon. *Object-Oriented Analysis.* Prentice-Hall, 1991.

[7] P. Demmert. *Strategien zur objektorientierten Modelierung von Aufgabenstellungen der numerischen Tragwerksanalyse.* PhD thesis, Ruhr-University Bochum, 1995.

[8] Y. Dubois-Pelèrin, T. Zimmermann, and P. Bomme. Object-oriented finite element programming: Ii. a prototype program in smalltalk. *Comp. Meth. in Appl. Mech. Engng.*, 98(3):361–397, 1992.

[9] G. J. Kiczales, M. J. Ashley, L. Rodriguez, A. Vahdat, and D. G. Bobrow. Metaobject protocols: Why we want them and what else they can do. *Object Oriented Programming: The CLOS Perspective*, 1993.

[10] A. Stepanov and M. Lee. The standard template library. Technical report, Hewlett Packard Company, 1995.

[11] B. Stroustrup. *The Design and Evolution of C++.* Addison-Wesley, 1994.

[12] T. Veldhuizen. Expression templates. *C++ Report*, 7(5), June 1995.

[13] T. Veldhuizen. Template metaprograms. *C++ Report*, 7(4), May 1995.

[14] O. C. Zienkiewicz and Taylor. *The Finite Element Method, fourth ed.*, volume 1. Prentice Hall, Inc., 1991.

8

Efficient Management of Parallelism in Object-Oriented Numerical Software Libraries

Satish Balay
William D. Gropp
Lois Curfman McInnes
Barry F. Smith[1]

ABSTRACT
Parallel numerical software based on the message passing model is enormously complicated. This paper introduces a set of techniques to manage the complexity, while maintaining high efficiency and ease of use. The PETSc 2.0 package uses object-oriented programming to conceal the details of the message passing, without concealing the parallelism, in a high-quality set of numerical software libraries. In fact, the programming model used by PETSc is also the most appropriate for NUMA shared-memory machines, since they require the same careful attention to memory hierarchies as do distributed-memory machines. Thus, the concepts discussed are appropriate for all scalable computing systems. The PETSc libraries provide many of the data structures and numerical kernels required for the scalable solution of PDEs, offering performance portability.

8.1 Introduction

Currently the only general-purpose, efficient, scalable approach to programming distributed-memory parallel systems is the message passing model. Other approaches, based on parallel languages or compiler directives, have worked well on shared-memory computers, particular hardware platforms (e.g., CM-5) [20], or specific problems but have never been able to demonstrate general applicability. The chief drawbacks to the message passing model have been

[1]Mathematics and Computer Science Division, Argonne National Laboratory, 9700 South Cass Ave., Argonne, IL 60439-4844, USA.
Email: {balay,gropp,curfman,bsmith}@mcs.anl.gov.

(1) lack of portability due to varying syntax for message passing or inefficient and poorly designed portable systems, and
(2) the difficulty experienced by end users in writing complicated message passing code.

Fortunately, with the development of the Message Passing Interface (MPI) [8], [15], [9], [19], drawback (1) is no longer a problem. MPI is an efficient, robust standard to which the major vendors are adhering. In addition, several high-quality implementations are freely available [6], [7]. Another advantage of MPI is that it is fully usable from Fortran 77, C, and C++; this feature allows programmers to use the language that is most appropriate for a particular task or with which they are most comfortable. Another important aspect of MPI is that it provides specific mechanisms to support the development of portable software libraries that most previous message passing systems did not provide.

Drawback (2) is far more challenging; however, the implementation of high-quality parallel numerical libraries using modern software engineering practices can ease this difficulty enormously. This paper discusses specific techniques used in PETSc 2.0 (the Portable, Extensible Toolkit for Scientific computation) to allow end users, with relative ease, to write sophisticated parallel application codes that involve the numerical solution of partial differential equations (PDEs) through the use of portable, efficient numerical libraries. Thus, we assert that the combination of

the message passing model
+
carefully designed and implemented parallel numerical libraries

is the solution to the problem of efficiently using large-scale distributed-memory, as well as clustered and NUMA (non-uniform memory access) shared-memory computers. This approach enables us to face the explicit tradeoffs that must be made to balance the code's performance (computational efficiency) and ease of use (programmer efficiency). Most important, this combination allows the gradual process of improving performance by the addition of new computational kernels, while retaining the remainder of the correctly working libraries and application code.

Over the past fifteen years, various government funding agencies worldwide have invested heavily to make parallel computing usable for more than very special high-budget custom projects. Aside from hardware development, much of the research funding has been directed toward (1) basic computer science research in parallel programming paradigms and languages and (2) basic mathematical research in parallel algorithm development and analysis. These two research thrusts have often been orthogonal, with neither leveraging the corresponding advances in the other field. Thus, the resources devoted to the high-performance communication and computation (HPCC) community unfortunately have not led to the advances in the usability of parallel computing that many had anticipated. Developing efficient applications for massively parallel processors (MPPs) and clusters is still difficult and requires a large commitment from application scientists.

The approach used in PETSc is to encapsulate mathematical algorithms using computer science developments in object-oriented programming and message passing. Object-oriented programming techniques enable us to manage the complexity of efficient numerical message passing codes. All the PETSc software is freely available and used around the world in a variety of application areas [4].

Our approach does not attempt to completely conceal parallelism from the application programmer. Rather, the user initiates combinations of sequential and parallel phases of computations, but the library handles the detailed (data-structure dependent) message passing required during the coordination of the computations. This provides a good balance between ease of use and efficiency of implementation. In this article we discuss six of our main guiding design principles; the first four focus on allowing the application programmer to achieve high performance, while the last two focus on ease of use of the libraries.

- Performance
  - overlapping communication and computation,
  - determining within the library the details of various repeated communications, and optimizing the resulting message passing code (similar to the inspector/executor model of PARTI/Chaos [1]),
  - allowing the user to dictate exactly when certain communication is to occur, and
  - allowing the user to aggregate data for subsequent communication
- Ease of use
  - allowing the user to work efficiently with parallel objects without specific regard for what portion of the data is stored on each processor, and
  - managing communication whenever possible within the context of higher-level operations on a parallel object or objects instead of working directly with lower-level message passing routines.

Note that the first four principles are chiefly related to reducing the number of messages, minimizing the amount of data that needs to be communicated, and hiding the latency and limitations of the bandwidth by sending data as soon as possible, *before* it is required by the receiving processor.

The six guiding principles, embedded in a carefully designed object-oriented library, enable the development of highly efficient application codes, without requiring a large effort from the application programmer. We note that PETSc is not intended to be a complete parallel mathematical software library like the Thinking Machines' Scientific Software Library; rather, PETSc focuses on components required for the solution of PDEs and related problems.

Another strength of the approach of message passing combined with numerical libraries is that application codes written with this model will also run well on shared-memory computers—often as well as codes custom written for a particular machine. This translation occurs because even shared-memory machines have a memory hierarchy that message passing programs must inherently respect. For the small number of code locations where taking explicit advantage of the shared memory can lead to improved performance, alternative library routines that bypass the message passing system may easily be provided, thus retaining a performance-portable library.

Other researchers are also investigating object-oriented techniques and programming strategies for large-scale numerical software. A few of the projects that are most closely related to PETSc in the problems they address include Diffpack [5] (a collection of uniprocessor libraries for solving PDEs), Aztec [11] (a library for iteratively solving linear systems in parallel), and POOMA [17] (a framework for parallel scientific simulation). The unique aspect of PETSc compared with other packages is the complete integration of the six guiding principles throughout its design.

This article is organized as follows. In Section 8.2 we introduce the message passing programming model and discuss why it can result in highly efficient programs but why programming with raw message passing is difficult for most numerical applications. Section 8.3 introduces the concept of a parallel distributed object (for example, a matrix) and explains how it is managed in PETSc. Section 8.4 briefly explains the importance of each of the six conceptual principles introduced above. Section 8.5 introduces the design of several fundamental PETSc objects and for each explains how the six guiding principles are related. This section discusses several important components and operations within numerical libraries for PDEs, namely

- vector assemblies,
- vector scatters and gathers,
- matrix assemblies,
- parallel matrix-vector products,
- parallel computation of Jacobian matrices, and
- linear and nonlinear solvers.

We discuss our implementation techniques for balancing efficiency and ease of use. We conclude the section by explaining how all three principles of object-oriented programming are crucial to the management of complexity in the PETSc design. Section 8.6 demonstrates the performance possibilities of an application code written using PETSc by presenting results for a three-dimensional, fully implicit Euler simulation.

This article is not intended as a users guide or introduction to the use of PETSc; for that information we refer readers to the PETSc users manual [3]. Rather, this article discusses in some technical detail several specific aspects that are important in the design of PETSc. In fact, users of PETSc do not have to understand the technical details discussed here in order to use PETSc effectively.

8.2 The Message Passing Model for Programming Distributed-Memory Parallel Systems

Hardware for parallel computers has been designed in many ways, which can be distinguished by memory layout and interconnection schemes. The main spectrum includes common memory and bus shared by all processors, common memory connected to all processors through a switch, separate memory "owned" by one processor but directly accessible to all processors, and separate memory accessible only to its local processor [10]. Each approach has both advantages and disadvantages. The common memory approach is limited by the ability of the memory banks to serve all processors efficiently, while the distributed-memory approach is limited by the need of all processors to share data. Thus, even moderately scalable systems (and single-processor systems as well) have a hierarchy of local and remote memory that is managed directly by the hardware.

At the programmer's level, of course, the details of the memory systems are well hidden. The programmer uses an abstract memory model (or parallel programming model) that is somehow related, through system software and hardware, to the physical machine.

8.2.1 Flat Address Space

In the simplest parallel programming model, the application programmer works with a *flat memory structure*; all processors share the same address space and are free to change data at any location in memory. This model's inherent simplicity is countered by two main drawbacks.

- The user must ensure that two (or more) different processes do not generate inconsistent values in memory. For example, two processes simultaneously incrementing the same memory location by one may actually increase the value by only one, rather than the intended two. While techniques for this are well understood in theory, including locks and monitors, it does require care on the part of the user to prevent programming errors and hot-spots (bottlenecks in the program where several processes are waiting on the same locks).

- A flat address space is not scalable; even with very sophisticated hardware only extremely carefully written code can completely utilize more than a few processors. To achieve good performance, even recent machines such as the SGI/Cray ORIGIN2000 will require libraries, such as PETSc, that truly acknowledge and respect the memory hierarchy of the machine. In fact, such carefully tuned shared-memory code strongly resembles message passing code in that chunks of data are moved among "local" memories in a very controlled way.

Parallelizing compilers have been postulated as the cure for these two problems, and on a small scale they have been quite successful. But even on systems for which parallelizing compilers work well, they are often limited to highly structured code for which the compiler can detect parallelism (e.g., double loops and rectangular array operations). Compilers that can handle general sparse matrices, for example, are only at the beginning research stages, while these are exactly the types of matrices that applications scientists need to use on a regular basis. Even if parallelizing compilers vastly improve, it seems highly unlikely that they will ever be able to compile complex sequential application codes into even moderately efficient parallel codes. In fact, few do well even in entirely sequential codes for sparse matrix operations.

8.2.2 Message Passing

The other standard parallel programming model is *message passing*; in this model each process can directly access only its own memory and must explicitly communicate with other processes to access the data in their memories. The communication is done through the *send* and *receive* operations. Thus, both the sending and receiving processors must be involved whenever a remote memory location is accessed. For example, if process 1 wanted to add to its local variable x the value y from processor 0, the code for the two processes could look like the following:

```
Process 0 code          Process 1 code
--------------          --------------
MPI_Send(y,......);     MPI_Recv(mess,....);
                        x += mess; /* Add the remote data to x */
```

The Message Passing Interface (MPI) standard contains a wide variety of basic communication routines, including reductions, scatters, and broadcasts [15], [9], [19]. But these routines are predicated on both the sending and receiving processors being aware of the data's origin and destination. Consequently, writing complicated message passing codes is tedious and prone to error. To illustrate this point, we consider a specific example:

sparse matrix-vector multiplication, $y = A * x$. This operation occurs in most iterative linear solvers, scalable eigenvalue solvers, etc.

8.2.2.1 *Why Writing Message Passing Code Is Tedious*

Given a parallel sparse matrix A and a parallel vector x, we wish to write a routine that scalably and efficiently computes $A * x$. We assume that A is distributed by rows among a collection of processors; that is, each processor contains (for simplicity) an adjacent series of rows of the matrix and the corresponding elements of the vector. See the matrix depicted below for an example division among three processors.

One could easily code a naive sparse matrix-vector product using MPI. Each processor could broadcast its elements of the vector x to all other processors with the command

```
MPI_Allgatherv(local,localsize,MPI_DOUBLE,global,localsizes,...);
```

Here each processor contributes its piece (of length `localsize`, called `local`) of the entire vector to a copy of the entire vector (called `global`), which lies in its entirety on each processor. The advantages of this approach are that the communication call is simple, and every processor knows exactly what messages it must send and receive. One disadvantage is that this code is not scalable; the amount of communication grows as $O(n)$, where n is the number of columns in the matrix. In addition, memory is wasted since each processor must store a complete copy of the vector x. Note that even in this simple case each processor must know the amount of data to expect from all other processors, as determined by prior communication (in a setup phase) and given by the array `localsizes`.

To discuss how we can take advantage of matrix sparsity, we consider the following matrix, which is partitioned by rows among three processors so that processors zero, one, and two "own" submatrices consisting of rows 0–2, 3–5, and 6–7, respectively. The corresponding vector is partitioned accordingly;

$$\left(\begin{array}{cccccccc} 1 & 2 & 0 & 0 & 0 & 0 & 0 & 0 \\ 0 & 5 & 6 & 7 & 0 & 0 & 8 & 0 \\ 9 & 0 & 10 & 11 & 0 & 0 & 12 & 0 \\ \hline 13 & 0 & 0 & 15 & 16 & 17 & 0 & 0 \\ 0 & 18 & 0 & 19 & 20 & 21 & 0 & 0 \\ 0 & 0 & 0 & 22 & 23 & 0 & 24 & 0 \\ \hline 25 & 26 & 27 & 0 & 0 & 28 & 29 & 0 \\ 30 & 0 & 0 & 0 & 0 & 33 & 0 & 34 \end{array}\right) \left(\begin{array}{c} x_0 \\ x_1 \\ x_2 \\ \hline x_3 \\ x_4 \\ x_5 \\ \hline x_6 \\ x_7 \end{array}\right)$$

Note that each processor's local submatrix contains certain columns with all zero elements, so that when computing a matrix-vector product, $y = A * x$, each processor requires only a subset of the entries of x. In this example, processor zero does not need x_4, x_5, and x_7; processor one does not need x_2 and x_7; and processor two does not need x_3 and x_4. To minimize communication, processor zero should distribute x_0, x_1, and x_2 to processor two but only x_0 and x_1 to processor one. Likewise, processor one need only distribute x_3 to processor zero and x_5 to processor two. Meanwhile, processor two needs to distribute x_6 to both processors zero and one. Clearly, for this small problem the communication reduction achieved by this organization is not worth the coding difficulties, but for large sparse problems (e.g., $n = 1,000,000$) for which the communication can potentially drop to 500 from $1,000,000$, such reduction is very important. What makes this problem nontrivial is that no processor knows *a priori* what components of the vector other processors will require.

Since sparse matrix-vector products are only a small part of a large application code, it is unrealistic to require an application programmer not highly trained or interested in message passing programming to write all the code required to perform efficient parallel sparse matrix-vector products. In later sections, we discuss how PETSc provides efficient implementations of these fundamental, low-level parallel routines, in a format immediately useful for an application code.

8.2.2.2 *Why Writing Correct Message Passing Code Is Difficult*

Not only is writing message passing code tedious, it also is technically difficult, since rather subtle issues become extremely important for guaranteeing robust, *correct* libraries. Writing such code requires expert knowledge that most application programmers have neither the time nor interest to master.

We now present a specific example for which a naive but quite reasonable implementation can result in unexplained deadlock, while a more sophisticated implementation will perform efficiently and correctly for all problem sizes and machines. Consider the case of dealing with finite sizes for the internal system buffers used in message passing. If two processors both initiate blocking sends to each other at the same time, the data to be transferred must be moved away from the sending processors' buffers before the sends can complete and return to the user's code. The data is usually placed in an intermediate buffer (the details of which vary among machines) until the receiving processor can accept the data. If the messages exceed the amount of available buffer space, then deadlock can occur. The likelihood of deadlock depends on the past history of the computation and even the order in which packets are switched through the computer. Thus, seemingly random deadlock can occur for different runs of the same code.

In more complicated applications that use blocking sends, it is not always easy to determine whether a possibility of deadlock exists. This situation arises because each processor may be running different portions of the code when the deadlock occurs. Also, locating the cause of the deadlock is often

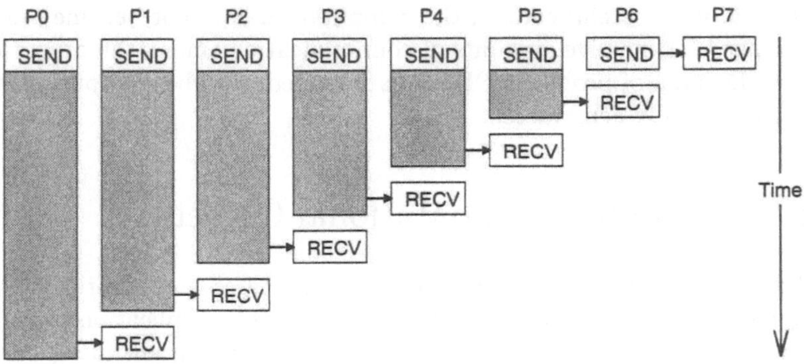

FIGURE 8.1. Blocking Sends Effect on Performance.

difficult due to its random appearance. There are, of course, techniques to prevent the occurrence of deadlock, including the use of nonblocking communications and explicit library control of buffering, but these require a more sophisticated and deeper understanding of parallel programming than most application programmers have time to master.

Another example illustrates degradation of a program's performance, due to a naive, yet seemingly reasonable, implementation. In this case a collection of processes all simultaneously send data to the processor to the right (the last processor does not perform a send). In Figure 8.1 this situation is depicted with eight processors. During the first stage, processor 6 actually transfers its message to processor 7 while processors 1 through 5 wait for their right-hand neighbors to post a receive. During the second stage, processor 5 transports its message to processor 6 while processors 1 through 4 continue to wait for their neighbors to post a receive. Thus, the entire communication requires seven stages, while the underlying hardware may have been able to perform the required communication in one or at most two stages. Again, the proper use of nonblocking techniques would alleviate this problem.

Understanding such details in message passing is similar to the necessity of understanding the numerical effects of working with finite-precision calculations when computing. Most of the time one can simply use common sense and generate the correct results, but catastrophic failure can occur if one is not grounded in the fundamentals of numerical analysis. Thus, for both message passing and numerical programming, we can encapsulate the needed functionality in software libraries that allow all users to take advantage of the experts' knowledge.

We remind the reader that the solution to PDEs at any point is determined mostly by input data that is geometrically near that point. Thus, for many application problems involving PDEs, a geometric decomposition of the solution domain among the processors is most appropriate. This leads immediately to data locality on the computer, an ideal situation for any NUMA parallel machine, including distributed-memory processors programmed with message passing. Since the bulk of the computation involves local data, with careful coding, the computation does not become limited by the need for massive amounts of communication among the processors. So, for the class of problems PETSc is intended, scalable computing is at least theoretically achievable.

8.3 Distributed Computational Objects

PETSc is built around a variety of data structures and algorithmic objects, some of which are depicted in Figure 8.2. The application programmer works directly with these objects rather than concentrating on the underlying (rather complicated) data structures.

The three basic abstract data objects in PETSc are index sets, vectors, and matrices. An index set is an abstraction of a list of integer indices, which is used for selecting, gathering, and scattering subsets of elements. A PETSc vector is an abstraction of an array of values (e.g., coefficients for the solution of a PDE), and a matrix represents a linear operator that maps between vector spaces. Each of these abstractions has several representations in PETSc. For example, PETSc currently provides three sequential sparse matrix data formats, four parallel sparse matrix data structures, and a dense representation. Each is appropriate for particular classes of problems. Several data distribution examples for particular PETSc objects are given in Section 8.5.

Built on top of this foundation are various classes of solver objects, including linear, nonlinear, and timestepping solvers. These solver objects encapsulate virtually all information regarding the solution procedure for a particular class of problems, including the local state and various options. Details are discussed in Section 8.5.

In general, the data for any PETSc object (vector, matrix, grid, linear solver, etc.) is distributed among several processors. The distribution is handled by an MPI communicator (called `MPI_Comm` in MPI syntax), which represents a group of processors. When a PETSc object is created, for example with the commands

```
VecCreate(MPI_Comm comm,int m,Vec* vector);
MatCreate(MPI_Comm comm,int m,int n,Mat *matrix);
SLESCreate(MPI_Comm comm,SLES *linear_solver);
```

the first argument specifies the communicator, thus indicating which processes share the object. The creation routines are collective over all processors in the communicator; thus, when creating a PETSc object, all processors in the communicator *must* call the creation routine.

The use of communicators in parallel software libraries is extremely important, since it enables all communication for a particular operation (e.g., a matrix-vector product) to be isolated from communication in other parts of code. Such encapsulation eliminates the problem of colliding tags (for example, when two libraries inadvertently use the same tag on different messages, one library may incorrectly receive a message intended for the

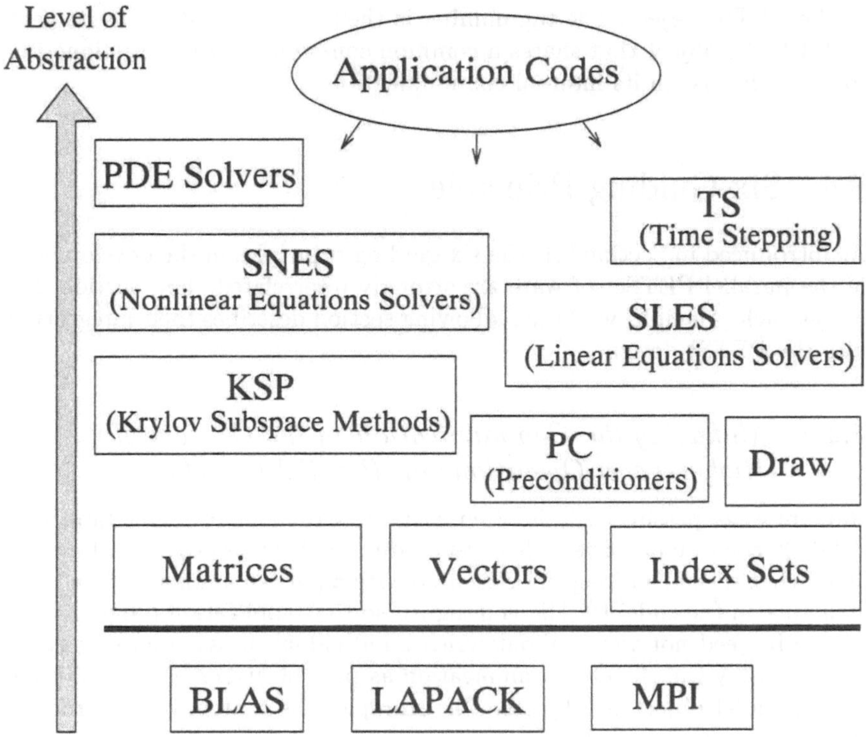

FIGURE 8.2. Organization of the PETSc Library.

other library), which was a serious limitation of older message passing systems.

The underlying communicators in PETSc objects ensure that communications for different computations are separate. We achieve this segregation upon object creation by immediately duplicating via `MPI_Comm_dup()` (an MPI function that makes a copy of a given communicator) any communicator that is not already a "PETSc communicator" and then denoting it as such by inserting an MPI attribute via `MPI_Attr_put()`. An MPI attribute is simply any collection of data a user chooses to attach to a communicator. This PETSc attribute essentially contains a tag number that is assigned to the PETSc object. The tag number is then decremented to ensure that each PETSc object that shares a common communicator has a unique tag (or tags) for use in its internal communication.

8.4 Six Guiding Principles

As introduced in Section 8.1, the six guiding principles in the development of the parallel PETSc software are strongly interrelated. This section discusses each principle, while the following section describes their integration into the PETSc design.

8.4.1 Managing the Communication in the Context of Higher-Level Operations on Parallel Objects

Raw message passing code is often extremely difficult to understand and debug because, unless the code is very carefully documented, it is often unclear what specific message is associated with a particular operation or data structure in the code. PETSc is designed so that application programmers generally need not worry about writing individual message passing calls. Instead, they can direct communication as part of higher-level operations on a parallel object or objects. For example, the matrix-vector product interface routine, given by

```
MatMult(Mat A,Vec x,Vec y);     /*  y = A*x  */
```

institutes a collection of MPI calls to manage the underlying communication required for the particular data structures being used (see Section 8.5.2 for details). Additional examples of encapsulating complicated lower-level communication sequences are the PETSc vector scatters/gathers as well as matrix and vector assembly routines.

The ability to encapsulate all details of communication inside a PETSc object/operation is vital for building a system that is relatively easy to use. Such organization also facilitates the design of higher-level modules, such as linear and nonlinear solvers, which can then focus more clearly on mathematical abstractions rather than being cluttered by excessive communication details.

8.4.2 Overlapping Communication and Computation

On virtually all modern computers, the data communication required in implementing an algorithm is often more of a limiting factor in its performance than the actual floating-point operations. To use a computer efficiently in numerical computations, it is important to limit data movement and, whenever possible, to perform numerical computation on one set of data while another set of data is in motion.

Within MPI, nonblocking operations support overlapping the communication of certain data with computation on other data. These nonblocking routines initiate communication but may return to the calling routine immediately, before the communication is complete. For example, a nonblocking send for double-precision data of length `count` can be handled with the code fragment

```
MPI_Request request;  int  count, proc, tag;
MPI_Status  status;   void *buffer;
MPI_Comm    comm;
MPI_Isend(buffer,count,MPI_DOUBLE,proc,tag,comm,&request);
... /* Do some other computation, etc. */
MPI_Wait(&request,&status);
```

Here `buffer` is the initial address of the send buffer, `proc` is the rank (number, where the processors are numbered from 0 to `size-1`) of the destination processor, `tag` is the message tag, `comm` is the communicator, and `request` is the communication request. Likewise, a basic nonblocking receive can be handled as follows:

```
MPI_Request request;  int  count, proc, tag;
MPI_Status  status;   void *buffer;
MPI_Comm    comm;
MPI_Irecv(buffer,count,MPI_DOUBLE,proc,tag,comm,&request);
... /* Do some other computation, etc. */
MPI_Wait(&request,&status);
... /* Now use the data in buffer */
```

Clearly, programmers working directly with message passing routines can themselves institute the overlap of computation and communication. More important, PETSc provides this opportunity within many of the higher-level operations mentioned in the preceding section for encapsulating complicated communication patterns. This makes all the optimizations in the communication transparent to the user.

8.4.3 Precomputing Communication Patterns

In many aspects of the numerical solution of PDEs (for example, iterative solution of linear systems and explicit timestepping), the same communication is repeated many times on essentially the same data structures. If, each time the communication had to occur, one redetermined what sends and receives had to be instituted (and just this determination requires its own communication), this process would be very inefficient. Fortunately, it is possible to precompute exactly what messages need to be sent and received and the amount of data that will be transferred. In fact, MPI even explicitly supports precomputing through *persistent* send and receive objects.

If one is aware that a particular receive is to occur multiple times, in MPI one can initialize the set of receives by calling

```
MPI_Request request;
MPI_Recv_init(buffer,count,MPI_DOUBLE,proc,tag,comm,&request);
```

In this case, we are receiving double-precision data of length `count`; `buffer` is the initial address of the receive buffer; `proc` is the rank of the receive processor; `tag` is the message tag; `comm` is the communicator; and `request` is the communication request. Then every time the receive is required, one can simply call

```
MPI_Start(&request);
/* Do other computations */
MPI_Wait(&request,&status);
/* Use the data in the buffer */
```

There is analogous syntax for repeated sends. In addition, MPI directly supports the use of persistent communication on a series of related messages.

As discussed in Section 8.2.2, it is often not known *a priori* by a receiver who is sending it data. Thus a set-up phase must be performed to even know what communication needs to take place. For similar, repeated communications, a clean mechanism is required to manage this set-up phase, so the same set-up need not be repeated. This is discussed in detail for vector scatters in Section 8.5.1.3.

It is natural to encapsulate information for potentially repeated communication patterns in the objects that define higher-level operations. As will be demonstrated in the following sections, we do exactly this for operations such as matrix-vector products and vector scatters/gathers.

8.4.4 Programmer Management of Communication

Certain parallel programming paradigms (for example, HPF) attempt to conceal completely from the user knowledge of when communication takes place. Although this approach to make parallel programming "easier" is very appealing, it can have serious effects on both efficiency and the user's knowledge of what is actually occurring in an application code. In PETSc, the user can explicitly initiate communication by calling specific PETSc routines. For example, to perform a parallel vector scatter/gather (discussed in Section 8.5.1), the user calls the routines

```
VecScatterBegin(Vec x,Vec y,InsertMode im,ScatterMode sm,
                VecScatter scattercontext);
VecScatterEnd(Vec x,Vec y,InsertMode im,ScatterMode sm,
                VecScatter scattercontext);
```

Thus, within an application code the user can dictate the time that communication takes place with respect to computations. In addition, the user often can arrange the overlap of communication and computation by placing code unrelated to the message passing between calls to multiphased routines such as the vector scatters/gathers. This situation is analogous to the use of prefetching for hierarchical memories.

8.4.5 Working Efficiently with Parallel Objects without Regard for Details of Data Location

When assembling a distributed object that requires a large amount of user-provided data (for example, a vector or matrix), one needs an efficient mechanism for transferring data from the application code to its correct location in the library data structures. This mechanism must be easy to use; otherwise, the application programmer will be tempted simply to bypass the library and assemble the data structures manually.

Within PETSc we allow the user to insert data into global PETSc objects without regard for the particular processor on which the data is destined ultimately to be stored. To require the user to generate all data on the "owner" processor is simply too much of a burden within most applications and is, in fact, unnecessary. Rather, PETSc internally retains (called stashing in PETSc) those values destined for another processor until an explicit communication phase is initiated by the user (an example of the concept of the application programmer determining when communication occurs, as discussed above). This topic is discussed in detail for vectors in Section 8.5.1 and for matrices in Section 8.5.2. This same stashing technique is commonly used and, in fact, is vital for obtaining good performance in I/O systems, where it is referred to as write caching.

8.4.6 Aggregation of Data for Communication

The overhead of initiating message transition is quite high relative to the cost of sending a segment of data. For efficient parallel programming, whenever possible one should pack relatively large amounts of data for a single subsequent aggregate transition instead of sending several smaller messages. PETSc incorporates this strategy in various communication phases throughout the library, for example in the transfer of user-supplied non-local data within parallel matrix and vector assembly routines. What is important in the PETSc design is that it allows this type of optimization without requiring an effort from the application programmer.

8.5 PETSc Design of Fundamental Objects

In this section we discuss several of the fundamental PETSc objects and their implementations in the context of the six guiding principles introduced in Section 8.4. We briefly summarize this information, which is given in full detail below.

- Vectors are perhaps the simplest PETSc objects and are used, for example, to store the coefficients of the solutions of PDEs. The main parallel operations that we perform on vectors are
  - vector assembly, which (1) allows the user to insert vector entries without regard to their processor ownership, and (2) aggregates the values until the user directs that communication can begin; and
  - vector scatters/gathers, which (1) allow overlap of the communication and computation, (2) reuse precomputed communication patterns, and (3) conceal all communication within the context of the higher-level abstraction of a scatter/gather.
- Matrices, another class of fundamental PETSc data objects, represent linear operators. Often, matrices are obtained as Jacobians of an underlying nonlinear algebraic system arising from the discretization of a PDE. We discuss
  - matrix assembly, which is similar to the vector assembly mentioned above; and
  - sparse matrix-vector products, which demonstrate (1) the overlap of communication and computation by dividing the calculation into sections that are purely local and that require off-processor data, (2) precomputation of a repeated communication pattern, and (3) management of communication within the context of a higher-level operation (the matrix-vector product).
- We discuss the efficient parallel computation of numerical Jacobians using coloring and finite differences. This topic demonstrates (1) the use of aggregation to reduce the amount of distinct communications required, (2) precomputation of communication patterns (including

the coloring), and (3) the encapsulation of the entire process within an abstract PETSc operation.

- Finally, we discuss the linear and nonlinear solvers. In this section we demonstrate how all communication required during the solution process is managed directly in the context of a higher-level "solver" object.

The remainder of this section discusses these operations in detail.

8.5.1 Vectors

In the numerical solution of PDEs, a vector may often be thought of as a set of coefficients representing a function on a grid. A vector may also be thought of as a set of elements of R^N, the usual Euclidean vector space. It is, however, too limiting to think of a vector as merely a one-dimensional array of floating-point numbers in memory, since the components of a vector in a parallel machine will generally not be stored in contiguous memory locations.

In PETSc one can create a parallel vector with the command

```
VecCreateMPI(MPI_Comm comm,int nlocal,int nglobal,Vec *vector);
```

As mentioned in Section 8.3, all processors in the communicator `comm` must call this routine, because vector creation is an aggregate operation. The arguments `nglobal` and `nlocal` indicate, respectively, the total size of the vector and the number of elements to be represented locally on a particular processor. Either one, but not both, of the arguments `nglobal` and `nlocal` may be set to `PETSC_DECIDE` to allow PETSc to determine the value. Upon this call, PETSc allocates memory to store the vector entries and sets up any data structures required for manipulating the vector.

8.5.1.1 *Vector Assembly*

Although the components of PETSc vectors are ultimately distributed among the various processors, it is highly desirable to allow users to assemble the vector components easily, without regard to the elements' final storage locations. That is, we would like to separate completely the vector data storage formats from the application codes. This capability simplifies the application codes and allows library writers to provide highly tuned data structures without imposing a burden on users. This facet of design is an example of data encapsulation, which is a very basic requirement for flexible libraries.

To illustrate the parallel vector assembly process, we consider three cases of constructing the right-hand side of a linear system: by the finite difference method, by the finite element method, and by a special case for a grid-based nonlinear system arising from a finite difference discretization of a nonlinear PDE. In all cases, the resulting parallel vector is distributed among the processors, so that each vector component is owned by exactly one processor.

Case 1: Simple finite difference discretization in one dimension. The user creates a parallel vector and partitions it among the processors by calling `VecCreateMPI()`, as discussed above. To evaluate the vector entries, one can use a routine such as the following:

```
Vec    F;                        /* global vector */
int    istart, iend, i, N;
double f;
  ...
VecCreateMPI(comm,PETSC_DECIDE,N,&F);
VecGetOwnershipRange(F,&istart,&iend);
/* Loop over local entries, inserting vector elements */
for ( i=istart; i<iend; i++ ) {
  /* Assign f to be some function of the grid node */
  VecSetValues(F,1,&i,&f,INSERTVALUES);
}
/* Complete the vector assembly process */
VecAssemblyBegin(F);
VecAssemblyEnd(F);
```

In this simple case each processor generates contributions only for its local part of the vector, `F`, and inserts these values one at a time with `VecSetValues()`; no parallel communication is required. In this example, the routines `VecAssemblyBegin()` and `VecAssemblyEnd()` do essentially nothing except verify that the user has completed inserting entries into the vector on all processors. However, in most serious applications the situation is not this simple, as demonstrated by Case 2.

Case 2: Finite element discretization of the right-hand side of a PDE. The vector is calculated by the integrals

$$F_i = \int_{\mathrm{supp}(\phi)} f(x)\phi(x)dx,$$

where x can be one-, two-, or three-dimensional, and supp(ϕ) denotes the support of ϕ. In most finite element codes these integrals are calculated numerically, element by element. Thus, certain vector components, F_i, receive contributions from more than one processor.

We demonstrate this situation with an example in two dimensions using piecewise linear finite elements. Assume that we have generated a parallel vector, `F`, and in addition have distributed the finite elements (in this case triangles) among the processors. That is, each processor has a list of its local elements (triangles) and, for simplicity, the locations of the element nodes. For demonstration purposes the data structures representing the grid in the following code fragment are very simple and are not necessarily the best choices for an actual code.

```
Vec    F;          /* global vector */
int    gidx[3];   /* work array of vector indices */
double f[3];       /* work array of vector contributions */
int    Nt, *tri, i;
double *X, *Y, x0, y0, x1, y1, x2, y2;
/* Loop over the local elements */
for ( i=0; i<Nt; i++ ) {
  /* Load the vertices of the element into local variables */
  x0 = X[3*i];    y0 = Y[3*i];
  x1 = X[3*i+1]; y1 = Y[3*i+1];
  x2 = X[3*i+2]; y2 = Y[3*i+2];
  /* Compute the local element vector in a work array, f,
     with an application-specific routine */
  CalculateElementIntegral(n,x0,x1,x2,y0,y1,y2,f);
  /* Load global indices for the nodes of the element into gidx */
  gidx[0] = tri[3*i]; gidx[1] = tri[3*i+1]; gidx[2] = tri[3*i+1];
  /* Add the local vector contribution, f, into the global vector,
     F, in the rows indicated by gidx */
  VecSetValues(F,3,gidx,f,ADD_VALUES);
}
/* Complete the vector assembly process */
VecAssemblyBegin(F);
VecAssemblyEnd(F);
```

Here we insert three elements into the parallel vector, `F`, with each call to `VecSetValues()`. Generally, it is desirable to set multiple values simultaneously; otherwise, the overhead for function calls increases. This is an example of aggregation. In most true applications, aggregating the insertions is natural and easy.

In this example, as for most problems discretized with parallel finite elements, some contributions to the vector components are generated by a different processor from their owner, because each finite element is assigned a unique processor and each vector element similarly is assigned a unique processor. Note that these contributions are added to the appropriate global vector component, as indicated by the flag `ADD_VALUES` within `VecSetValues()`. The vector assembly routines perform the required message passing to move the values to their correct locations. As discussed below, separating the assembly process into two stages, `VecAssemblyBegin()` and `VecAssemblyEnd()`, is critical for allowing application programmers to overlap communication and computation. Although the procedure was not done in this example, a careful user will attempt to generate the majority of the vector entries on the correct processor. (Note that there are tricks for reducing the cost of integral evaluation that also may exploit locality.) This example indicates that trying to generate all vector entries on the correct processor can lead to a unnecessarily messy application code, because certain finite elements would have to be dealt with several times on multiple processors.

Application programmers can, in fact, manage such message passing themselves. However, this task should be done only by experienced parallel programmers who wish to develop a custom implementation that optimizes performance for a particular problem. Since most users have neither the time nor the experience to finely tune their assembly routines, their special-purpose codes generally would not perform as efficiently as well-written library routines.

Since the application code is free to insert components into a vector from any processor, regardless of their ultimate storage location, there must be a mechanism to store the values locally until they are transferred to the processor that owns them. In PETSc, these values are stored in a stash until they are transferred to the correct processor. The stash can be implemented in many ways; the one tricky point is correctly handling values that are inserted multiple times for the same component. Our current implementation is fairly naive; we simply retain an array of values and their indices. A simple optimization might be to sort the list by indices; more sophisticated implementations might use a hash table.

The final movement of the stashed values to their proper home processor is done in several stages:

- By making calls to `MPI_Allreduce()`, each processor determines the maximum size delivery to expect, as well as a count of the number of deliveries.
- Each processor allocates enough buffers to accommodate the arrival of all stashed values and posts corresponding nonblocking receives.
- Each processor posts nonblocking sends of all its elements to be distributed.
- Each processor waits for its receives and inserts the received values into the correct locations as indicated by the received indices.
- Finally, all processors wait for their nonblocking sends.

The first three steps are done during the `VecAssemblyBegin()` stage and the last two during the `VecAssemblyEnd()` phase.

The process of setting values into a vector demonstrates several of our six guiding principles:

- Aggregation:
  - The user aggregates several vector values into a single call to `VecSetValues()`.
  - The "off-processor" values are aggregated in the stash before being moved *en masse* to the correct processor.
- Values are stashed locally until the user explicitly initiates communication by calling `VecAssemblyBegin()` and `VecAssemblyEnd()`.

- The communication is managed within a vector object, not as a series of seemingly unrelated calls to `MPI_Send()` and `MPI_Recv()`.

- Communication and computation can be overlapped between the two assembly calls.

- The user can insert values into the vector without specific regard for the processor on which the data will ultimately reside. But through the use of `VecGetOwnershipRange()`, the user can ensure that most vector entries are generated on the "correct" processor, thus allowing efficiency without requiring complicated application code including application-specific message passing.

8.5.1.2 *Distributed Arrays: Managing Ghost Points*

Next, we consider a more complicated scenario, requiring vector communication before local function evaluation can occur. In this example, we introduce another PETSc construct, the distributed array (called `DA` in PETSc). These data structures manage the communication required for *ghost points*, which are the bordering portions of the vector that are owned by neighboring processors. Figure 8.3 illustrates the ghost points for processor six of a two-dimensional, regular, parallel grid. Each box represents a processor; the ghost points for processor six's local part of a parallel array are shown in gray.

Two types of distributed arrays can be created: one based on star-type stencils and one based on box stencils. In both cases the ghost points are identical, the only difference being that with star-type stencils certain ghost points are ignored, potentially decreasing substantially the number of messages sent. In two dimensions a `DA` object of stencil type `st` and periodicity type `pt` can be created with the command

```
DACreate2d(MPI_Comm comm,DAPeriodicType pt,DAStencilType st,
           int M,int N,int m,int n,int w,int s,DA *da);
```

Here `w` indicates the number of degrees of freedom per node, and `s` is the stencil width. The arguments `M` and `N` are the global dimensions of the array, while `m` and `n` indicate the partition of the array among the processors. One should either use `PETSC_DECIDE` for `m` and `n` or ensure that `m*n` equals the number of processors in the communicator. We employ two-dimensional distributed arrays in the following example.

Case 3: Consider the classic Bratu problem [2],

$$-\Delta u - \lambda e^u = 0, \qquad 0 < x, y < 1,$$

with boundary conditions

$$u = 0 \text{ for } x = 0, x = 1, y = 0, y = 1,$$

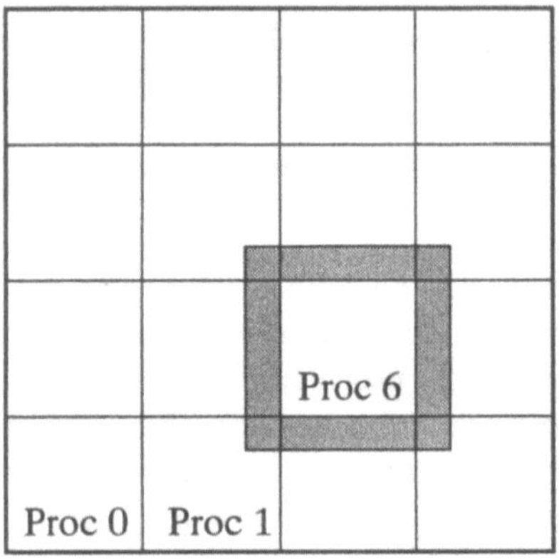

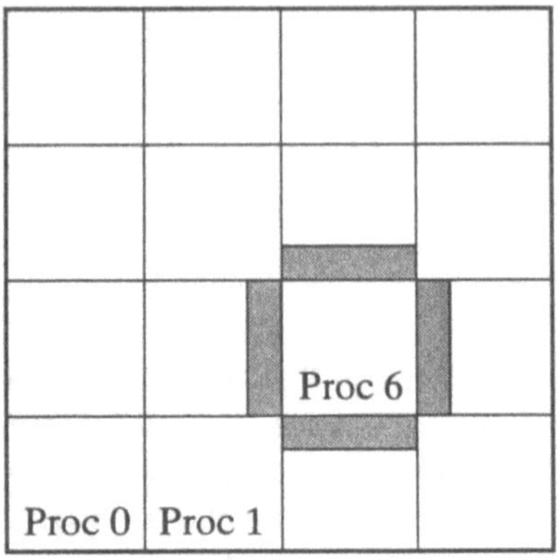

FIGURE 8.3. Ghost Points for Two Stencil Types on Processor Six.

which is discretized by finite differences on a uniform m by m grid. The resulting nonlinear algebraic equation for an interior node is given by

$$4u_{i,j} - u_{i+1,j} - u_{i-1,j} - u_{i,j+1} - u_{i,j-1} - h^2 \lambda e^{u_{i,j}} = 0,$$

where $h = 1/(m-1)$ is the grid size. The PETSc code for evaluating this function (`F(X)`) in parallel, for use in a Newton-based method within SNES (the nonlinear solvers component), is given by the following routine:

```
/* SNES - nonlinear solver context */
/* X    - input vector */
/* F    - output vector (nonlinear function) */
/* user - user-defined context with application-specific data */
int EvaluateFunction(SNES snes,Vec X,Vec F,ApplicationCtx user)
{
  DA      da = user->da;         /* distributed array */
  Vec     localF = user->localF; /* local work vector */
  Vec     localX = user->localX; /* local work vector */
  int     i,j,row,xs,ys,xm,ym,Xs,Ys,Xm,Ym,m = user->m;
  double  h,hh,ut,ub,ul,ur,u,uxx,uyy,*x,*f,lambda = user->param;
  h      = 1.0 / (double)(m-1);
  hh     = h*h;
  /* Transfer local portion (including ghost points) of global
     vector, X, into a local work vector, localX */
  DAGlobalToLocalBegin(da,X,INSERTVALUES,localX);
  DAGlobalToLocalEnd(da,X,INSERTVALUES,localX);
  /* Get the indices of the lower left corner of the region
     owned by this processor; get the corresponding indices
     for the region including ghost points. */
  DAGetCorners(da,&xs,&ys,0,&xm,&ym,0);
  DAGetGhostCorners(da,&Xs,&Ys,0,&Xm,&Ym,0);
  /* Directly access arrays (x and f) containing vector data */
  VecGetArray(localX,&x);
  VecGetArray(localF,&f);
```

```
  /* Loop over local region owned by processor, performing
     application operations */
  for (j=ys; j<ys+ym; j++) {
    for (i=xs; i<xs+xm; i++) {
      row = i - Xs + (j - Ys)*Xm;
      if (i == 0 || j == 0 || i == m-1 || j == m-1 ) {
        f[row] = x[row]; continue;
      }
      u = x[row]; ub = x[row - Xm]; ul = x[row - 1];
      ut = x[row + Xm]; ur = x[row + 1];
      uxx = (-ur + 2.0*u - ul); uyy = (-ut + 2.0*u - ub);
      f[row] = uxx + uyy - hh*lambda*exp(u);
    }
  }
  /* Place values in global vector */
  DALocalToGlobal(da,localF,INSERTVALUES,F);
  return 0;
}
```

The application portion of the routine (within the double `for` loop) appears exactly like a sequential code, except that the loop indices range over only the local portion of the distributed array. The indices of the lower left corner of the local portion of the array as well as the local array size are obtained with the commands `DAGetCorners()` and `DAGetGhostCorners()`. The first version excludes any ghost points, while the second version includes them and deals with the fact that subarrays along boundaries of the problem domain have ghost points only on their interior edges, but not on their boundary edges.

To evaluate its local function, each processor requires its portion of the vector `x` as well as its ghost points. The calls to `DAGlobalToLocalBegin()` and `DAGlobalToLocalEnd()` handle all of the detailed message passing to retrieve the needed ghost points from neighboring processors.

The distributed arrays in PETSc demonstrate several of our guiding design principles:

- Overlapping communication and computation.
- Precomputing repeated communication patterns.
- Managing the communication in the context of a PETSc object, in this case the `DA`.
- Allowing the user full control of the time that communication occurs, via calls to `DAGlobalToLocalBegin()` and `DAGlobalToLocalEnd()`.
- Aggregation, in that from the user's point of view a single communication of all the ghost values occurs, while internally a variety of message passing activities (which have been organized for efficiency) take place.

8.5.1.3 *Index Sets and Vector Scatters*

The previous example illustrates a very special type of vector scatter and gather. PETSc also includes general-purpose scatters and gathers, which we illustrate below with a contrived example.

To facilitate vector scatters and gathers, PETSc employs the concept of an *index set.* An index set, which is a generalization of a set of integer indices, is used to define various operations on vectors and matrices. The following command creates a sequential index set, `is`, based on a list of integers of length `n` in the array `indices`:

```
ISCreateGeneral(MPI_Comm comm,int n,int *indices,IS *is);
```

Another standard index set, which is defined by a starting point (`first`), a stride (`step`), and a length (`n`), can be created with the command

```
ISCreateStride(MPI_Comm comm,int n,int first,int step,IS *is);
```

We use strided index sets in the following vector scatter/gather example. In this case, one vector is half the length of the other; we wish to gather every second element of the longer vector, `x_from`, into the shorter, `x_to`.

```
VecScatter ctx;
IS         is_from, is_to;
ISCreateStride(MPI_Comm comm,n/2,0,1,&is_to);
ISCreateStride(MPI_Comm comm,n/2,0,2,&is_from);
VecScatterCreate(x_from,is_from,x_to,is_to,&ctx);
ISDestroy(is_to); ISDestroy(is_from);
VecScatterBegin(x_from,x_to,INSERT_VALUES,SCATTER_ALL,ctx);
  /* Do other useful work */
VecScatterEnd(x_from,x_to,INSERT_VALUES,SCATTER_ALL,ctx);
```

The scatter context can be saved and reused whenever the same scatter is required.

As discussed below, the vector scatters are performed in several stages to enable reusing communication patterns and overlapping communication and calculation. In the first scattering stage, which is initiated by calling `VecScatterCreate()`, the destination and source of each component are determined. First, each processor decides the number of components it needs from the other processors; it then calls `MPI_Allreduce()` with these quantities to determine the number of requests to expect and the maximum length of any request. Next, each processor posts a suitable number of nonblocking receives with suitable buffer sizes, followed by posts of all of its requests with nonblocking sends. Each processor then waits for its requests. As the requests arrive, the processor creates the `VecScatter` data structure that contains information regarding all the sends it must perform during the scatter. Once this has been completed, the processor creates an analogous data structure with information about all of its expected receives and the storage location of the values. Finally, the processor waits for the posted nonblocking sends. Note that the construction of the `VecScatter` data structure is modeled on the inspector/executor ideas in PARTI/Chaos [1].

We have decided to hide the creation of the `VecScatter` data structure in a single stage. To allow even more overlap of application computation and communication with the nonblocking sends and receives, one could break the context creation into two or three stages. In the interest of simplicity we have not done so, believing that it is an overoptimization, since the scatter context is created only once and then used repeatedly.

The actual vector scattering begins with the `VecScatterBegin()` stage, when the processor posts all of its nonblocking receives, followed by posting its nonblocking sends. The `VecScatter` contains the required MPI communicator, all information regarding what must be posted, and locations to store the `MPI_Request` data structure.

The `VecScatterEnd()` phase waits for the receives, inserts the arriving data into the correct vector locations, and waits for the posted sends. In our implementation, we do not use the MPI scatter construct to do the scattering. Instead, we manually gather the data into a buffer (hidden in `VecScatter`) and send the buffer with a persistent nonblocking send. Similarly, we receive the data into a buffer and then scatter the buffer into the actual vector. We wrote the code this way to support a scatter-add operation, which MPI does not directly support.

This implementation facilitates the overlap of calculation and communication by allowing the application code to perform work unrelated to the scatters while the messages are in transition (that is, between the calls to `VecScatterBegin()` and `VecScatterEnd()`). However, we ignore several additional possibilities for overlap. For instance, the final wait-on-sends could constitute a separate stage at the scatter's conclusion. Such modifications could provide additional efficiency, though at the cost of a steeper learning curve and more complicated application and library codes. We have tried to strike a healthy balance between the ability to achieve the important overlap of communication and calculation with little added complexity.

An observant reader may have noticed that since we keep MPI request objects as well as temporary buffers in the `VecScatter`, it may be used for only a single scatter at a time. To prevent programming errors resulting from accidental concurrent reuse, the scattering context contains a flag that indicates whether it is currently involved in a scatter.

The vector scatter construction in PETSc demonstrates several of our six guiding principles:

- Overlapping communication and computation.
- Precomputing repeated communication patterns during the routine `VecScatterCreate()`, which are then reused for several scatters.

- Managing the communication in the context of a PETSc object, in this case the `VecScatter`.
- Allowing the user complete control of the time that communication takes place, via calls to `VecScatterBegin()` and `VecScatterEnd()`.

In a later section we discuss the use of the basic vector scatter to implement some of our parallel sparse matrix-vector products.

8.5.2 Matrices

PETSc provides a variety of matrix data structures and many of the matrix operations required for the solution of PDEs. In this section we focus on two specific facets of our matrix implementation (matrix assembly and matrix-vector products) that reflect our guiding design principles.

8.5.2.1 *Matrix Assembly*

Matrix assembly is organized similarly to vector assembly, as demonstrated by the following routine for forming a finite element stiffness matrix. The call to `MatSetValues()` adds a 3×3 block to the stiffness matrix. Again, as for the vector assembly, it is more efficient to insert elements by blocks, rather than individually.

```
Mat    K; /* Global stiffness matrix */
int    Nt, *tri, i, gidx[3];
double *X, *Y, x0, y0, x1, y1, x2, y2, k[9];
/* Loop over local elements */
for ( i=0; i<Nt; i++ ) {
  /* Load the vertices of the element into local variables */
  x0 = X[3*i];   y0 = Y[3*i];
  x1 = X[3*i+1]; y1 = Y[3*i+1];
  x2 = X[3*i+2]; y2 = Y[3*i+2];
  /* Compute the element stiffness matrix with an application-
     specific routine that stores it in k */
  CalculateElementStiffness(x0,x1,x2,y0,y1,y2,k);
  /* Load global indices for nodes of the element into gidx */
  gidx[0] = tri[3*i]; gidx[1] = tri[3*i+1]; gidx[2] = tri[3*i+1];
  /* Add the local stiffness matrix, k, to the global stiffness
     matrix, K, in the rows and columns indicated by gidx */
  MatSetValues(K,3,gidx,3,gidx,k,ADD_VALUES);
}
MatAssemblyBegin(F,FLUSH_ASSEMBLY);
MatAssemblyEnd(F,FLUSH_ASSEMBLY);
```

Certain internal data structures (for instance, data structures needed for a matrix-vector product) must be created once during sparse matrix assembly, after all of the matrix elements have been inserted. The second argument of `MatAssemblyBegin()` and `MatAssemblyEnd()` indicates whether the assembly is complete, `FINAL_ASSEMBLY`, or only partial (so that values will be added/changed later), `FLUSH_ASSEMBLY`. Only after calling the `FINAL_ASSEMBLY` variant of the assembly routines is the parallel matrix ready for use. The `FLUSH_ASSEMBLY` is to allow the local stashes to be unloaded to the correct processor, without setting up the rest of the data structures required to make the matrix ready for use.

PETSc provides a variety of parallel matrix data structures that are distributed by rows, including (block) compressed sparse row, (block) diagonal, and dense formats. The matrix assembly process for all of these implementations is analogous to that for vectors, discussed in Section 8.5.1. Specifically, stashing is employed to enable the application code to insert any components into a matrix, regardless of their ultimate storage location. The assembly phase automatically handles the movement of the elements into their final locations. The matrix assembly design reflects the same guiding principles as the vector assembly.

8.5.2.2 *Matrix-Vector Products*

One of the most computationally expensive operations within iterative methods for sparse systems of linear equations is the matrix-vector product. We next describe the use of the vector scatters introduced above to compute the matrix-vector product for one of our simple matrix implementations. This section continues the discussion at the end of Section 8.2.2.

The default sparse matrix representation supported in PETSc is the compressed row storage format (also called the Yale sparse matrix format or AIJ storage). This format employs three arrays: the matrix nonzeros (ordered by rows), their respective column indices, and pointers to the beginning of each row. The internal data representation is compatible with standard Fortran 77 storage. In PETSc these three arrays, as well as other useful information, are encapsulated in the internal part of the `Mat` datatype.

In our implementation, parallel AIJ matrices are distributed by rows among the processors, so that each processor owns a submatrix consisting of all of the nonzeros within its assigned rows. Each processor's local matrix is divided into diagonal and off-diagonal parts. For a square global matrix we define each processor's diagonal portion to be its local (owned) rows and the corresponding columns (a square submatrix); each processor's off-diagonal portion encompasses the remainder of the local matrix (a rectangular submatrix). As described in the following paragraph, this subdivision enables the overlap of computation and communication during the matrix-vector products. We indicate below the pertinent parts of the parallel AIJ data structure for our discussion:

```
typedef struct {
Mat        A;     /* diagonal submatrix */
Mat        B;     /* off-diagonal submatrix */
Vec        lvec;  /* local vector to receive scattered elements */
```

```
VecScatter Mvctx; /* scatter context for matrix-vector product */
....              /* additional data */
} Mat_MPIAIJ;
```

After all elements have been inserted into a parallel matrix, we create the data structures needed for the matrix-vector product within the MatEndAssembly() phase. These data structures are generally formed a single time for a matrix; however, they must be recomputed when the nonzero structure of the matrix changes.

Each processor first compresses its off-diagonal submatrix, renumbering the column numbers from one through *nc*, where *nc* denotes the number of nonzero columns in the entire off-diagonal submatrix. Each processor also creates a mapping from the compressed to the actual column numbers (needed for such actions as printing a matrix) and forms a local vector of dimension *nc* that will be used to receive scatters of nonlocal vector elements during the matrix-vector product. Then each processor forms a vector scatter context with VecScatterCreate(), so that *only* the vector elements needed by the off-diagonal part of the local matrix will be scattered to that processor during a matrix-vector product.

After this setup phase, the matrix-vector product computations are then extremely simple, as indicated by the routine given below for computing $y = A * x$:

```
int MatMult_MPIAIJ(Mat A,Vec x,Vec y)
{
Mat_MPIAIJ *aij = (Mat_MPIAIJ *) A->data;
VecScatterBegin(x,aij->lvec,INSERT_VALUES,SCATTER_ALL,aij->Mvctx);
MatMult(aij->A,x,y);
VecScatterEnd(x,aij->lvec,INSERT_VALUES,SCATTER_ALL,aij->Mvctx);
MatMultAdd(aij->B,aij->lvec,y,y);
return 0;
}
```

We initiate scattering with VecScatterBegin() and compute the local part of the matrix-vector product, thus overlapping calculation and communication. We then complete the scattering with VecScatterEnd(). Finally, we perform the off-diagonal part of the matrix-vector product.

The matrix-vector product exemplifies several of the guiding principles:

- Overlapping communication and computation.
- Managing the computation in the context of a PETSc object, in this case a matrix.
- Precomputing communication patterns inside the vector scatter.

8.5.3 Jacobian Computation via Finite Differences

PETSc provides a variety of scalable nonlinear solvers. At the heart of these is a need to provide (approximate) Jacobian information. In the context of finite element and finite difference methods, these Jacobians are usually extremely large sparse matrices that for realistic applications are difficult to compute analytically. PETSc provides the infrastructure to compute Jacobians via finite differences efficiently and scalably. Since this is an important example of the use of our six guiding principles, we discuss it in detail.

The Jacobian, J, of a function $F(x)$ is a matrix whose columns are the derivatives of F with respect to each of the components of x. Thus,

$$\text{column}_j(J) = \nabla_{x_j} F \approx \frac{F(x + dx_j) - F(x)}{dx_j},$$

where dx_j represents a small perturbation to the jth component of x. Once $F(x)$ has been evaluated, one can compute each column of the Jacobian at the cost of one function evaluation. In addition, one can determine the nonzero pattern of each column by locating the nonzeros generated during this computational process.

Unfortunately, this approach is completely nonscalable because, if the Jacobian is n by n, this method requires n function evaluations and n^2 checks to locate the nonzeros in the Jacobian. Fortunately, for the class of problems that PETSc tackles, both of these problems can be eliminated.

Determining the nonzero pattern: For PDEs discretized on finite element or finite difference grids, it is generally straightforward to determine the nonzero structure of the Jacobian matrices. The nonzero structure arises from the underlying coupling between nodes used on the grid (which corresponds to the stencil for finite difference methods), and in general this is determined by cell or node neighbors. Most important, it is determined locally and thus is easily parallelized.

Efficient computation of the Jacobian entries: Consider the function

$$F(x_0, x_1, x_2, x_3, x_4) = \begin{pmatrix} x_0 + x_1 * x_0 \\ x_1 + x_4^2 \\ 3 * x_2 + x_0 \\ x_3 + x_0 \\ x_4 + 2 * x_1 \end{pmatrix}.$$

Its Jacobian is easily computed to be

$$\begin{pmatrix} 1+x_1 & x_0 & & & \\ & 1 & & & 2x_4 \\ 1 & & 3 & & \\ 1 & & & 1 & \\ & 2 & & & 1 \end{pmatrix}.$$

Note that columns 1, 2, and 3 share no common rows. Consequently, one could compute all three columns with a single function call

$$\text{column}_1(J) + \text{column}_2(J) + \text{column}_3(J)$$
$$\approx$$
$$\frac{F(x+dx_1)-F(x)}{dx_1} + \frac{F(x+dx_2)-F(x)}{dx_2} + \frac{F(x+dx_3)-F(x)}{dx_3}$$
$$=$$
$$\begin{pmatrix} \frac{F_0(x+dx_1)-F_0(x)}{dx_1} \\ \frac{F_1(x+dx_1)-F_1(x)}{dx_1} \\ \frac{F_2(x+dx_2)-F_2(x)}{dx_2} \\ \frac{F_3(x+dx_3)-F_3(x)}{dx_3} \\ \frac{F_4(x+dx_1)-F_4(x)}{dx_1} \end{pmatrix}$$
$$\approx$$
$$\begin{pmatrix} \frac{F_0(x+dx_1+dx_2+dx_3)-F_0(x)}{dx_1} \\ \frac{F_1(x+dx_1+dx_2+dx_3)-F_1(x)}{dx_1} \\ \frac{F_2(x+dx_1+dx_2+dx_3)-F_2(x)}{dx_2} \\ \frac{F_3(x+dx_1+dx_2+dx_3)-F_3(x)}{dx_3} \\ \frac{F_4(x+dx_1+dx_2+dx_3)-F_4(x)}{dx_1} \end{pmatrix}.$$

Similarly, columns 0 and 4 may be computed simultaneously. From this example, one sees a very general mechanism for computing sparse Jacobians efficiently.

First, one *colors* the columns of the sparse matrix so that no two columns of the same color share a common row. Then, the number of function evaluations required to approximate the Jacobian drops from n to the number of colors. In most applications the number of colors required can vary between 4 and 50.

The use of coloring to compute Jacobians efficiently has been known for many years and is widely used on sequential machines, for example in the library MINPACK [14]. We discuss briefly how efficient Jacobian computation is performed in PETSc.

- Compute the nonzero structure of the Jacobian; as indicated above, this is essentially a local computation.

- Generate a coloring of the resulting matrix. This may be done in one of two ways:
  - color the underlying grid directly (this again is essentially a local operation), or
  - use a parallel graph coloring algorithm (for example, [13]), to color the parallel sparse matrix directly.
- Determine all the local computation and the communication that will be required for each color in the actual Jacobian approximation. This is an inspector step, and in PETSc it generates a data structure called `MatFDColoring`. Note that the actual Jacobian calculations will (generally) require communication of the scaling factors dx_j, since a column change on one processor will (in general) affect rows on other processors (though usually only a few).

A code fragment that demonstrates this process is given below.

```
ISColoring    iscoloring;
MatFDColoring fdcoloring;
Mat           J;

DAGetColoring2dBox(da,&iscoloring,J);
MatFDColoringCreate(J,iscoloring,&fdcoloring);
```

The PETSc data structure `ISColoring` contains the information about the column coloring, and the routine `DAGetColoring2dBox()` is a PETSc utility routine that colors two-dimensional grids using a nine-point stencil.

The routine `SNESDefaultComputeJacobianWithColoring()` that actually computes the approximate Jacobian can be expressed in pseudo-code as follows:

```
Loop over all colors {
  For each local column, j, of that color {
    Compute an appropriate dx[j] perturbation and add it to x
  }
  Evaluate the function at the new x
  Communicate the required scaling factors dx[j]
  For each local column of that color {
    Loop over known nonzero rows {
      Scale Jacobian entry by appropriate dx[j]; insert in matrix
    }
  }
}
```

The data structure `MatFDColoring` contains all the precomputed information required in implementing this pseudo-code.

The efficient computation of Jacobians via coloring as implemented in PETSc exemplifies several of our guiding principles:

- Managing the computation in the context of a PETSc object, in this case a `MatFDColoring` object.
- Precomputing communication and computation patterns, required for efficiently determining nonzero rows and scalings.
- Aggregating the computation, by updating all columns of the same color simultaneously.

8.5.4 Linear and Nonlinear Solvers

We next provide an overview of the techniques used in PETSc for coordinating various types of algorithms (e.g., linear, nonlinear, and timestepping solvers) by focusing on the the simplest case—linear solvers. We build these solvers on the foundation of the previously discussed matrix and vector data structures.

Most modern linear iterative solvers may be viewed as the combination of a preconditioner (simple stationary iterative solver) and a Krylov subspace accelerator. (Even multigrid methods may be categorized in this manner if we view the Richardson iteration as playing the role of a Krylov subspace method.) Since its earliest days PETSc has provided uniform and simple access to a variety of parallel Krylov subspace accelerators. These methods are relatively easy to code (since they merely require a matrix-vector product and several vector operations), and the data-structure-neutral implementations within PETSc are independent of underlying matrix and vector data structures [18]. We emphasize that this data-structure-neutral approach exposes mathematical details in a uniform fashion, without making unnecessary assumptions about the representation of mathematical objects. In particular, the parallelism within the Krylov subspace methods is handled completely within the vector and matrix modules of PETSc. Such organization is crucial in managing the complexity of the parallel software.

In contrast to the Krylov subspace techniques, most preconditioner implementations must be data-structure-dependent, so that each preconditioner must be explicitly coded for use with each new (parallel, sparse) matrix data structure. In addition, the source code and parallelism for matrix-based preconditioners are generally much more complicated than those for Krylov subspace methods. Thus, the numerical kernel of most preconditioners within PETSc is within the matrix module of the library, where such actions as incomplete factorization, triangular solves, and extraction of matrix subblocks occur.

We coordinate the solvers by using a context data type (called `SLES` for linear solvers and `SNES` for nonlinear solvers), to store all information about the solution process, including the right-hand side, convergence tolerances, options, parameters, etc. All of the computations and communications related to a particular solution process are managed in the solver context variable. These solvers, as well as the lower levels of code on which they are built, of course also employ the previously discussed software design principles.

To demonstrate the easy-to-use interface that this approach enables, we consider a simple example of solving a linear system. We first create and assemble the matrix A and right-hand-side vector b; then we create a solver context, set some solver parameters, and finally solve the system.

```
Vec      x, b;
Mat      A;
SLES     sles;
MPI_Comm comm;
...
MatCreate(...,&A);
/* ... Assemble the matrix ... */
VecCreate(...,&b);
/* ... Assemble the vector ... */
SLESCreate(comm,&sles);   /* Create a linear solver context */
SLESSetFromOptions(sles); /* Set solver parameters at runtime */
SLESSetOperators(sles,A,A,SAME_NONZERO_PATTERN);
/* Set the matrix that defines the linear system and an
   optionally different preconditioning matrix.  The flag
   SAME_NONZERO_PATTERN indicates the matrix A will have
   the same nonzero structure for several distinct linear
   solves; this flag allows reuse of information during
   successive solves to reduce the computation required
   in constructing the preconditioner for the next linear
   system. */
SLESSolve(sles,b,x);      /* Solve the system */
```

All communication required by the linear solver is handled internally within the SLES routines (and the routines that they call) and uses the MPI communicator set in the call to `SLESCreate()`. This communication includes

- passing matrix information during the preconditioner construction,
- computing vector inner products and norms within Krylov subspace codes, and
- performing vector scatters in parallel matrix-vector products.

The advantage of this organization is that the communication is always intrinsically associated with a higher-level object or objects and the operations performed on them, rather than using simply a jumble of sends, receives, and other MPI calls for which it is difficult to understand the connection between a particular message passing call and the actual mathematical operations on the data. Consequently, parallel application codes, including the linear, nonlinear, and timestepping solvers within PETSc, are much easier to develop, maintain, and understand.

Through its design, PETSc provides a uniform interface to all of its linear solvers. These solvers include a variety of Krylov subspace accelerators such as

- the conjugate gradient method,
- GMRES,
- CGS,
- Bi-CG-stab, and
- transpose-free QMR,

and a growing family of preconditioners, such as

- block Jacobi,
- overlapping additive Schwarz,
- ILU(0) and ICC(0) (through an interface to BlockSolve95 [12]), and
- ILU(k) (currently sequential only).

There are three main parallel sparse matrix formats:

- CSR (compressed sparse row),
- block CSR, and
- storage by block diagonals.

PETSc provides two main approaches to solving nonlinear algebraic systems (in parallel): line search techniques and trust region methods. Both support the use of any of the linear solvers in a truncated Newton algorithm. Recently, we have added support for using backward Euler timestepping schemes. We plan to include additional higher order timestepping schemes in the future. All the solvers are introduced in the PETSc User's Manual [3].

8.5.5 Object-Oriented Features of PETSc

This article has touched on several object-oriented features of PETSc without providing a complete overview of the object-oriented design. In this section we focus on how object-oriented techniques are used to organize the overall software package.

Object-oriented programming is often defined by the three complementary principles:

- data encapsulation,
- polymorphism, and
- inheritance.

Data encapsulation (also sometimes called data hiding) refers to writing objects (data structures) so that application code does not directly access the underlying data in the object. Rather, the application code can affect the data only by making subroutine calls that change the data. In C++, these routines are called member functions, and they are the only part of the code that can change data that is "private" to an object.

Polymorphism refers to techniques that allow one to call the same function from the application level of the code to perform a specific operation regardless of the underlying data structure used internally to store the data. An example of polymorphism in PETSc is the use of the routine `MatMult()` to perform a matrix-vector product regardless of the particular matrix and vector formats being used. In C++, polymorphism is usually accomplished by using virtual functions.

Inheritance is the process of defining new objects by either combining properties of several different types of objects or adding properties to an object that is already defined.

PETSc uses all three of the defining principles of object-oriented programming. PETSc uses strong data encapsulation in both the vector and matrix data objects. Application access to the vector and matrix data is obtained through function calls such as `VecSetValues()` and `MatMult()`. All of the PETSc operations are supported via polymorphism. The user need not call routines specific to a particular data structure from the application code. Instead, the user calls a generic interface routine that correctly calls the underlying data-specific routine. This process is handled in PETSc through the use of structures of function pointers, quite similar to the way virtual functions are handled in C++.

Surprisingly, PETSc also uses inheritance in its design. All PETSc objects are derived from an abstract base object. From this fundamental object an abstract base object is defined for each PETSc object; examples include the `IS` (index set) object, the `Vec` (vector) object, the `Mat` (matrix) object, and the `PC` (preconditioner) object. Each of these abstract base objects then has a variety of instantiations. For example, there are at least five different matrix storage formats. No attempt is made to use inheritance directly inside these instantiations. For example, the CSR and block CSR objects do not share any common code. For specific details on how the inheritance is managed (e.g., for vectors), one may refer to the files `petsc/include/phead.h` and `petsc/src/vec/vecimpl.h` as well as `petsc/src/vec/impls/dvecimpl.h`.

Another level of inheritance has been incorporated in the newest component of PETSc, the `GVec` (Grid Vector) package. Grid vectors are vectors with additional information about an underlying grid and discretization. Since PETSc is coded in C, which does not provide direct support for inheritance, we manage inheritance by using techniques similar to those used for handling attributes in MPI.

Thus, although PETSc is written in C, which is not an object-oriented programming language, it is still possible to take advantage of the three underlying principles of object-oriented programming to manage the complexity of such a large package. Clearly, a software package similar in design to PETSc could be implemented in C++ with virtually a one-to-one mapping between the underlying data structures and routines in the two packages. We have chosen to develop PETSc in C to ensure complete portability across a wide range of machines, while C++ has continued to evolve over the years.

8.6 Sample Performance Results

To demonstrate the type of parallel performance one can expect using PETSc, we consider here a three-dimensional structured Euler simulation on a C-type grid using a finite volume discretization. The case for which the following data was generated is transonic flow over an ONERA M6 wing, a standard test case originally coded (for a serial computer) by David Whitfield of Mississippi State University in a Newton-iterative formulation with low-CFL, pseudo-transient continuation and explicit enforcement of boundary conditions [22]. As part of the conversion to PETSc, we have replaced the explicit boundary conditions with a fully implicit variant [21].

The solution technique used for this problem is a Newton-Krylov approach with pseudo-transient continuation and adaptive advancement of the CFL number. Much of the approximate Jacobian matrix has block-band structure corresponding to the three-dimensional, seven-point stencil, with five degrees of freedom per node (three-dimensional momentum, internal energy, and density). The fully implicit boundary conditions spoil the strictly banded structure due to the C-type grid wrap-around of the mapped grid in the wake region. We use the PETSc matrix format for block, compressed, sparse rows (block CSR) to exploit this structure.

We present timings on an IBM SP2 for the matrix-vector product, the entire linear solve, and a fixed number of iterations (29) of the nonlinear solve. The times for the matrix-vector products and linear solves are for the entire solution process. Thus, for each matrix-vector product in the table, the times are given for between 110 and 150 products, while the linear solve times are for 29 complete linear solves. To put these numbers in perspective, we note that the peak performance of one processor of the IBM SP2 is 264 Mflops, the LINPACK 100 benchmark produces 130 Mflops, and a sparse matrix-vector product that uses the standard compressed sparse row format (CSR) attains 27 Mflops.

Tables 8.1 and 8.2 are for a fixed grid size of $98 \times 18 \times 18$, so that the Jacobian matrix has a dimension of 158,760 with 4,636,200 nonzero elements. The linear Newton systems are solved using GMRES(30) and block Jacobi preconditioning, where each processor has one block that is solved with ILU(0). The speedup over two processors is given in parentheses in the tables. Note that the entire problem does not fit on one processor when explicitly forming the Jacobian matrix and using ILU preconditioning.

We next consider similar runs for the same problem on a refined mesh of size $194 \times 34 \times 34$, which produces a system that is roughly eight times as

Number of Processors	Mat-Vec Products		Linear Solves		Nonlinear Solve	
	Time	Speedup	Time	Speedup	Time	Speedup
2	5.67	–	17.46	–	73.63	–
4	3.03	(1.9)	9.02	(1.9)	39.22	(1.9)
8	1.65	(3.4)	4.74	(3.7)	21.17	(3.5)
16	0.92	(6.2)	2.62	(6.7)	11.50	(6.4)
32	0.58	(9.8)	1.62	(10.8)	6.70	(11.0)
64	0.40	(14.2)	1.14	(15.3)	4.87	(15.1)

TABLE 8.1. Performance on Small Grid Problem: Computation Time (Seconds).

Number of Processors	Mat-Vec Products		Linear Solves	
	Mflops	Speedup	Mflops	Speedup
2	179	–	147	–
4	337	(1.9)	286	(2.0)
8	620	(3.5)	540	(3.7)
16	1137	(6.4)	994	(6.8)
32	2038	(11.4)	1785	(12.1)
64	3003	(16.8)	2546	(17.3)

TABLE 8.2. Performance on Small Grid Problem: Computation Rate (Mflops).

large as the previous one, having 1,121,320 rows and 35,742,760 nonzeros. These results are given in Tables 8.3 and 8.4, where the speedup over sixteen processors for each phase is given in parentheses.

For the final problem having 1,121,320 unknowns, the computation rate on sixteen processors for the matrix-vector product was 1,281 Mflops, while the complete linear solve achieved 1,011 Mflops. On sixty-four processors the matrix-vector product ran at 4,217 Mflops, while the complete linear solve achieved 3,744 Mflops. Note that for this problem we solved a single linear system in .18 seconds on sixty-four processors using a general-purpose object-oriented numerical library.

For representative parallel computation rates for a similar problem, we consider use of the block tridiagonal (BT) method in a three-dimensional Navier-Stokes simulation within category two of the NAS Parallel Benchmarks (NPB 2) [16]. This method solves three sets of uncoupled systems of equations, each of which is block tridiagonal with 5×5 blocks. For a problem size of 102^3 (which is comparable to our large grid problem mentioned above), the BT code achieves 924 Mflops on sixteen processors of an IBM SP2; the sixty-four processor case achieves 3,487 Mflops. While the numerical methods used in this benchmark differ from those employed in our Euler code, this comparison does illustrate that the PETSc libraries achieve computation rates in line with those of similar sparse applications.

Number of Processors	Mat-Vec Products		Linear Solves		Nonlinear Solve	
	Time	Speedup	Time	Speedup	Time	Speedup
16	6.38	–	20.04	–	83.08	–
32	3.29	(1.9)	9.38	(2.1)	42.15	(2.0)
64	1.94	(3.3)	5.34	(3.8)	25.44	(3.3)

TABLE 8.3. Performance on Large Grid Problem: Computation Time (Seconds).

Number of Processors	Mat-Vec Products		Linear Solves	
	Mflops	Speedup	Mflops	Speedup
16	1281	–	1011	–
32	2483	(1.9)	2154	(2.1)
64	4217	(3.3)	3744	(3.7)

TABLE 8.4. Performance on Large Grid Problem: Computation Rate (Mflops).

8.7 Conclusion

We have outlined six of the guiding principles that are used to efficiently manage the communication and computation in a parallel, large-scale, numerical object-oriented software library. These techniques are appropriate for distributed as well as NUMA-based shared-memory computers. In fact, the underlying PETSc libraries can deliver high performance on all modern scalable machines. The guiding principles are appropriate for all scalable computers, not merely those requiring message passing programming.

We have shown how it is possible to organize a library around two conflicting goals—*ease of use* and *high efficiency*—while maintaining a good balance between the two and allowing users to move along the spectrum in either direction depending on their needs.

The complete PETSc distribution is freely available via our home page, `http://www.mcs.anl.gov/petsc/petsc.html`.

Acknowledgments: The work of the first author was supported by the Applications Technology Research Division subprogram of the Office of Computational and Technology Research, U.S. Department of Energy, under Contract W-31-109-Eng-38. The work of the third author was supported by the National Science Foundation under ECS-9527169, through the Old Dominion University Research Foundation. The second and fourth authors were supported by the Mathematical, Information, and Computational Sciences Division subprogram of the Office of Computational and Technology Research, U.S. Department of Energy, under Contract W-31-109-Eng-38.

8.8 REFERENCES

[1] Gagan Agrawal, Alan Sussman, and Joel Saltz. Compiler and runtime support for unstructured and block structured problems. In *Proceedings of Supercomputing '93*, pages 578–587, 1993.

[2] Brett M. Averick, Richard G. Carter, and Jorge J. Moré. The MINPACK-2 test problem collection. Technical Report ANL/MCS-TM-150, Argonne National Laboratory, 1991.

[3] Satish Balay, William Gropp, Lois Curfman McInnes and Barry Smith. PETSc 2.0 User's Manual. Technical Report ANL-95/11, Argonne National Laboratory, November 1995.

[4] Satish Balay, William Gropp, Lois Curfman McInnes, and Barry Smith. PETSc home page. http://www.mcs.anl.gov/petsc/petsc.html, December 1996.

[5] A. M. Bruaset and H. P. Langtangen. A Comprehensive set of Tools for Solving Partial Differential Equations: Diffpack. In M. Dæhlen and A. Tveito, editors, *Numerical Methods and Software Tools in Industrial Mathematics*, pages 63–92. Birkhäuser, 1997.

[6] Greg Burns, Raja Daoud and James Vaigl. LAM: An open cluster environment for MPI. In John W. Ross, editor, *Proceedings of Supercomputing Symposium '94*, pages 379–386. University of Toronto, 1994.

[7] William Gropp, Ewing Lusk, Nathan Doss and Anthony Skjellum. A high-performace, portable implementation of the MPI message passing interface standard. *Parallel Computing*, 22:789–828, 1996.

[8] William Gropp, Ewing Lusk, Nathan Doss and Anthony Skjellum. MPICH home page. http://www.mcs.anl.gov/mpi/mpich/index.html, December 1996.

[9] William Gropp, Ewing Lusk and Anthony Skjellum. *Using MPI: Portable Parallel Programming with the Message Passing Interface.* MIT Press, 1994.

[10] R. W. Hockney and C. R. Jesshope. *Parallel Computers 2.* Adam Hilger, 1988.

[11] Scott A. Hutchinson, John N. Shadid and Ray S. Tuminaro. Aztec User's Guide Version 1.1. Technical Report SAND95/1559, Sandia National Laboratories, October 1995.

[12] Mark T. Jones and Paul E. Plassmann. BlockSolve v1.1: Scalable library software for the parallel solution of sparse linear systems. Technical Report ANL-92/46, Argonne National Laboratory, 1992.

[13] Mark T. Jones and Paul E. Plassmann. A parallel graph coloring heuristic. *SIAM J. Sci. Comput.*, 14(3):654–669, 1993.

[14] Jorge J. Moré, Danny C. Sorenson, Burton S. Garbow, and Kenneth E. Hillstrom. The MINPACK project. In Wayne R. Cowell, editor, *Sources and Development of Mathematical Software*, pages 88–111, 1984.

[15] MPI: A message-passing interface standard. *International J. Supercomputing Applications*, 8(3/4), 1994.

[16] NAS Parallel Benchmarks home page. http://www.nas.nasa.gov/NAS/NPB/index.html, December 1996.

[17] J. V. W. Reynders, J. C. Cummings, P. J. Hinker, M. Tholburn, M. Srikant S. Banerjee, S. Karmesin, S. Atlas, K. Keahey and W. F. Humphrey. *POOMA: A FrameWork for Scientific Computing Applications on Parallel Architectures*, chapter 14. 1996.

[18] Barry F. Smith and William D. Gropp. The design of data-structure-neutral libraries for the iterative solution of sparse linear systems. *Scientific Programming*, 5:329–336, 1996.

[19] Marc Snir, Steve Otto, Steven Huss-Lederman, David Walker and Jack Dongarra. *MPI: The Complete Reference.* MIT Press, 1995.

[20] Thinking Machines Corporation. *Users Manual for CM-Fortran.* Thinking Machines Corporation, 1993.

[21] M. D. Tidriri. Krylov methods for compressible flows. Technical Report 95-48, ICASE, June 1995.

[22] D. Whitfield and L. Taylor. Discretized Newton-relaxation solution of high resolution flux-difference split schemes. In *Proceedings of the AIAA Tenth Computational Fluid Dynamics Conference*, pages 134–145, 1991.

9

Object-Oriented Construction of Parallel PDE Solvers

Michael Thuné
Eva Mossberg
Peter Olsson
Jarmo Rantakokko
Krister Åhlander
Kurt Otto[1]

ABSTRACT
An object-oriented approach is taken to the problem of formulating portable, easy-to-modify PDE solvers for realistic problems in three space dimensions. The resulting software library, Cogito, contains tools for writing programs to be executed on MIMD computers with distributed memory. Difference methods on composite, structured grids are supported. Most of the Cogito classes have been implemented in Fortran 77, in such a way that the object-oriented design is visible. With respect to parallel performance, these tools yield code that is comparable to parallel solvers written in plain Fortran 77. The resulting programs can be executed without modifications on a large number of multicomputer platforms, and also on serial computers.

The uppermost level of abstraction in Cogito decouples the numerical method from the PDE problem. The validity of these tools has been preliminarily demonstrated with a C++ implementation for one-dimensional problems.

9.1 Introduction

The problem to be discussed in this report is how to write computer programs for the numerical solution of time-dependent partial differential equations (PDEs). Many significant real-life applications, e.g., air-craft design and environmental modelling, involve simulations based on the numerical solution of systems of time-dependent partial differential equations. In general, the computational domain has a complicated geometry, and the computations demand high-performance computers.

[1]Department of Scientific Computing, Uppsala University, Box 120, S-751 04 Uppsala, Sweden. Email: `{michael, evam, peter, jarmo, krister, kurt}@tdb.uu.se`.

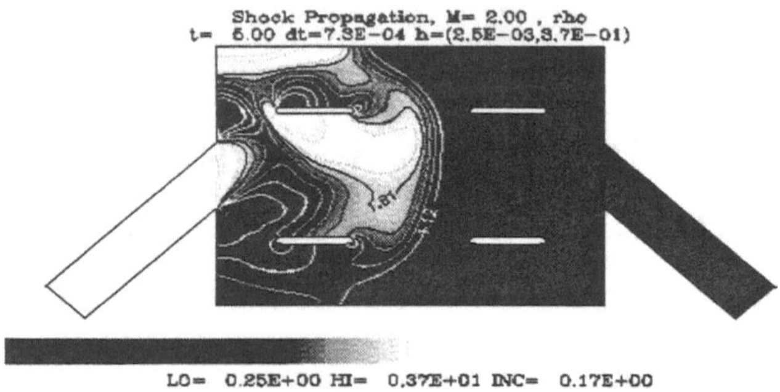

FIGURE 9.1. Simulation of airflow through an expansion pipe.

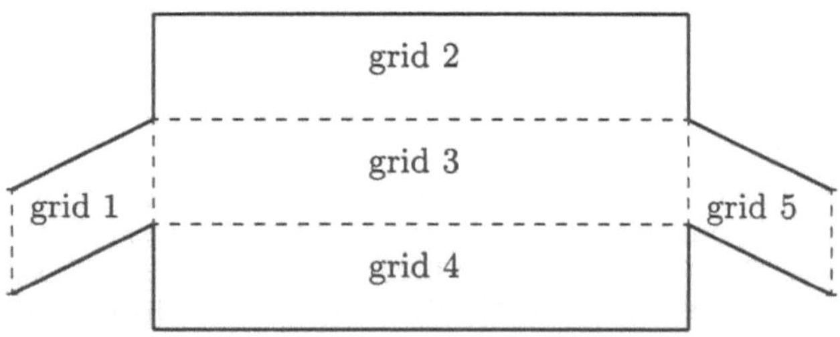

FIGURE 9.2. A schematic view of the multiblock grid used in the airflow simulations.

9.1.1 Motivation

As a model problem—simplified but nevertheless containing all the relevant features of a real-life problem—consider a 2D simulation of airflow through an expansion pipe. Figure 9.1 shows simulated flow in a two-dimensional cross-section of the pipe. There are four plates in the pipe.

We focus on finite difference methods. They are common in PDE solvers, and require structured computational grids. Since the geometry is too complicated to be covered by a single, structured grid, we use a multiblock grid with five blocks, one for the inlet, one for the outlet, and three for the main section of the pipe, where the plates make it impossible to use a single block (see Figure 9.2).

The airflow is governed by the compressible Navier–Stokes equations:

$$\frac{\partial Q}{\partial t} + \frac{\partial}{\partial x}\left(F - \frac{1}{Re}F_\nu\right) + \frac{\partial}{\partial y}\left(G - \frac{1}{Re}G_\nu\right) = 0,$$

where

$$Q = \begin{pmatrix} \rho \\ \rho u \\ \rho v \\ E_s \end{pmatrix}, \quad F = \begin{pmatrix} \rho u \\ \rho u^2 + p \\ \rho uv \\ u\,(E_s + p) \end{pmatrix}, \quad G = \begin{pmatrix} \rho v \\ \rho uv \\ \rho v^2 + p \\ v\,(E_s + p) \end{pmatrix}.$$

The viscous parts of the equations are gathered in F_ν and G_ν:

$$F_\nu = \begin{pmatrix} 0 \\ \tau_{xx} \\ \tau_{xy} \\ f_4 \end{pmatrix}, \quad G_\nu = \begin{pmatrix} 0 \\ \tau_{yx} \\ \tau_{yy} \\ g_4 \end{pmatrix}$$

with

$$\begin{aligned}
\tau_{xx} &= 2\mu\frac{\partial u}{\partial x} + \lambda\left(\frac{\partial u}{\partial x} + \frac{\partial v}{\partial y}\right), \\
\tau_{xy} &= \mu\left(\frac{\partial u}{\partial y} + \frac{\partial v}{\partial x}\right), \\
\tau_{yx} &= \tau_{xy}, \\
\tau_{yy} &= 2\mu\frac{\partial v}{\partial y} + \lambda\left(\frac{\partial u}{\partial x} + \frac{\partial v}{\partial y}\right),
\end{aligned}$$

and

$$\begin{aligned}
f_4 &= u\tau_{xx} + v\tau_{xy} + \frac{k}{Pr(\gamma-1)}\left(a^2\right)_x, \\
g_4 &= u\tau_{yx} + v\tau_{yy} + \frac{k}{Pr(\gamma-1)}\left(a^2\right)_y.
\end{aligned}$$

Here u and v are the components of the velocity in the x and y directions respectively, ρ is the density, E_s the total energy per unit volume, p is the hydrostatic pressure, a is the speed of sound, Re is the Reynolds number, k is the heat conductivity, Pr is the Prandtl number, $\gamma = c_p/c_v$ is the ratio between the heat capacities at constant pressure and constant volume, μ and λ are the viscosity coefficients. As the geometry of the tube is not rectangular, we must make a transformation of the coordinates

$$\begin{cases} \xi = \xi(x, y) \\ \eta = \eta(x, y) \end{cases}$$

and rewrite the Navier–Stokes equations in this new coordinate system.

The Navier–Stokes equations are valid in the interior of the pipe. At the inlet and outlet boundaries, at the plates, and at the pipe walls, boundary conditions are prescribed. For a detailed description of these, see [5]. Finally, initial conditions describe the airflow at the beginning of the simulation.

We solve this problem by an explicit finite difference method, using second order, centered finite differences in space, and a Runge–Kutta type time-marching scheme. The centered finite differences are too wide to be applied at the boundaries, so there we need additional, numerical boundary conditions. Moreover, we need to connect the solutions between the different blocks of the multiblock grid. This is accomplished through interpolation.

On a parallel computer with distributed memory, the exchange of interpolation data between grids leads to communication between processors. The communication pattern for this interaction is irregular (in contrast to the communication pattern for the interaction between different partitions of a single element grid).

Traditionally, PDE solvers have been written in low-level Fortran. For example, a "Navier–Stokes solver" is a huge Fortran program which contains the mathematical model, boundary conditions, grids, grid functions, and specifications of the numerical method to be used. The expression "Navier–Stokes solver" reflects that all the components of the program are tightly linked to each other, making it difficult to replace one component without rewriting major parts of the code. Moreover, the initial construction of such a program is a significant task.

The advent of parallel computers has made the drawbacks of the traditional approach even more apparent. In traditional PDE solvers, the data structures are visible at the top level of the program. This approach is unsatisfactory in the context of parallel programming, for portability reasons. Also, the increase in execution speed, through parallel processing, has opened possibilities for more realistic simulations with more complicated data structures, which are not supported in plain Fortran. One example of this is multiblock grids.

A conclusion of the discussion above is that there is need for a software tools and languages that provide a raised level of abstraction in scientific computing. The established way of reaching this goal is to use mathematical subroutine libraries, such as NAG and IMSL. A subroutine from such a library encapsulates a big chunk of work. Typically, it solves a complete problem of a certain type, using a predefined method. The user can influence the result by setting certain parameters, such as tolerances, but the basic method is fixed.

The traditional subroutine libraries are appropriate for a user interested in solving a certain application problem, without bothering about the numerical methods. However, they are too inflexible for a user who wishes to experiment with different choices of numerical algorithms.

An ideal set of software tools for scientific computing should make it possible to express computational problems

- flexibly (e.g., not restricted to predefined numerical methods or mathematical models)
- generally (i.e., not restricted to model problems)
- briefly
- with concepts that belong to the application area, thus making the program text immediately meaningful to an application specialist.

9.1.2 Our Contribution

An object-oriented approach is suitable for reaching these goals. Our contribution is a set of object-oriented software tools, *Cogito*, for composite-grid based PDE solvers. Programs using these tools will be portable over a wide range of computers, serial and parallel (MIMD with distributed memory). Moreover, there is support for composite grids as an abstract data type. (Here, composite grids can be multiblock grids, but also more complicated structures with overlapping grids.) Finally, care has been taken to decouple different components of the program, for easier program construction, modification and maintenance.

The majority of our tools are implemented in Fortran, with the object-oriented design visible in the code. For example, we use dynamic memory management to create objects dynamically. On the highest level of abstraction, we use C++ for the implementation.

As a model problem for testing our ideas, we have been using the application presented above, i.e., the 2D compressible Navier–Stokes equations on a multi-block grid. This code has been executed—without modification—on several platforms: single SUN workstations, IBM SP1 and SP2 with MPL, Parsytec parallel computers with Parix, Intel parallel computers with NX, and a network of SUN SPARC 10 with PVM. The performance of the code is comparable to that of a corresponding implementation in plain Fortran 77.

9.2 The Object-Oriented Approach

In recent years there has been a rapidly growing interest in applying object-oriented ideas to the field of numerical software. Most of these efforts consist in enriching the programming language (normally C++) with some abstract data types (ADTs) suitable for scientific computing. Typical examples are arrays, matrices, grids, and stencils [12, 13, 22]. The *overall* style of programming remains procedure-oriented. There is a main program and procedures implementing the numerical method and the mathematical problem to be solved. These ADTs are essentially data stores with associated operations. They do not initiate interaction with other objects of other classes.

This is a big leap forwards compared with the traditional programming style in scientific computing. The new ADTs hide low-level implementation details, and thus make the programs easier to modify and to port.

However, with the type of approach discussed above, it is still difficult to change to a new numerical method, or to change the mathematical problem to solve with a given method. In order to achieve this kind of flexibility, it is necessary with *program decoupling.* Program decoupling is a combination of data decoupling and algorithm decoupling. By data decoupling, we mean that if our data set consists of l conceptually separate categories of data, each of these categories is represented by a separate abstract data type, $\mathcal{D}_i$. Data decoupling is what we achieve with the ADT-oriented approach discussed above. By algorithm decoupling, we mean that if our algorithm has m options (in our context, the options are: choice of discrete space derivatives, choice of time-marching scheme, choice of numerical and physical boundary conditions, choice of PDE problem, etc.) then each option is isolated in a separate subalgorithm. This can typically be achieved through appropriate modularization of a program. By *program decoupling*, we mean that the modules coincide with the ADTs, so that the separate subalgorithms are encapsulated in different objects, in such a way that the complete algorithm can be composed of different objects in a "plug-and-play" manner.

Our goal is a fully object-oriented approach, leading to program decoupling, as described above. Then, there is no main program in the traditional sense. Instead, the program consists of active objects that interact with each other. The construction of a program consists in creating objects, composing objects, and finally initiating the execution of one object, which will subsequently activate other objects.

Note that there is no restriction to predefined numerical methods. The objects can provide general operators that can be flexibly combined, to allow (almost) the same freedom as if the program was written in a traditional way. Ideally, there should be an object data base, with a user-friendly interface, through which this composition can be accomplished. However, even without such an interface, the indicated style of programming can be achieved. There will then be a main program, but the purpose of the main program will simply be the creation and composition of objects.

In the remainder of this chapter, we will develop the ideas outlined above, and demonstrate our realization of them in the context of finite difference methods for time-dependent PDE.

Other efforts similar to ours in spirit are presented in other chapters of this book. In particular, we want to mention Diffpack [3], and ELEMD [15], since these projects, like ours, emphasize the object-oriented *analysis*. The models resulting from the object-oriented analysis are considered as research results *per se*.

Let us also mention some related efforts not represented in this book. The POOMA framework [18] addresses the same general issues as Cogito, and with a similar approach, but with differences in details. Especially, our top level of abstraction, Cogito/Solver, has no apparent counterpart in POOMA.

A group at Los Alamos National Laboratory has developed a C++ class library, Overture [2], which is similar in scope to Cogito/Grid (see below). Fritzson et al. [7] have an interesting approach, combining symbolic algebraic manipulations and numerical computations in the framework of an object-oriented environment, ObjectMath. So far, their emphasis has been on mathematical modelling. Recently, a cooperation has been established between the ObjectMath and Cogito groups [11], and between the Cogito and Overture groups.

The list of related work could be made even longer, including also approaches that are not object-oriented. Suffice it here to note that the general issue of raising the level of abstraction in software for scientific computing is currently being addressed by a large number of research groups around the world. Many of these efforts are so successful that it is appropriate to say that we are currently seeing a breakthrough in the area of numerical software.

9.3 Overview of Cogito

For a set of software tools to be of interest to the intended audience of scientific computing specialists, it is necessary that programs based on the tools are comparable with corresponding plain Fortran programs with respect to *efficiency*. Experiments indicate that this can be achieved if the fully object-oriented design is restricted to the highest level of abstraction, whereas the computational kernels are treated in a Fortran-like manner (even if the implementation would be made in a language such as C++) [1].

These recent experiments support the design strategy that was chosen at the outset of the Cogito project, some years ago. Cogito is structured into three layers: one subsystem for the computer related classes (Cogito/Parallel), one for the grid and grid function classes (Cogito/Grid), and one for the classes related to numerical solution methods and PDE problems (Cogito/Solver). The two lower layers are implemented in Fortran. These layers are based on an object-oriented design, and care has been taken to make the design visible in the code. For example, objects are created dynamically. However, important object-oriented features such as polymorphism are lost in this implementation. Thus, it is proper to say that the implementation of the two lower layers is *object-based.*

The Cogito/Solver layer is fully object-oriented, and provides for the program decoupling discussed above. Cogito/Solver is implemented in C++.

The three layers meet the following goals:

- (Cogito/Parallel). To abstract away from computer dependencies, for portability.

- (Cogito/Grid). To abstract away from low-level representations of the data structures. This has two motivations. One is that advanced data structures, such as composite grids, are not supported in existing programming languages. The other motivation is that we should not need to rewrite the PDE solver if the low-level representation of data is changed. For example, by expressing grids in terms of an abstract data type *Grid*, we make the PDE solver independent of the low-level grid representation being used by a particular grid generation package.

- (Cogito/Solver). To decouple the different components of the numerical method and the PDE problem from each other, and from the grid and grid function. The purpose is to make program construction and modification easier.

A fourth issue is to make tools flexible enough to address the individual needs of different users. In our context, there are three typical user categories: the application engineer (who is mainly interested in the application, and needs easy-to-use, predefined numerical methods), the numerical analyst (who is interested in experimenting with new finite difference methods, new boundary conditions, etc.), and the parallel programmer (who wants to govern the parallel implementation in detail). Roughly speaking, Cogito/Parallel is intended for the latter user category. Cogito/Grid is suitable for the numerical analyst. Finally, Cogito/Solver can be used by the application engineer, but has also the flexibility required by the numerical analyst.

Cogito/Parallel [16] handles three types of tasks: message passing, data distribution, and management of trees of arrays in distributed memory. Portability is achieved by providing different implementations of the class *Message*. Currently, the following platforms are supported: IBM SP1 and SP2 with MPL, Parsytec parallel computers with Parix, Intel parallel computers with NX, and any platform with PVM or MPI.

Cogito/Grid deals with the management of composite structured grids and grid functions defined on such grids [17]. The task in focus is how to use an *existing* grid. It is assumed that the grid has been generated in advance, by some grid generation package, and is stored on disk. The classes in this layer are based on Cogito/Parallel. Cogito/Grid is portable over the same platforms. Moreover, Cogito/Grid can be executed without modification on serial computers, since it uses an SPMD programming style, hiding all message passing.

The Cogito/Parallel and Cogito/Grid layers have been implemented in Fortran 77. (Cogito/Grid has also been implemented in Fortran 90. This version does not depend on Cogito/Parallel, and the intention is to use HPF for the parallelization.) By simulation of dynamic memory, using a large array `memory`, objects can be created dynamically. The operations on the classes are implemented as subroutines, having a reference (integer) to the object as an argument. The name of the subroutine is a concatenation of the operator and class names (abbreviated in order to avoid extremely long subroutine names).

On the Cogito/Grid level of abstraction, the PDE solver is expressed by writing the equations explicitly in the code, replacing derivatives by finite difference operators. Thus, there is a close coupling between the mathematical problem and the numerical method. Moreover, the boundary conditions (both mathematical and numerical ones) are hardcoded, in the same way as the equations.

In the Cogito/Solver layer [21, 23], these aspects are decoupled as well. The implementation is in C++. Currently, this subsystem exists only in an implementation for 1D problems, and is not linked to the other two subsystems. However, the intention is that a future implementation will use the classes in Cogito/Grid (and, through them, Cogito/Parallel).

9.4 Case Study 1: Application of Cogito/Grid

To indicate the style of a code based on Cogito/Grid, consider the model problem discussed in Section 9.1.1.

The class *Composite Grid* contains a number of blocks. Each block is an object of the class *Rectangular Grid.* (The blocks can be curvilinear, and are assumed to be logically rectangular.) Moreover, the composite grid contains an object of the class *Interpolation Table*, i.e., a description of how the different blocks relate to each other. The user will normally not make explicit references to the interpolation table.

As mentioned, Cogito does not contain tools for grid generation. It is assumed that the grid has been generated in advance, and is available in a file. The following code sequence creates a composite-grid object cg (reads the structure of the grid from file—number of grids, interpolation table, etc.), then decides how to partition the grid, and finally reads the grid.

```
c -- Create and distribute the grid
       call Create_CG(memory,cg,'myFile')
       call Distribute_CG(memory,cg,'rectangle')
       call Read_CG(memory,cg,'myFile')
```

When an object of the class *Grid Function* is created, the corresponding grid is given as a parameter. In our model application, the solution has four components at each grid point. We represent the solution by a grid function u defined on the entire composite grid:

```
       call Create_GF(memory,'u',4,cg,u)
```

Here cg and u are integers, acting as pointers to the grid and grid function object, respectively.

A PDE solver on the Cogito/Grid level of abstraction will mainly operate on grid functions. There are operations for inspecting properties of the grid function, and for performing arithmetic operations. In particular, there are methods for applying finite difference operators to the grid function. For example, the central part of the PDE solver for our model application is the following code sequence, which performs the timestepping. Here, u1, u0, and du are grid functions. Note that this is a *parallel* program!

```
c -- Runge-Kutta stages
      do 10 r=1,nstages
c --    Space discretization
```

```
         call spdisc(memory,cg,du)
c --     u1=u0+alpha*du
         call Saxpy_GF(memory,u1,alpha(r),du,u0,all,arb)
c --     Boundary conditions
         call bound(memory,cg)
 10   continue
```

In the subroutine spdisc, the finite difference operators in space are applied. Some of the most common finite difference operators are predefined as operators on the class *Grid Function*. Moreover, the class has a general operation *ApplyDiff*, which allows the user to specify the coefficients of the finite difference operator, and get the operator applied to the grid function.

The subroutine bound contains the application of boundary conditions, and, which is important in the composite-grid context, the interpolation needed to connect the grid functions between the different grids.

For the implementation of boundary conditions, we use subdomains. It is possible to define subdomains of a grid. To specify a subdomain, the corresponding index ranges are given. Also, a subdomain identity (integer id) and a boundary condition label (integer bc) are associated to each subdomain:

```
call DefSD_G(memory,g,indexRange,id,bc)
```

The arithmetic operations on the class *Grid Function* take a bc label as input. The operation is performed only on the subdomains which have this label. By this mechanism, boundary conditions can be conveniently implemented.

It is also possible to specify the subdomain identity, so that the operation is performed on a single subdomain. (In this case the bc label should be specified as arb, for "arbitrary".) If the subdomain identity all is given, the operation should be performed on the entire grid.

Interpolation is performed through the operation *Interpolate* (class *Grid Function*). Assume that a grid function u is to be interpolated from grid G1 to grid G2. The interpolation consists of two phases. In the first phase, the processors owning relevant partitions of G1 compute interpolation values. These values are sent to the processors who have the corresponding partitions of G2. The second phase consists in the latter processors receiving the interpolation values, and updating u on G2. The operation *Interpolate* has a *mode* parameter, which makes it possible to execute each phase in a separate call. With *mode* = *pre* the first phase is executed, with *mode* = *post* the second one. This allows for overlap of communication and computation:

```
c -- Interpolate u and apply boundary conditions
     call Interpolate_GF(memory,u,pre)
     ...computations not needing interpolation data...
     call Interpolate_GF(memory,u,post)
```

9.5 Case Study 2: Application of Cogito/Solver

As mentioned, Cogito/Solver exists only in a pilot implementation for 1D problems. Thus, we cannot show an executable Cogito/Solver based program for the model application. Instead, we illustrate Cogito/Solver by the following program for the solution of a one-dimensional nonlinear PDE problem similar to the (two-dimensional) Shallow-Water equations (see [6]). The example is executable, using the present pilot implementation of Cogito/Solver. A program for the model application will have the same structure, once Cogito/Solver has been fully implemented, using the lower level layers of Cogito as a basis.

Note, that this program simply performs "plug-and-play": objects are created, then they are connected to each other (plug), and finally, one of the objects is activated (play).

```
int main()
{
 // -- Declare the main objects. --
    NumExp myExp;
    NonConsProblem *problem = new ShallowWater(); // An heir to
                                                  // NonConsProblem
    SpaceHandler theSH;
    TimeHandler  timehandler;
    const int N = 100;          // Grid point indices from 0 to N
    Grid grid1(0.0, 1.0, 0, N); // This is a simple class Grid used for
                                // the pilot implementation.
 // -- Prepare the problem. --
 //   "ShallowWater_*" are previously declared functions.
    BCDirichlet bcleft(0,1,ShallowWater_BC);  // Same BC function
    BCDirichlet bcright(1,1,ShallowWater_BC); // at both boundaries
    problem->use( &ic); problem->add( &bcleft); problem->add( &bcright);
 // -- Set up the space handler. --
    SpaceDisc * interiorSD = new Dzero;
    SpaceDisc * boundarySD1 = new Dplus;
    SpaceDisc * boundarySD2 = new Dminus;
    DirichletHandler bchd;
    NeumannHandler bchn;
    theSH.add(interiorSD);
    theSH.add(boundarySD1); theSH.add(boundarySD2);
    theSH.add(&bchd); theSH.add(&bchn);
 // -- Set up the time handler. --
    double dt = 0.4/N ; // Set time step.
    TimeDisc * timedisc = new RungeKutta(dt);
    timehandler.add(timedisc);
 // -- Prepare the experiment. --
    myExp.use(&grid); myExp.use(problem);
    myExp.use(&theSH); myExp.use(&timehandler);
 // -- Run until time 1.0, and store solution on file! --
    myExp.solveUntil(1.0,"exShallow.m");
    return 0;
}
```

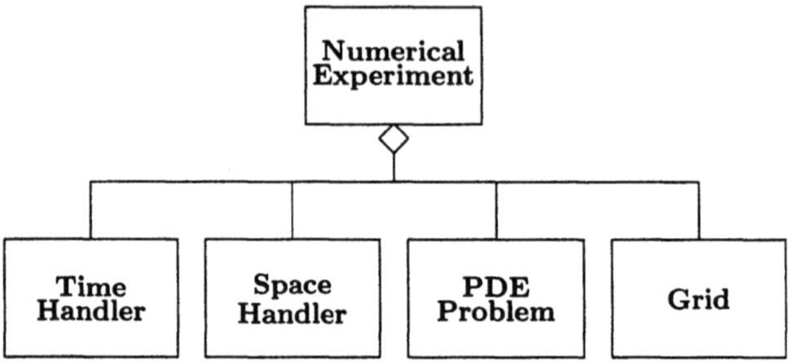

FIGURE 9.3. The base classes of the object model for Cogito/Solver. The diagram uses the OMT notation. Note that the diagram gives a top level *overview* of the object model. Component classes of aggregates are not shown, neither are inheritors.

The object `myExp` is an object of the class *Numerical Experiment.* The critical reader might suspect that `myExp.solveUntil` is a traditional PDE solver in disguise. However, this is not the case. Disregarding some code for correctness checks and for controlling intermediate outputs from the experiment, `myExp.solveUntil` simply reads:

```
int NumExp::solveUntil(double endTime,...)
{
  gf = tHandler->advanceUntil(endTime);
  return 1;
}
```

So, `myExp.solveUntil` activates the time handler object, which in turn activates the space handler object, etc. There is no single object that contains the complete solver.

Our approach to program decoupling is based on a complete object-oriented analysis, in which attention is paid not only to the static object model, but also to the dynamic model [21]. Program decoupling has been the *goal* of our work. Consequently, we have formulated scenarios for the dynamic model in such a way that this goal should be ensured. Essentially, this is achieved through a careful definition of base classes *and* associations between them, and a careful division of responsibilities between classes.

The central classes of the object model on base class level are shown in Figure 9.3. (The diagram uses the Object Modelling Technique (OMT) notation [19].) The program is composed by connecting a *Time Handler* object, a *Space Handler* object, and a *PDE problem*, into a *Numerical Experiment.* The *Numerical Experiment* knows nothing about the details of the numerical method or PDE problem. It is a control object, with responsibility for storing the details belonging to a certain numerical experiment. The user can store results through operations on this object. Also, the user can change the state of the object, by ordering continuation of the experiment further in time.

When the numerical experiment is asked to continue in time, it delegates this responsibility to the time handler. This, as well, is a control object. It knows the present state (and the states for required previous time steps) of the computed solution to the PDE problem. It also knows how to proceed the solution one or more steps in time. However, it does *not* know how the space-dependent parts should be computed. This task is delegated to the space handler. The space handler knows how to compute the discrete space derivatives involved, and turns to the PDE problem object for information about the problem dependent coefficients, etc.

Thus, within the framework provided by the object model in Figure 9.3, we get a decoupling of the space handler, the time handler, the PDE problem, and the administration of the experiment (intermediate output etc.).

However, further decoupling is achieved by means of classes not visible in Figure 9.3. For example, the space handler is an aggregate of *Space Discretization* objects. Each such object knows what finite difference operators to apply for discrete differentiation—for derivatives of various orders—in a subdomain of the grid.

Also not visible in Figure 9.3 are the classes for management of boundary conditions (BC). There are *BC descriptor* objects associated with the PDE problem, and *BC handler* objects associated with the space handler. In this way, we can change the way a certain BC is *implemented*, by replacing the corresponding BC handler object, while the BC descriptor remains unchanged.

We emphasize that Figure 9.3 shows *base* classes. There is an inheritance structure in the proposed object model, which makes it straightforward to extend the system with new solution methods, PDE problems, etc. For example, the compressible Navier–Stokes equations of conservative form would be an inheritor of the abstract class *Conservative PDE*, which in turn inherits the abstract base class *PDE problem.*

Finally, note that the object model above is *not* restricted to finite difference methods. Space handlers based on, e.g., finite volume discretizations and unstructured grids would also fit into this framework. The concepts we have chosen on the top level of abstraction are, we believe, general enough for any type of time-dependent PDE solver. The adaptation to special choices, such as structured grids or finite difference methods, is made through inheritance from the general concepts.

9.6 Cogito for Implicit Methods

At the outset of the Cogito project, the focus was on *explicit* finite difference methods. Recently, we have taken the first steps towards also handling implicit finite difference methods [14]. So far, this has been implemented as an extension of the Cogito/Grid layer. (Later on, the corresponding extensions will be made also on the Cogito/Solver level).

Cogito/Grid, in the version for explicit finite difference methods, contains two main classes: *Grid* and *Grid Function* (where *Grid* has two inheritors, *Composite Grid* and *Rectangular Grid*). Figure 9.4 shows the base classes of the extended object model, which allows for the treatment of implicit methods. The broken line marks the old model, for explicit methods only.

In the implicit case, there will be an algebraic system to solve at each time step. For a multistep method, this system can be expressed as $Qv^{n+1} = g(v^n, \dots, v^{n-s+1})$. The problem is to invert the operator Q in order to get an explicit expression for the grid function v^{n+1}. The right-hand side g is a grid function, which can be computed through operations on the known grid functions $v^n, \dots, v^{n-s+1}$. Thus, the *System* is defined by the *Operator* Q and the *Grid Function* g. The operator Q depends on the coefficients of the PDE problem (class *Coefficient*), on the finite difference stencils chosen for the discretization of space derivatives (class *Stencil*), and on the boundary conditions (class *Boundary Condition*).

The user treats the *Operator* object as an abstract operator. Whether this operator is internally implemented as a matrix, or whether it is represented by its components (coefficients, stencils, and boundary conditions) is hidden to the user. This allows for internal optimization of the code. Certain subclasses of operators may be stored as sparse matrices, whereas others are not. In the latter cases, the application of the operator to a grid function is achieved by the successive application of the individual components (coefficients, stencils, and boundary conditions). Many state-of-the art solvers for discretized PDEs are based on preconditioned Krylov subspace methods. Thus, it is important to extend the object model also with preconditioners. On the top level of abstraction, we have any kind of preconditioner. Particular *types* of preconditioners are introduced as subclasses. A Krylov-based solver would be associated to a preconditioner and would operate on a system. In [4], an object model along these lines is presented. This type of structure will be appropriate also for Cogito.

As an initial effort, we have concentrated on a particular *family* (P) of preconditioners, proposed by Holmgren and Otto [9, 10]. These preconditioners are based on fast transforms, and can be efficiently executed on high-performance computers. Figure 9.5 shows a preconditioner class for the family (P). (NB.: In the general model indicated in the previous paragraph, this class would be one of the subclasses to the abstract base class *Preconditioner*. However, for simplicity, since we presently have only one preconditioner class, we have chosen to let it have the name of the abstract base class.)

A specific preconditioner belonging to the family (P) is defined by an operator $\tilde{Q}$ and a transform. The operator is related to but not identical to Q (see above). For example, $\tilde{Q}$ may have other boundary conditions than Q. Our object model allows the user to construct any preconditioner belonging to (P), by plug-and-play. First, compose an operator $\tilde{Q}$ by replacing some stencil and boundary condition objects in Q by other stencils and boundary conditions. Then select a transform object $\mathcal{T}$. Finally, the

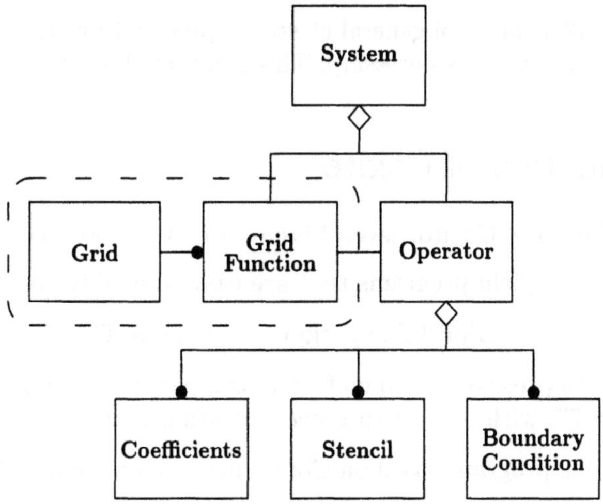

FIGURE 9.4. Object model for the treatment of implicit finite difference methods.

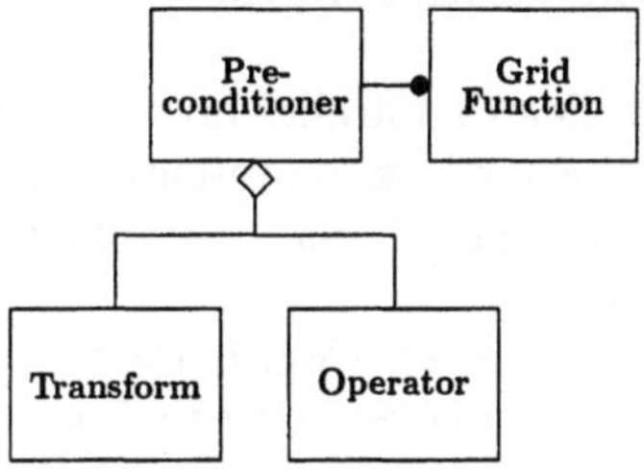

FIGURE 9.5. Object model for the the preconditioner.

constructor operation of our preconditioner class takes $\tilde{Q}$ and $\mathcal{T}$ as input, and creates the preconditioner. The creation of the preconditioner is a nontrivial programming task, which the user is thus relieved of. In addition, as explained above, our framework makes it trivial to change to another preconditioner of the same family. The classes for implicit methods have been implemented in Fortran 90, and so far only for serial platforms. The parallelization of these tools is a topic for future work. Also, the present implementation is for noncomposite grids.

It should be noted that the parallelization of the implicit methods discussed above can be carried out within the framework of Cogito/Grid. The parallelization of these methods has been studied in detail by Holmgren and Otto [8]. The iterative methods for solving the algebraic system perform sparse matrix vector multiplication, which is essentially the same kind of operation as appears in an explicit finite difference method. The preconditioners of family (P) can also be expressed as operations on grid functions, where the corresponding grids are related to the grid used for the solution of the PDE.

For the parallelization of general classes of preconditioners, extensions to our object models may be necessary. This is a topic for continued research.

9.7 Validation of Cogito

For the validation of Cogito, several issues are of importance:

1. Will Cogito yield programs that are easy to modify, as claimed?
2. Will Cogito yield portable programs, as claimed?
3. How will a program based on Cogito compare with a program in plain Fortran 77, with respect to *serial* execution time?
4. How will a program based on Cogito perform on a parallel computer?

Issue 1 relates in particular to the Cogito/Solver layer, which has been designed with regard to the program decoupling discussed in Section 9.2. The discussion of Section 9.5 leads to the conclusion that Cogito/Solver achieves its goals.

The Cogito/Grid and Cogito/Parallel layers, as well, simplify program modification, by hiding the low level representations of, e.g., grids and grid functions. In particular, the new tools for implicit methods are based on the plug-and-play principle.

As for the portability (Issue 2), Cogito/Parallel exists in an MPI version and a PVM version. Moreover, Cogito/Parallel supports a number of vendor-specific message-passing libraries: Intel's NX, IBM's MPL, and Parsytec's PARIX. Thus, Cogito/Parallel (and the Cogito/Parallel based version of Cogito/Grid) is portable over a wide range of platforms.

Issues 3 and 4 relate mainly to the two lower layers of Cogito, where the heavy computations take place. (The results reported in [1] indicate that code based on a future, reimplemented version of Cogito/Solver can be fast enough, provided that the underlying layers are efficient.) The remainder of this section presents some experiments intended to study the efficiency of Cogito/Parallel and Cogito/Grid, in terms of execution speed and parallel efficiency.

9.7.1 Cogito/Grid for the Compressible Navier–Stokes Equations

In the first experiment, we solve the model problem discussed in Section 9.1.1. For the serial execution speed, we have compared two codes for this problem: a serial code written in plain Fortran 77, and a code based on Cogito/Grid. Both solvers use the same numerical method: centered, second order accurate finite differences in space, and Runge–Kutta time-marching [17].

Computer	*Model*	*OS*	*CPU/MHz*	*Memory*
Sun	Sparc 10	SunOS 5.3	Sparc 10/40	64 MB
IBM	SP2	AIX 3.2.5	POWER2	128 MB
Intel	Paragon XP	OSF/1 1.3.3	i860 XP/50	32 MB

TABLE 9.1. The hardware platforms used for numerical experiments with the Navier-Stokes equations and the wave equation.

Computer	*Compiler*	*Flags*
Sun	f77 SparcCompilers 3.0	-O3
IBM	xlf 2.2	-O3 -Q -qhot -qipa
Intel	if77/Paragon SUN4/Solaris (cross compiler) R5.0.1	-O4

TABLE 9.2. The compiler versions and compilation flags used for the numerical experiments with the Navier-Stokes equations and the wave equation.

In our experiments, the solvers were executed on a single node on three different platforms, see Table 9.1 and Table 9.2. The results are summarized in Figures 9.6 to 9.8. All grids are quadratic and of the same size.

The results in Figures 9.6 to 9.8 show no clear tendency in difference between the Cogito and the plain Fortran 77 version. The differences on IBM SP2 and Intel Paragon are clearly acceptable.[2] On the SP2, the Cogito version was even slightly better. However, the result for the Sun Sparc 10 shows that improvements are possible. More attention should be paid to processor type and memory management issues. One possibility is to modify the Cogito/Grid tools so that they can adapt automatically to such details. Information about processor type, etc., could be incorporated into the (already existing, but not very elaborated) class *Computer Model*, which is part of the Cogito/Parallel layer. There is an additional source of difference that has to do with fundamental differences in programming style. The object-oriented approach tends to yield more subroutine calls. Further optimization could be achieved by adding compound operations to the class *Grid Function*, in order to decrease the number of Cogito subroutine calls (and related loops).

[2]The drop in performance seen in Figure 9.8 for the Cogito version has nothing to do with the Cogito tools. The low flop rate was caused by excessive swapping, and reflects the fact that the Cogito version does not reuse arrays to the same extent as the plain Fortran version. During one time step of the finite difference method, the plain Fortran version uses the same array for several different grid functions, which do not appear simultaneously. The same effect can be achieved in the Cogito version by using one grid function object to store the data of these. Alternatively, separate objects could be used, being created *and* deleted within each time step, so that they do not exist simultaneously.

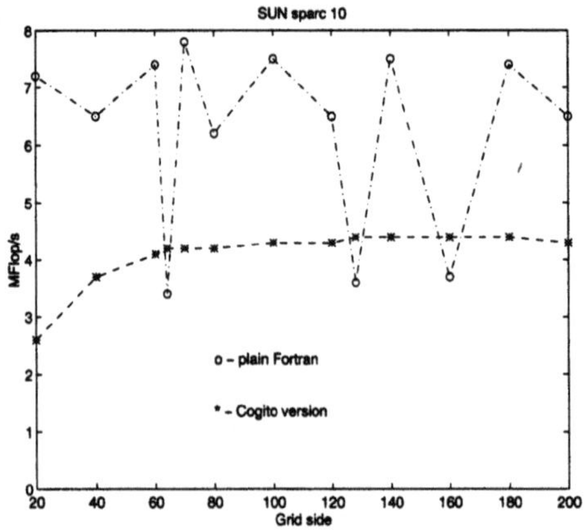

FIGURE 9.6. Performance, SUN Sparc 10.

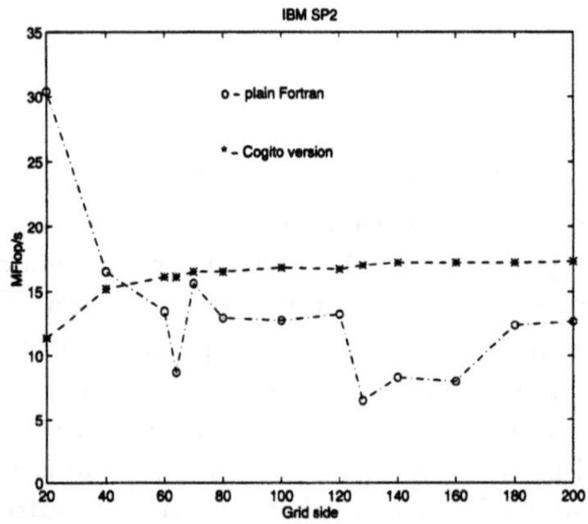

FIGURE 9.7. Single node performance, IBM SP2.

For an evaluation of the parallel performance of the tools, the performance of the Cogito based Navier–Stokes solver was studied. However, there was no parallel version of the corresponding plain Fortran 77 code. Thus, we made a second set of experiments, on the same platforms, comparing parallel codes for the wave equation on a single grid, one version being based on Cogito/Grid, the other written in plain Fortran 77.

As a measure of parallel performance, we used sizeup. This means increasing the problem size, to maintain *fixed execution time* when the number of processors grows. Sizeup is defined as parallel work divided by serial work. Sizeup has advantages over the traditional speedup or scaled speedup measures, which are sensitive to computer properties and code optimization [20].

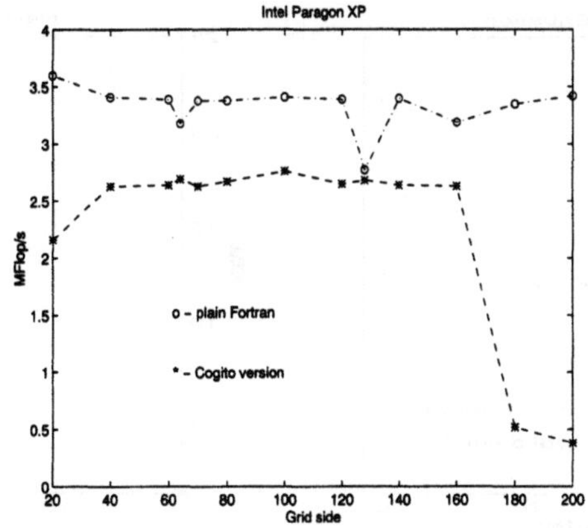

FIGURE 9.8. Single node performance, Intel Paragon XP.

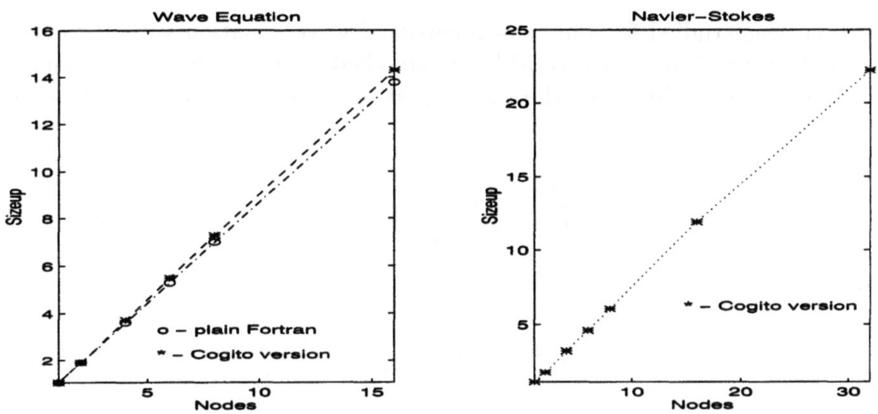

FIGURE 9.9. Sizeup, IBM SP2.

In our experiments, the one-processor problem size is 200 × 200 grid points for the wave equation, and 50 × 50 points in each block for the Navier–Stokes equation.

The results show that the Cogito-based codes scale very well. The comparison for the wave equation indicates that Cogito gives a parallel performance comparable to that of a hardcoded solver.

9.7.2 The Tools for Implicit Methods

As mentioned in Section 9.6, our tools for implicit methods have so far only been implemented for serial platforms. The serial performance of code based on these tools, compared with a corresponding plain Fortran 90 implemen-

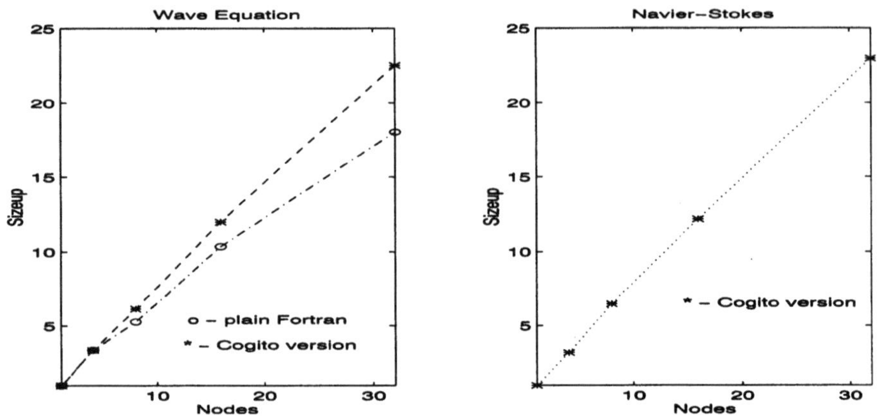

FIGURE 9.10. Sizeup, Intel Paragon XP.

tation, was studied for the two-dimensional wave equation with Dirichlet boundary conditions. We consider a stretched grid, which is transformed into a uniform grid. Then the wave equation transforms into a PDE with variable coefficients:

$$\frac{\partial u}{\partial t} + \sigma_1(\xi_1)\frac{\partial u}{\partial \xi_1} + \sigma_2(\xi_2)\frac{\partial u}{\partial \xi_2} = f'(x_1 + x_2 - t)$$

where

$$\sigma_k(\xi_k) = a_k^{-1}\cosh^2(s_k(2\xi_k - 1))\frac{\tanh(s_k)}{s_k}.$$

In the experiment, this problem was solved with the trapezoidal rule for implicit time discretization. The spatial derivatives were approximated by second order centered finite differences in the interior of the domain, and with a first order finite difference at the outflow boundary.

The corresponding system of equations was solved with the GMRES method. We implemented this method using the tools indicated in Section 9.6 (for details, see [14]). The Cogito-based GMRES procedure was compared with a traditional hardcoded Fortran 90 routine on a Dec Alpha 8200. (See Table 9.3 for details.) From Figure 9.11, we can see that the new version has approximately twice the execution time of the old one for different problem sizes N (total number of grid points) and convergence tolerances n (the tolerance limit is 10^{-n}). The figure shows the results for GMRES(10). The same behavior was also observed for other restarting lengths. Moreover, it can be noted that the relative performance of the Cogito-based solver is better for the larger problem size, where there are more computations. This indicates, as expected, that the overhead related

Model	*OS*	*CPU/MHz*	*Memory*
Alphaserver 8200	Dec Unix V3.2	EV5/300MHz	1 GB
Compiler	*Flags*		
Dec Fortran 90 V4.0	none		

TABLE 9.3. Data concerning the Dec Alpha 8200 platform used in the GMRES experiment. In other experiments, the optimization flag -O was set. That gave faster execution, but did not change the relative behavior of the two codes.

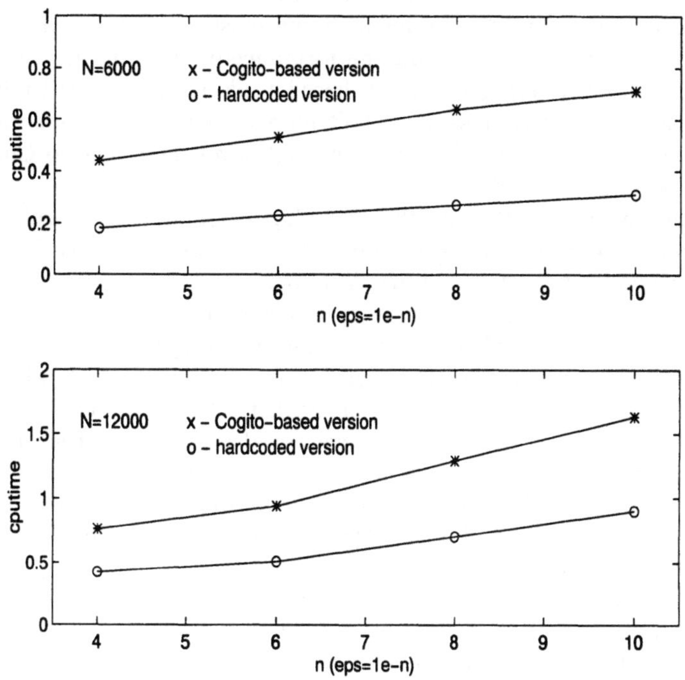

FIGURE 9.11. Performance for two different GMRES-solvers, for different problem sizes N (total number of grid points) and convergence tolerances n (the tolerance limit is 10^{-n}). The figure shows the results for GMRES(10). The same behavior was observed also for other restarting lengths.

to the new tools is constant. Thus, when amortized over a sufficiently large problem size (large computations) the overhead will be negligible.

The new software tools make it *much* easier to set up new preconditioners (within the family of preconditioners we currently support). In this perspective a performance degradation of a factor two is acceptable. Moreover, we are convinced that this factor can be decreased by further optimization of Cogito/Grid, along the lines discussed in Section 9.7.1.

9.8 Concluding Remarks

We have presented ideas for the formulation of portable, easy-to-modify PDE solvers. Moreover, we have described implementations showing that the ideas are valid for realistic problems. The resulting software library, Cogito, contains tools for writing programs to be executed on MIMD computers with distributed memory. Difference methods on composite, structured grids are supported. Finally, the ability to decouple the numerical method from the PDE problem has been demonstrated with an implementation of C++ tools for one-dimensional problems.

Cogito is still under development. Some existing parts need to be improved: increasing the single node performance of the Fortran-based parts of Cogito, and extending the C++ based tools to handle multiple space dimensions. Moreover, classes for I/O management and out-of-core data storage need to be added. In addition, high-level interfaces to Cogito remain to be implemented.

In order to validate the usefulness of Cogito for industrial problems, we are currently working on using some of our tools in connection with a multiblock code at SAAB Military Aircraft.

In collaboration with the Overture group at Los Alamos [2], Overture is being used as a platform for implementing Cogito/Solver in a "real-life" version (as opposed to the existing pilot version), which will be able to handle multidimensional problems, realistic boundary conditions, and realistic geometries. In this context, we are also modifying the underlying object model to be able to treat more complex problems, and to be able to treat simple problems more straightforwardly than is allowed by the present *Space Handler* and *PDE Problem* classes. A detailed discussion of these issues will be the topic of a future report.

The new version of Cogito/Solver can later be modified into a version that adapts to Cogito/Grid instead of Overture.

The work on implicit methods, of which the first results have been reported above, will be continued. Cogito will be extended to fully support implicit finite difference methods.

Finally, Cogito provides a framework for expressing PDE solvers. This framework is applicable beyond the special cases covered by the present implementation (i.e., multicomputers, composite grids, and finite difference methods). To demonstrate this, it would be of interest to extend Cogito to other classes of parallel computers, and other numerical techniques.

Acknowledgments: The Cogito project is supported by grants from the Swedish National Board for Industrial and Technical Development, and from the Swedish Research Council for Engineering Sciences.

9.9 References

[1] E. Arge, et al. On the numerical efficiency of C++ in Scientific Computing. In M. Dæhlen and A. Tveito, editors, *Mathematical Models and Software Tools in Industrial Mathematics*, pages 93–119. Birkhäuser, 1997.

[2] D. L. Brown and W. D. Henshaw. Overture: An advanced object-oriented software system for moving overlapping grid computations. Technical Report LA-UR-96-2931, Los Alamos National Laboratory, Los Alamos, NM, 1996.

[3] A. M. Bruaset, E. Holm, and H. P. Langtangen. Increasing the efficiency and reliability of software development for systems of PDEs. In E. Arge, A. M. Bruaset, and H. P. Langtangen, editors, *Modern Software Tools for Scientific Computing*, pages 247–268. Birkhäuser, 1997.

[4] A. M. Bruaset and H. P. Langtangen. Object-oriented design of preconditioned iterative methods in Diffpack. *To appear in Transactions on Mathematical Software*, 1997.

[5] R. Enander. *Grid patching and residual smoothing for computations of steady state solutions of first order hyperbolic systems.* PhD thesis, Dept. of Scientific Computing, Uppsala University, Uppsala, Sweden, 1991.

[6] B. Engquist. A sample of time dependent partial differential equations. Technical Report Internal Report 85-02, Dept. of Scientific Computing, Uppsala University, Uppsala, Sweden, 1985.

[7] P. Fritzson, et al. Industrial application of object-oriented mathematical modeling and computer algebra in mechanical analysis. In *Proceedings of the 7:th International Conference on Technology of Object-Oriented Languages and Systems: TOOLS EUROPE'92.* Prentice-Hall, 1992.

[8] S. Holmgren. *Fast Solvers for First-Order PDE.* PhD thesis, Dept. of Scientific Computing, Uppsala University, Uppsala, Sweden, 1993.

[9] S. Holmgren and K. Otto. A framework for polynomial preconditioners based on fast transforms I: Theory. Technical Report 167, Dept. of Scientific Computing, Uppsala University, Uppsala, Sweden, 1995.

[10] S. Holmgren and K. Otto. A framework for polynomial preconditioners based on fast transforms II: PDE applications. Technical Report 168, Dept. of Scientific Computing, Uppsala University, Uppsala, Sweden, 1995.

[11] P. Hägglund, et al. Implementation of a PDE application in ObjectMath and Cogito. Technical Report 5, Parallel and Scientific Computing Institute, Uppsala, Sweden, 1996.

[12] J. F. Karpovich, et al. A parallel object-oriented framework for stencil algorithms. In *Proceedings of the Second International Symposium on High-Performance Distributed Computing*, pages 34–41, 1993.

[13] M. Lemke and D. Quinlan. P++, a parallel C++ array class library for architecture-independent development of structured grid applications. *ACM SIGPLAN Notes*, 28:21–23, 1993.

[14] E. Mossberg. Object-oriented software tools for the construction of preconditioners. Report in preparation.

[15] G. Nelissen and P. F. Vankeirsbilck. Electrochemical modelling and software genericity. In E. Arge, A. M. Bruaset, and H. P. Langtangen, editors, *Modern Software Tools for Scientific Computing*, pages 81–104. Birkhäuser, 1997.

[16] P. Olsson. Object-oriented software tools for parallel computing with distributed memory. Report in preparation.

[17] J. Rantakokko. Object-oriented software tools for composite-grid methods on parallel computers. Technical Report 165, Dept. of Scientific Computing, Uppsala University, Uppsala, Sweden, 1995.

[18] J. V. W. Reynders, et al. POOMA: A framework for scientific simulation on parallel architectures. Available on Internet, http://www.acl.lanl.gov/PoomaFramework/.

[19] J. Rumbaugh, et al. *Object-Oriented Modeling and Design*. Prentice-Hall, Englewood Cliffs, NJ, 1991.

[20] X. Sun and J. L. Gustafson. Towards a better parallel performance metric. *Parallel Computing*, 17:1093–1109, 1991.

[21] M. Thuné and K. Åhlander. Towards an expressive language for PDE solvers. In H. Nielsen, editor, *Programming Languages and Systems—ESOP '96*, pages 373–386, Berlin, 1996. Springer-Verlag.

[22] R. D. Williams. DIME++: A language for parallel PDE solvers. Technical Report CCSF-29-92, Caltech, Pasadena, CA, 1993.

[23] K. Åhlander. An object-oriented approach to construct PDE solvers. Technical Report 180, Dept. of Scientific Computing, Uppsala University, Uppsala, Sweden, 1996.

10

Modern Software Techniques in Computational Finance

Konstantinos N. Pantazopoulos
Elias N. Houstis[1]

ABSTRACT
Option management is among the most important and interesting problems in computational finance where modern techniques can be used to improve the design and implementation of software systems and supporting tools. The purpose of option valuation is to compute the value of a financial instrument, the *option*, which is governed by economic laws and is a function of a number of financial variables. The option valuation problem has many similarities to classical problems in physics and applied mathematics such as boundary value problems, free boundary problems, and optimal stopping problems. Knowhow from these areas has been proven useful in the mathematical treatment of options. The treatment of options in a mathematical framework facilitates their classification and provides the basis for an object–oriented abstraction. The object–orientation of the option valuation problem is used in FINANZIA, a library for option valuation, hedging, and implied parameter calculation. Introducing object–oriented techniques in the design and implementation of option management software helps in mastering the complexity and dynamic nature of the application domain, as well as in meeting the strict requirements for software maintainability, robustness, and performance.

10.1 Introduction

Option valuation is among the most important and interesting problems in the emerging area of computational finance. The option valuation problem is related to classical mathematical problems such as boundary value problems, optimal stopping problems, and free boundary problems, and in that sense, has gained a lot from technology transfer from these areas [9]. The interaction with mathematics has been very fruitful and has resulted in a vast literature of results. However, modern software techniques for the

[1]Computer Science Dept., Purdue University, W. Lafayette, IN47907-1398, USA, Email: {kp,enh}@cs.purdue.edu.

implementation and management of option valuation algorithms have attracted relatively little attention so far. In this direction we discuss how object–orientation can be used to improve the handling of the option valuation problem in a mathematical framework.

Options come in many different flavors: the most basic options are *calls* and *puts*, often referred to as *vanilla* options. A call is a contract among two parties, with the buyer of the contract having the right, but not the obligation, to *buy* the *underlying* asset for the *strike* price. A put allows the buyer to *sell* the underlying asset for the strike price. The duration of an option is the time period between the date of initiation and the *expiration* date. The underlying asset is usually a stock (e.g. company XYZ), an index (e.g. S&P 500), a commodity (e.g. gold) or an interest rate (e.g. Libor). For example, an *XYZ Sep $150 call* option, could be bought in the Chicago Board of Options Exchange. This option would give the right to the owner to purchase 100 shares of XYZ for $150 each, any time on, or before, the third Friday of September (the coming September). In this case the underlying asset is the XYZ stock, the strike price is $150 and the expiration date is the third Friday of the coming September. Notice that the XYZ stock itself is traded independently in the world financial markets, and its price fluctuates over time based on various economic and other parameters. Because of the uncertainty involved, the price of XYZ on the third Friday of September is not known at the time the option contract is initiated (in fact, it is not known for any future date). Depending on the actual realization of the XYZ stock price, the call owner stands to gain or lose. If XYZ shares on the third Friday of September cost $170 each, the call owner can *exercise* the right and buy them for $150, gaining $20 on each of the 100 shares that the option covers (the owner can realize the gain by immediately selling the shares for their market value). If, on the other hand, XYZ shares on the expiration date sell for $105 each, the call option is worthless, (the owner can buy the shares in the market at a lower price than the option's strike price) and consequently the owner loses the amount paid up-front for the option. Thus, the *payoff* of a call option (i.e. the amount the option worths at expiration) can be described as $\max(AssetPrice - StrikePrice, 0)$.

The primary task is to value an option, i.e. to come up with a "fair" price that the buyer of an option must pay to the seller of the option (the seller of an option is referred to as the *writer* of the option). This problem is termed *option valuation.* In the case of the XYZ example given above, we are looking for the price to pay in order to buy the specified XYZ call option. "Fair" in the context of option valuation refers to the price of the option that all agents in an economy, i.e., buyers and sellers, would agree, independent of their risk preferences. That is, independent of what they think of the underlying asset itself. The "fair" option price is defined as the risk–neutral price.

Mathematically, an option is a function V of the underlying asset S, the strike price K, the risk–free interest rate r (the risk–free interest rate is the rate earned by an investment that has no risk, such as a government security), the volatility of the underlying asset's return σ, the expiration date of the option T, the time to expiration t, the dividend yield of the underlying asset δ (for stocks or indices) and the payoff of the option V_T. Customarily, only S and t are explicitly included in the value function, while the rest are considered implied parameters and are inputs to the option valuation problem. The option value is denoted as $V(S,t)$. The uncertainty for the future values of S is modeled by taking S to be a stochastic process. Usually S is modeled as a diffusion, log-normally distributed, process. In simple models σ and r are constants.

There are numerous characteristics that differentiate the various option products that are encountered in practice. If the buyer of an option can only *exercise* the right at the *expiration* of the option, then the option is *European.* If the right can be exercised at any time on, or before, the *expiration* of the option, then the option is *American.* Another important characteristic of an option is the *payoff* at expiration, i.e. the distribution of the option value with respect to the price of the underlying asset at expiration. For a call it is given as $max(S-K,0)$ and for a put as $max(K-S,0)$, where S is the price of the underlying asset and K is the option's strike price.

The valuation of an option involves solving a mathematical problem in discrete or in continuous time, depending on the modeling of the process followed by the underlying asset. In the case of European options, the mathematical problem involved is usually an *initial/boundary value* problem. In the case of American options one has to solve a *free boundary value* problem (which has an equivalent formulation as an *optimal stopping* problem in a probabilistic framework). A discussion of free boundary problems and a survey of numerical methods to deal with them is given in [7].

Apart from the value of an option, one is interested in the risk that incurs from an option transaction. Knowing the risk allows for protective actions (hedging). Risk analysis and hedging of an option are as important as the valuation of the option. Consider the writer of an option that is in the business of selling options and making a profit from the commissions and transaction costs involved; usually such writers are financial institutions and investment banks which are interested only in the profit and do not wish to risk any funds. If they write an option that results in a profit to the buyer they stand to lose the corresponding amount. In this case they need to know the risk that is inherent in the option they sell and be able to take protective actions, i.e., *hedge* their position. In a mathematical framework, the risk and the appropriate hedging parameters are translated to *partial derivatives* of the option value V function.

Often, one knows the solution to the valuation problem, for instance in the case of publicly traded options with quoted prices in the financial

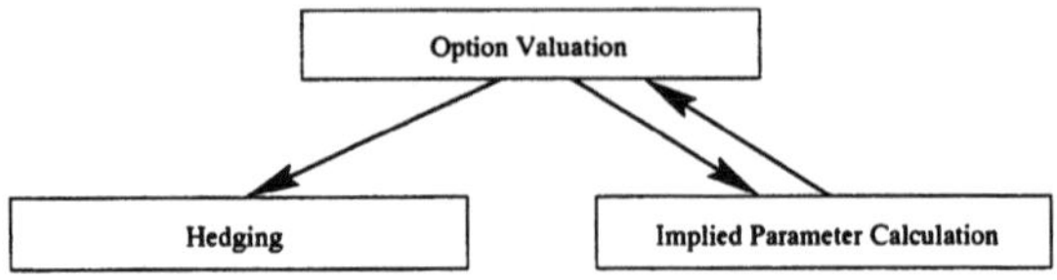

FIGURE 10.1. The three basic tasks in option management.

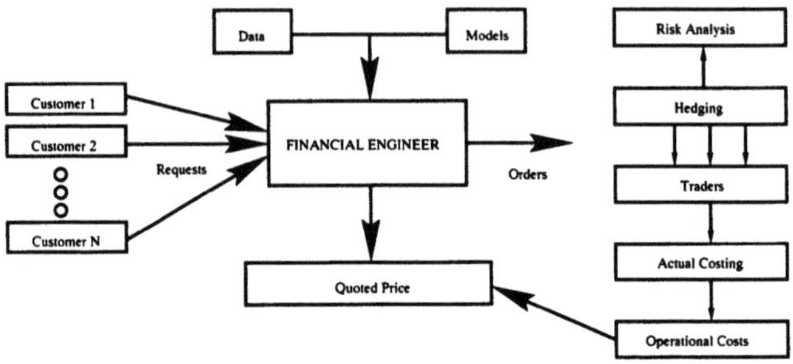

FIGURE 10.2. The financial engineering process.

markets, and is interested in obtaining values for various parameters that lead to the particular price, for example calculating the implied volatility. The implied parameter estimation is accomplished by solving an *inverse problem* based on the mathematical model that provides the option value. The inter-relationship among valuation, hedging and implied parameter calculation is depicted in Figure 10.1. We collectively refer to these three tasks as *option management.* Once the option value has been found, the hedging actions can be decided. This is indicated by the one–way connection between the two tasks. The implied parameter calculation and the valuation interact in both directions. For instance, the publicly quoted option prices can be used to calculate the implied volatility σ, which is then used to value a new option. The two–way connection between the corresponding tasks indicates this interaction.

The main objective of using modern software techniques, and in particular object–orientation, in the option management problem is to assist in the development of computational tools, capable of handling the complexity and requirements of the *financial engineering process.* A general model of the financial engineering process is described in Figure 10.2. The financial engineer receives a number of requests from customers to provide price quotes of options. Using existing data and the appropriate mathematical models the request is decomposed into specific items (orders) for hedging the position. Hedging generates a number of transactions which incur costs. Combining the operational and actual costs, and the fair price which has been calculated for the option, the quote is generated.

In order to provide practical support for the option management problem, it is necessary to address the dynamics of the financial engineering process. New options are designed on the fly in order to meet customer needs. Participants' competitiveness depends on the ability to value, hedge and manage options as fast as possible. Valuations of options are performed millions of times every minute and speed is crucial in order to promptly provide quotations or analyze new options.

Object–orientation is needed in order to manage the complexity and expose the natural properties of the problem. The complexity arises primarily because of the multitude of option products and the available algorithms for handling their management. The challenge is to properly assess the applicability and identify adequate abstractions for the option valuation problem. An object–oriented software system for option computations and the financial engineering process would be quite useful in providing support for the problem specification, the problem solution and the solution management and interpretation. The applicability of such a system could be wide, including

- education and academic research
- banking, financial industry, and corporations
- regulatory bodies and government agencies
- individual investors, market researchers, and analysts

From an operational point of view a software system for the option management problem needs to be:

- portable, maintainable, expandable
- user friendly, configurable, mobile
- robust, efficient, accurate

Maintainability and expandability cannot be overstated in view of the dynamic nature of the problem domain and the diversity in computing support. Portability is of primary importance mainly due to the vast heterogeneity of real world computing environments. The context of applications is also important in this direction. For example, investment banks in their majority use workstation technology for back–office and real time support while they incorporate personal computer technology for the human interaction with the applications. The customizability is also important in order to address the issue of many synthetic option products whose management can be decomposed in more primitive blocks. Robustness, efficiency, and accuracy are mandated by the nature of the application where failure is directly translated into financial loss.

In summary, there are four main dimensions on which the software design challenge for option management systems presents itself: the variety of option products, the variety of algorithmic approaches, the variety of implementations, and the environment of operation. Object–orientation comes as a natural framework for managing the complexity, providing efficient design, and guaranteeing easy maintainence of an option management system.

The rest of the chapter is organized as follows. In Section 10.2 we discuss the mathematical formulations and models that are used by financial institutions, academics, and practitioners, in order to handle the option management problem. We also give an overview of the available algorithmic infrastructure. In Section 10.3 we describe the software design issues in the context of the option management problem and in Section 10.4 we present the architecture and implementation of the FINANZIA library, a software library for option management that incorporates object–oriented techniques. We discuss the implementation of the library and give some examples of how it can be used. In Section 10.5 we outline possible extensions and we conclude in Section 10.6.

10.2 Option Computations

The mathematical modeling of options was pioneered in the seminal work of Black and Scholes [4]. Since then, tremendous effort has been devoted in extending and improving the option valuation models and their computational solutions. Partial differential equation (PDE) based models and their corresponding numerical solutions have been widely applied to the option valuation problem with considerable success [13], along with the more popular discrete time approximations and Monte Carlo methods.

In broad terms the option valuation problem is attacked by

- using analytic approximations to the fundamental PDE (Equation (10.2)) for specific cases,
- simulating the stochastic process of the underlying asset in discrete time,
- or, simulating the stochastic process in continuous time.

The majority of option valuation algorithms are based on the risk–neutral valuation approach which has been introduced in [5].

10.2.1 Mathematical Models for the Option Valuation

In the case of European options, the option value is the expected value of the payoffs, over all possible realizations of S, discounted with the risk–free interest rate r (risk-neutral valuation)

$$V(S,t) = e^{-r(T-t)} E^Q(V_T(S)). \tag{10.1}$$

The expectation $E^Q(\cdot)$ is taken using the risk–neutral density of S, which is the probability density under the risk–neutral measure Q. The notion of risk–neutrality (i.e., all discounting is done using the risk–free interest rate) is central and very essential in option valuation. Notice that in principal one could discount the future expectation of the option payoff using a discount rate, say μ, that would reflect its individual expectations. Since each agent would have different μ, there would be no way to come up with a unique option value (it would depend on the particular choice of the discount factor). The basis for risk–neutrality is the financial requirement to eliminate *arbitrage* opportunities, i.e., to avoid "free lunches". In a probabilistic framework, the risk-neutrality is justified by the existence of an *equivalent martingale measure* Q, under which the discounting can be done using r.

Equation (10.1) is the Feynmann-Kac solution to the fundamental partial differential equation (PDE)

$$V_t + \frac{1}{2}\sigma^2 S^2 V_{SS} + (r-\delta) S V_S - rV = 0 \tag{10.2}$$

with appropriate boundary conditions for the option product that V describes. The partial derivatives of the value function V are denoted with a subscript using the corresponding variable (second order partial derivatives are denoted using double subscript). For European options the problem is then an *initial/boundary value problem.* Recall that a put and a call differ in their payoff functions. The payoff serves as a *final* condition (which can be transformed to an initial condition by an appropriate variable transformation). The value of the option at the two extremes of the underlying asset's price, i.e., for $S = 0$ and $S = \infty$, is also different and serves as the boundary condition. However, the equation is the same for both and it is described by Equation (10.2). The boundary conditions for the put and call are given as

$$\begin{aligned} V(S,T) &= \max(S-K,0), \\ V(0,t) &= 0, \\ \lim_{S\to\infty} V(S,t) &= S - e^{-r(T-t)}K, \end{aligned}$$

for the call, and

$$V(S,T) = \max(K-S,0),$$

$$\begin{aligned} V(0,t) &= e^{-r(T-t)}K, \\ \lim_{S\to\infty} V(S,t) &= 0, \end{aligned}$$

for the put. Equation (10.2) is known as the Black–Scholes equation. It is valid for options on a single underlying asset under constant dividend yield δ, risk–free interest rate r and volatility σ. Many other types of options, such as barriers, can be valued by simply applying the appropriate initial and boundary conditions to Equation (10.2). In several cases the resulting PDE problem has a closed form solution as in the case of the put and the call demonstrated in [4]. For multi–asset European options, the fundamental PDE takes the form

$$V_t + \sum_{i=1}^{n}(r-\delta_i)\sigma_i S_i V_{S_i} + \frac{1}{2}\sum_{i=1}^{n}\sum_{j=1}^{n}\rho_{ij}\sigma_i^2\sigma_j^2 S_i^2 S_j^2 V_{S_iS_j} - rV = 0, \qquad (10.3)$$

where n is the number of underlying assets and ρ_{ij} is the correlation between asset i with asset j.

A number of options such as *lookback* options, or *Asian* options, have payoff V_T which is dependent on the path that the underlying asset follows. For example, the average strike option is defined using a strike price which is the average of the prices attained by the underlying asset during the life of the option. This translates into substituting K with a quantity like

$$K = \frac{1}{T}\int_0^T S dt.$$

Depending on the sampling rate and the average that is used (e.g. geometric or arithmetic) the problem might, or might not, have a closed form solution. In the PDE formulation of the problem, when there are no similarity reductions, an additional independent variable that keeps track of the path dependent quantity is introduced.

The *early exercise* feature, e.g. for American or Bermudan options, results in an optimal stopping problem which, in a probabilistic framework, is treated as an optimization problem. In the PDE framework it is treated as a free boundary problem.

The value of an American style option for $t \leq \tau$ is given by

$$V(S,t) = \max_{\tau} E^Q(e^{-r(\tau-t)}V_T(S)), \qquad (10.4)$$

where again Q denotes the expectation under the risk–neutral measure, and $\tau \leq T$ all possible stopping times.

Equation (10.4) implies the existence of an optimal exercise boundary, denoted with $f_S(t)$. The optimal exercise boundary is a time path of critical underlying asset prices at which, early exercise of the option is optimal (intuitively, the optimal exercise boundary can be thought of as the point where water and ice separate in a melting ice bar). In the case of a call option, it is optimal to exercise at t, when $S \geq f_S(t)$ at time t, and in the case of a put when $S \leq f_S(t)$ at time t. The optimal exercise boundary of an American option is not known ex ante, and must be determined as part of the solution to the valuation problem. Unlike some European options, for the majority of American options it is difficult to obtain an analytic solution. The principal difficulty arises from the absence of a simple expression for the optimal exercise boundary.

In the PDE framework the early exercise feature results in the free boundary problem

$$\begin{aligned} &V_t + \frac{1}{2}\sigma^2 S^2 V_{SS} + (r-\delta)SV_S - rV = 0, \\ &S \in [0, f_S(t)), \quad t \in [0,T] \end{aligned}$$

with boundary conditions

$$\begin{aligned} V(S,T) &= \max(S-K,0), \quad S \in [0, f_S(T)) \\ V(0,t) &= 0, \\ V(f_S(t),t) &= f_S(t) - K, \\ V_S(f_S(t),t) &= 1 \end{aligned} \tag{10.5}$$

in the case of a call option, and

$$\begin{aligned} V(S,T) &= \max(K-S,0), \quad S \in (f_S(T), \infty) \\ V(0,t) &= K, \\ V(f_S(t),t) &= K - f_S(t), \\ V_S(f_S(t),t) &= -1 \end{aligned} \tag{10.6}$$

in the case of a put option. The complete option value is given as

$$V_{complete}(S,t) = \begin{cases} V(S,t), & S \in [0, f_S(t)), \\ \max(S-K,0), & S \in [f_S(t), \infty) \end{cases} \tag{10.7}$$

in the case of the American call and

$$V_{complete}(S,t) = \begin{cases} \max(K-S,0), & S \in [0, f_S(t)], \\ V(S,t), & S \in (f_S(t), \infty) \end{cases} \tag{10.8}$$

in the case of the American put option. Equations (10.7) and (10.8) make explicit that the American option has an optimal exercise boundary, $f_S(t)$, which indicates whether the option should be held or exercised at time t.

There are many similarities between the free boundary problem encountered in the American option valuation and the *Stefan problem* for the melting or solidification of a pure material by heat transfer. One notable difference is the *smooth pasting* condition (Equations (10.5) and (10.6)) which makes difficult the use of numerical techniques such as front–tracking.

Apart from the features and characteristics of options discussed above, further extensions to the basic valuation model are used in practice, including stochastic interest rates, stochastic volatility, and varying underlying assets (e.g. options on interest rates).

10.2.2 Valuation Algorithms

Closed form solutions of the PDE problem are available in specific cases, such as the European option, the American call option with dividends, and the American put. However, most valuation problems of practical interest do not have closed form solutions and numerical methods must be employed.

Discrete time approximations were first introduced in [6] and include variations of what are known as *binomial* methods. Extensions for multiple assets (*multinomial* methods) are described in [2]. Binomial and multinomial methods simulate a discrete *random walk* on a lattice that represents the evolution of the prices of the underlying asset. The option value is computed based on the realization of the prices in the lattice. In general, such methods are not very accurate and are time and space consuming. They are, however, good tools for prototyping since they are easy to implement; for this reason they are used extensively in practice.

Monte Carlo methods for the option valuation problem are discussed in [1]. Monte Carlo methods are used in order to evaluate the integral in Equations (10.1) and (10.4). They are very useful in handling general path dependent option products in the European case. They are widely used in practice, mainly because of their powerful simulation capabilities and their usefulness in scenario analysis. However, they are rather inefficient and difficult to implement for the American option problem since they are "forward" looking algorithms, i.e., paths of asset variables are simulated forward in time. An average over independent samples of path prices gives an unbiased estimate of the option value. By contrast, valuation procedures for assets such as American options with early exercise features are generally "backward" looking algorithms. That is, the optimal exercise strategy at the expiration of the option is easily determined via some sort of dynamic programming. The problem of using Monte Carlo methods to valuate American options stems from the difficulty of applying inherently forward–looking methods to a problem that requires a backward procedure to solve.

	European	American	Path Dependent	Multi–Asset
Multinomial	1D lattice	1D lattice	2D lattice	nD lattice
Monte Carlo	Integral Computation	Optimal Stopping	Integral Computation	Integral Computation
PDE	Boundary Value	Free Boundary	Additional Dimension	Multi–Dimensional

TABLE 10.1. Option Characteristics and Algorithms.

PDE methods for the option valuation problem had been originally proposed for European options in [4]. In the context of the American option problem, finite difference techniques were first presented in [3]. The convergence of finite difference methods for the American option problem was first discussed in the context of the variational inequality approach to the problem in [11]. A general discussion of variational inequalities in the context of moving boundary problems is given in [8]. Finite difference discretization of the variational inequality formulation of the problem leads to the linear complementarity solutions presented extensively in [13]. Front–tracking methods for the solution of the free boundary formulation of the problem are discussed in [12]. PDE methods are very efficient and accurate for single–asset options but are inefficient and difficult to generalize for the case of multi–asset options.

Table 10.1 summarizes the main characteristics of an option valuation problem and the corresponding numerical algorithms to deal with it. Closed form approximations are usually option specific and are not considered as a general algorithm to solve a class of problems.

There are many other features that one needs to consider in practice. For an extensive guide to option products and financial derivatives the interested reader can look in [10]. In our discussion so far, we have described the various models assuming that interest rates are constant, i.e., r is constant. Most of the models described can be naturally extended for time-dependent, deterministic interest rates, i.e., the interest is described by a given function of time $r(t)$. However, when interest rates are assumed to be stochastic there are several things that change, both from a modeling perspective, and from an algorithmic point of view.

10.3 Software Design Issues in Option Valuation

In order to design and implement modern software solutions for the option management problem (i.e. for valuation, hedging, and implied parameter calculation of options), we need to take into account the multiplicity of options, as well as the numerous algorithmic approaches. In principal, all features and characteristics of options described in Section 10.2 (and many more which we have not discussed here for brevity) can be arbitrarily combined to create a new option and a corresponding valuation problem.

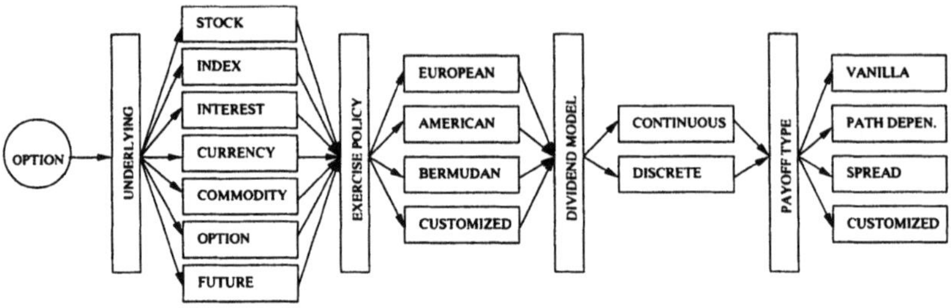

FIGURE 10.3. A hierarchy of options based on key characteristics.

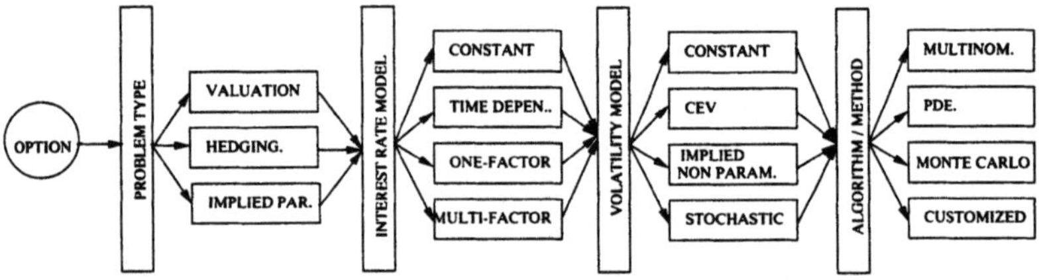

FIGURE 10.4. A hierarchy of valuation problems and models.

Systematic treatment of option problems can be achieved by identifying the key characteristics and features which can be used to classify existing options, and options that will be created by combination of these key elements. Figure 10.3 describes a possible hierarchy of options according to basic characteristics such as the type of underlying asset (S), the dividend model (δ), the payoff (V_T) and the exercise policy. Every possible path in Figure 10.3, going from left to right, describes a different option.

Figure 10.4 describes a hierarchy of the valuation models and schemes that can be used to valuate a particular option. The problem type indicates which of the three main tasks is to be performed: valuation (i.e., estimation of V), hedging (i.e., estimation of θV), or implied parameter calculation (i.e., estimation of V^{-1}). The interest rate model selection (i.e., model for r) results in a different mathematical problem. In Section 10.2 we described the PDE formulations for r constant, which remain pretty much the same for deterministic, time–dependent choice for interest rates. We hinted though, that stochastic modeling of r changes the mathematical formulation and results in one–factor or multi–factor models. Similar variations are based on the volatility (σ) model selection. The models described in Section 10.2 are valid for constant σ and hold with minor changes for the constant elasticity of variance (CEV) model. Non-parametric or stochastic volatility models result in different formulations. In both cases of stochastic interest rates and volatility, the underlying mathematical problems are still of the type discussed in Section 10.2. However, the form of the differential operator, the dimension of the problem, and the boundary conditions change. For each valuation model there are more than one algorithmic approach that one can take to solve it. We have suggested that the most widely used in practice are binomial simulation, Monte Carlo simulation

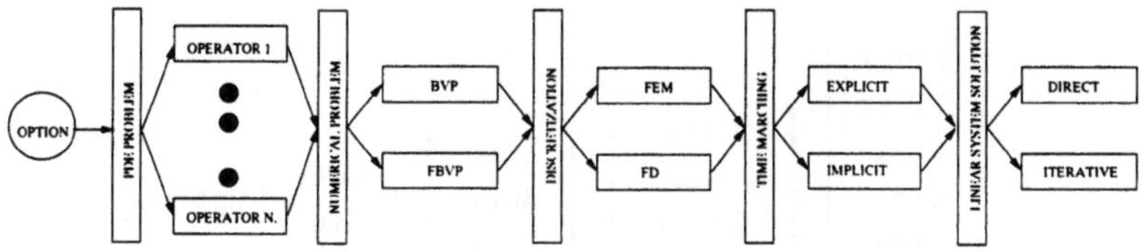

FIGURE 10.5. Hierarchy of PDE methods for valuation problems.

and numerical solution of PDEs. The basis on which selection of the algorithmic approach is made might be applicability, convenience, speed, or some other requirement.

Systematic treatment of the option management problem, gets even more complicated if we consider that for each algorithmic approach there are more than one implementation choices to make. Figure 10.5 gives a hierarchy of possible choices for a PDE type of algorithm approach. For example, a boundary value problem can be solved numerically using a number of different techniques (e.g. finite differences, finite elements, etc.) and the same holds for the free boundary problems (e.g. linear complementarity, front–tracking, etc.). Binomial (or multinomial) methods can also be implemented in a number of ways. Monte Carlo methods can have different implementations using, for example, different variance reduction techniques.

The hierarchies outlined in Figures 10.3, 10.4, and 10.5 can provide a starting point for an object–oriented abstraction of the option management problem. In the design of FINANZIA we use three basic abstractions for options, valuation models, and for the numerical algorithms that can be used. These three basic abstractions are naturally interpreted into classes.

The `Option` class serves as a container of information related to an option product. The various options are defined through inheritance from the base `Option` class. The use of the hierarchy described in Figure 10.3 simplifies the problem specification by identifying the characteristics which are important. As suggested in Figure 10.3 the main points of derivation from the base `Option` class are the *payoff*, the *exercise policy*, the *underlying*, and the *dividend model*. The advantage that a derivation tree of options based on key features has, is that it facilitates overloading of methods for the different options and simplifies the necessary data structures and code.

The `Valuator` Class serves as a container of information for the type of valuation problem, as described in Figure 10.1. The other points of derivation depend on the interest rate model (i.e. *term structure*), the *volatility model* and the actual valuation algorithm to be used. As discussed in Section 10.2 there are four main categories, the multinomial, numerical, Monte Carlo and customized. The customized family includes closed formulas that have been suggested and other specific algorithms for a given problem.

The `Calculator` Class serves as an abstraction for the actual valuation algorithm that is going to be used.

A number of secondary abstractions are also encapsulated into class definitions. Such classes include specific numerical algorithms such as root find-

$$\begin{bmatrix} \texttt{Stock} \\ \texttt{Index} \\ \texttt{Rate} \\ \texttt{Curr} \\ \texttt{Comm} \\ \texttt{Opt} \\ \texttt{Fut} \end{bmatrix} \left\{ \begin{matrix} \texttt{Call} \\ \texttt{Put} \\ \texttt{Asian} \\ \texttt{Exotic} \end{matrix} \right\} \left\{ \begin{matrix} \texttt{Euro} \\ \texttt{Amer} \\ \texttt{Berm} \end{matrix} \right\} \begin{bmatrix} \texttt{Cont} \\ \texttt{Disc} \end{bmatrix} \texttt{Option}$$

FIGURE 10.6. Derivation tree for the `Option` Class used in FINANZIA.

ing routines, arrays, vectors, and interpolation methods. *Assistant* classes are defined as wrappers to main classes in order to assist in the faster instantiation of a particular object. For example the option assistant simplifies the instantiation of an object from the `Option` class. Analogous assistant classes are available for `Calculator` and `Valuator` classes.

10.4 FINANZIA Implementation and Examples

The current implementation of FINANZIA uses C++ to implement object–orientation and is available in UNIX and Windows platforms. We use the following notation for representing class derivations: a name enclosed in $< * >$ indicates a variable, i.e. it can be substituted for a specific entity. A method argument is written as $< \texttt{argument} >$ and is understood to describe the actual arguments of the method. A name enclosed in $[*]$ indicates an optional part that may or may not be used. If the square brackets include more than one name, any of those names can be exclusively selected. A name enclosed in $\{*\}$ is mandatory. If more than one name are enclosed in the curly brackets, then any of them can be exclusively selected. A name appearing without any surrounding notation is understood to be a fixed part of the expression.

Option Classes are available for a number of option products. The derivation tree for the `Option` class is shown in Figure 10.6. An American call option can be derived as `CallAmerOption`. In this case the underlying asset and the dividend model are taken to be the defaults, `Stock` and `Cont` respectively for stock and continuous yield. The rest of the classes in Figure 10.6 are defined as follows: `Disc` stands for discrete dividend model, `Euro` for a European type of option, `Berm` for a Bermudan type of option, `Put` for a put option, `Asian` for an Asian type of option such as the average strike option, and `Exotic` for an exotic such as a barrier option. The various types of underlying assets are derived as `Index` for index options, `Rate` for interest rate options, `Curr` for currency options, `Comm` for commodity options, `Opt` for options on options, and `Fut` for options on futures.

The base `Option` class provides methods for setting up the various characteristics of the option, such as strike, spot price, etc. The general format

$$\begin{Bmatrix} \left.\begin{bmatrix}\texttt{EFD}\\ \texttt{IFD}\\ \texttt{EFM}\\ \texttt{IFM}\end{bmatrix}\texttt{Pde}\right. \\ \texttt{MoC}\\ \texttt{Bin}\\ \texttt{BS}\end{Bmatrix}\texttt{Calc}$$

FIGURE 10.7. Derivation tree for the `Calc` Class used in FINANZIA.

of these methods is defined as

$$<\texttt{class}>.\texttt{Set}\begin{Bmatrix}\texttt{S}\\ \texttt{K}\\ \texttt{T}\\ \texttt{D}\\ \texttt{R}\\ \texttt{Vol}\\ <\texttt{other}>\end{Bmatrix}(<\texttt{argument}>)$$

For example, in order to set the spot price (i.e. the current price) of the underlying asset `S` the method `<class>.SetS(double S)` is used. This method is overloaded when a derived class for a specific underlying asset is instantiated. The base `Option` class also defines the payoff function with the method `<option>.Payoff(<spot>)`; this method is overloaded in the derived classes `Call`, `Put`, etc., accordingly. For example, a `PutEuroOption` overloads `Payoff()` by defining

```
double PutEuroOption::Payoff( double S ) {
      return( max(this.K() - S, 0 );
}
```

There is a large number of methods that deal with specific issues such as boundary conditions necessary for each of the PDE solvers. In this case, for example, `Option` defines methods `LeftB()` and `RightB()` which return the left and right boundary conditions depending on the option and its characteristics.

Calculator Classes are available for several types of algorithms. The derivation tree for the `Calc` class is shown in Figure 10.7. The classes in Figure 10.7 are as follows: `MoC` for Monte Carlo methods, `Bin` for binomial methods, and `BS` for closed Black–Scholes formulas applicable in simple options. There are several classes for PDE based calculators; `EFD` stands for explicit finite differences and `IFD` for implicit finite differences, `IFM` for implicit finite elements and `EFM` for explicit finite elements.

Vanilla European or other options with closed formulas can be priced using a Black–Scholes calculator of type `BSCalc(<option>)` or using the

$$\left\{\begin{array}{c}\texttt{Price}\\ \texttt{Hedge}\\ \texttt{Implied}\left[\begin{array}{c}\texttt{Vol}\\ \texttt{Rt}\\ \texttt{Dv}\end{array}\right]\end{array}\right\}\texttt{Val}$$

FIGURE 10.8. Derivation tree for the `Val` Class used in FINANZIA.

binomial method `BinCalc(<option>)`. Exotic options can be priced using either binomial (`Bin`), Monte Carlo (`MoC`) or PDE (`Pde`) type calculators. American options are necessarily handled by either binomial or PDE calculators.

Overloading and dynamic binding are used extensively by the methods in the `Calc` class and its derived classes. For example, the main method `Calc::Solve(Option O)` is overloaded by an explicit finite difference PDE calculator as

```
EFDPdeCalc::Solve(EuroOption& EO)
EFDPdeCalc::Solve(AmerOption& AO)
```

Depending on whether the option is European or American, the appropriate method will be invoked and a boundary value problem, or a free boundary value problem, will be solved. Within the code of the methods, the appropriate boundary conditions and other model specific information can be accessed using dynamic binding. If for example the option is of type `CallEuroOption`, the invocation of `EO.LeftB()` will return the appropriate boundary condition on the left end of the domain.

Valuator Classes are available for the three main tasks. The derivation tree is given in Figure 10.8. An instantiation of a valuator that will perform option valuation is derived as `PriceVal`. Similarly, `HedgeVal` and `ImpliedVal` are derived for hedging and implied parameter calculation. The implied parameter calculation is currently supported for volatility σ (`Vol`), interest rates r (`Rt`) and dividends δ (`Dv`). The default derivation, e.g. `ImpliedVal` will work for volatility. A derivation of a valuator as `ImpliedDvVal` will perform implied calculation of δ.

Apart from the three main classes a number of **Utility Classes** are included to provide support and secondary operations. A list of the most important ones is given in Table 10.2.

10.4.1 Examples

In the following we give some examples to illustrate the use of FINANZIA. In order to price a particular option, the option must be defined, the valuator must be instantiated for pricing and the calculator must be selected for the specific algorithm to use.

Class	Services
Normal Distribution	Calculation of normally distributed random variables
Matrix, Vector	Abstractions and management of arrays and vectors
Root Finding	Bisection, Secant and Newton methods used in implied calculations
Interpolation	Linear and spline interpolation algorithms for one and two dimensions
Linear Solvers	Direct and Iterative linear solvers (tridiagonal LU, LU, SOR, Projected SOR)

TABLE 10.2. Main utility classes used in FINANZIA.

```
#include <Finanzia.h>

// Instantiate an option of the particular type
CallAmerOption  myOption;

// Initialize the option
myOption.SetS(100).SetK(120).SetR(0.05).SetD(0.02).SetT(0.5);

// Instantiate a calculator of binomial type
BinCalc         myCalculator;

// Instantiate a valuator
PriceVal                myValuator;

// Solve the problem
double myOptionValue = myValuator(myOption, myCalculator);
```

FIGURE 10.9. Pricing an American Option.

Pricing an American Option. Assume that we need to price an American call option on an underlying stock whose spot price is \$100 (i.e., $S = \$100$), the strike price is \$120 (i.e., $K = \$120$), with constant interest rate 5% (i.e., $r = 0.05$), constant dividend yield 2% (i.e., $\delta = 0.02$ and time to expiration six months ($T = 0.5$ years). A possible implementation is shown in Figure 10.9.

Using the Assistant Classes. In case a more complicated option with a more general algorithm needs to be priced an assistant can be used in order to simplify the necessary instantiation sequences. A generic example using the assistant classes is shown in Figure 10.10.

In general the assistant will return error information in case the request is invalid. In a robust use the appropriate error handling must also be considered. Instantiating other types of problems such as hedging and implied parameter calculation is straightforward.

```
#include <Finanzia.h>

// Instantiate the assistants
OptionAssistant                myOptionAssistant;
ValuatorAssistant              myValuatorAssistant;
CalculatorAssistant       myCalculatorAssistant;
Option                    myOption;
Valuator                       myValuator;
Calculator                     myCalculator;

// Set up the problem
myOptionAssistant.Asset(TYPE).Exercise(TYPE).DModel(TYPE).Payoff(TYPE);
myValuatorAssistant.ProblemType(TYPE).RateModel(TYPE);
myCalculatorAssistant.Method(TYPE).Algorithm(TYPE).Accuracy(#number);

// Create the appropriate instances
myOption = myOptionAssistant.newOption();
myValuator = myValuatorAssistant.newValuator();
myCalculator = myCalculatorAssistant.newCalculator();

// Initialize myOption, myValuator, myCalculator

// Solve the problem
```

FIGURE 10.10. Using the Assistant Classes.

Adding a new Component. The extension of FINANZIA to include new options, valuation problems or calculation algorithms can be done in a natural way. In order to add a new option problem the appropriate class must be derived from the hierarchy and the new type must be registered with the system. The same procedure is followed for new valuators and calculators.

10.5 Future Extensions

A number of extensions are considered useful for future implementations of FINANZIA. The wide expansion on the use of Internet, particularly in the financial industry, points to the direction of "calculator on demand" implementation that would offer the possibility to the user to download a specific portion of the library, or to use a service of the library over the network. An Internet enable implementation of FINANZIA could be done using Java. Java is among the most intriguing candidates for such an extension because the class hierarchy and methods can be interpreted almost directly. A Java implementation will allow network computing to be used in conjunction with FINANZIA.

The object hierarchy of FINANZIA should be extended to include stochastic volatility and interested rate models. Also existing object–oriented packages for numerical computing such as LAPACK++ and DIFFPACK (both are available at `http://www.netlib.org`) should be considered for extensions to the numerical algorithms included in FINANZIA. Such extensions would require redesign of parts of the `Calculator` class but would not affect the other main classes. Parts of LAPACK++ have already been used in FINANZIA (e.g., matrix and vector classes).

Furthermore, in order to provide an integrated interface for FINANZIA, a link with a problem solving environment of some sort is necessary. We are considering the integration of the library into MATLAB and EXCEL. For the multi–dimensional PDE calculators we are considering the integration with //ELLPACK[2] which provides a number of tools including mesh generators, visualization tools, and parallel numerical libraries.

10.6 Conclusions

We have discussed and presented the option valuation problem from a mathematical and computational perspective, and demonstrated how a systematic treatment of the problem can assist in the development of modern software tools for this computational finance application area. The need for modern software techniques in order to manage the complexity and meet the operational requirements of an option management system has been underlined in Section 10.1. The close relation of the option management problem to traditional problems and applications in applied mathematics and physics has been discussed in Section 10.2. Systematic treatment of options can be done based on those characteristics that affect the mathematical models, for example, in the case of early exercise, or their implementation. This hierarchical classification and its use for the development of an object–oriented abstraction for the option management problem has been discussed in Section 10.3. In Section 10.4 we have briefly presented the design and implementation overview of FINANZIA, an option valuation library which addresses the problem in an integrated, incremental way. The design and implementation of FINANZIA follows the hierarchical abstraction of the option valuation problem.

Acknowledgments: The authors acknowledge the support from NSF grants 9123502-CDA, 92022536-CCR, 620-92-J-0069, and ARPA grant DAAH04-94-G-0010. Also, the support of the Purdue Special Initiative Fellowship and the Purdue Research Foundation is gratefully acknowledged.

[2]//ELLPACK is a parallelized version of ELLPACK.

10.7 REFERENCES

[1] P. Boyle, M. Broadie, and P. Glasserman. Monte Carlo methods for security pricing. Technical report, Columbia University, 1995.

[2] P. Boyle, J. Evnine, and S. Gibbs. Numerical evaluation of multivariate contingent claims. *Review of Financial Studies*, 2:241–250, 1989.

[3] M. Brennan and E. Schwartz. Finite difference methods and jump processes arising in the pricing of contigent claims: A synthesis. *Journal of Financial and Quantitative Analysis*, 13:461–474, Sep 1978.

[4] F. Black and M. Scholes. The pricing of options and corporate liabilities. *Journal of Political Economy*, 81:637–659, 1973.

[5] J. Cox and S. Ross. The valuation of options for alternative stochastic processes. *Journal of Financial Economics*, 3:145–166, Mar 1976.

[6] J. Cox, S. Ross, and M. Rubinstein. Option pricing: A simplified approach. *Journal of Financial Economics*, 7:229–264, Oct 1979.

[7] J. Crank. *Free and Moving Boundary Problems.* Oxford University Press, 1984.

[8] C. M. Elliot and J. R. Ockedon. *Weak and Variational Methods for Moving Boundary Problems.* Pitman Publishing, Inc., 1982.

[9] S. D. Howison, F. P. Kelly, and P. Wilmott. *Mathematical Models in Finance.* Chapman & Hall, London, 1995.

[10] J. Hull. *Futures, Options and Other Derivatives.* Prentice Hall, third edition, 1996.

[11] P. Jaillet, D. Lemberton, and B. Lapeyre. Variational inequalities and the pricing of american options. *Acta Applicandae Mathematicae*, 21:263–289, 1990.

[12] K. N. Pantazopoulos, E. N. Houstis, and S. Zhang. Front-tracking finite difference methods for the american option valuation problem. Technical Report CSD TR-96-033, Computer Science Dept., Purdue University, May 1996.

[13] P. Wilmott, J. Dewynne, and S. Howison. *Option Pricing: Mathematical Models and Computation.* Oxford Financial Press, 1993.

11

Increasing the Efficiency and Reliability of Software Development for Systems of PDEs

Are Magnus Bruaset[1]
Erik Jarl Holm[2]
Hans Petter Langtangen[3]

ABSTRACT
In this chapter we address an object-oriented strategy for the development of software solving systems of partial differential equations. The proposed development strategy encourages heavy reuse of modules capable of solving the involved subproblems. Using class inheritance for successive refinement of the involved solvers, the complexity of the overall model is increased stepwise, layer by layer. In addition to the obvious advantage of code reuse and modular testing, this approach allows the developer to pull the pieces apart at any time for individual verification.

11.1 Introduction

The development of large codes for scientific computing is known to be a comprehensive and time consuming process. Moreover, large stand-alone FORTRAN codes dominate the field of scientific computing. Long-term evolvement of such codes is usually an error-prone and expensive process, unless the original software is carefully designed for future extensions. Turning to the field of computer science, years of experience indicate that human efficiency and software reliability can be significantly improved by a modular design that encourages reuse of code. This is also the basic principle underlying the ongoing development of *Problem Solving Environments* in various branches of scientific computing, see [16] and references therein.

---

[1]SINTEF Applied Mathematics, P.O. Box 124 Blindern, N-0314 Oslo, Norway. Email: `Are.Magnus.Bruaset@math.sintef.no`.

[2]Institute for Energy Technology, N-2007 Kjeller, Norway. Email: `erikh@ife.no`.

[3]Dept. of Mathematics, University of Oslo, P.O. Box 1053 Blindern, N-0316 Oslo, Norway. Email: `hpl@math.uio.no`.

A wide range of phenomena in science and technology is modeled by systems of partial differential equations (PDEs). The numerical solution of such systems is a demanding task and has therefore been one of the major research activities in scientific computing. The purpose of this chapter is to show how modular design and code reuse can be applied to the development of software for the numerical solution of systems of PDEs. Related mathematical models for a class of physical phenomena often exhibit a common basic structure, e.g., the fundamental differential operators in a system of PDEs might be the same. This property advocates the development of general modules that can represent the basic structure of a certain type of equations, preferably in a way that makes it easy to specify the details that differ from application to application. In this type of software environment, one can utilize previously debugged modules and thereby achieve increased software reliability and higher human efficiency in the coding process. Although the basic idea is simple and attractive, the design and implementation are far from straightforward. However, in this chapter we will address a particular approach to such software designs based on experiences with an implementation in Diffpack [5,7]. It should be mentioned that there are also other software libraries for PDEs available, e.g. Cogito [18], ELEMD [14], Kaskade [4], PETSc [2,12], and FEMLAB [9].

Initially, the software design discussed in this chapter was motivated by the development of a simulator for a particular type of plastic forming process. In this process, hot polymer is injected in a thin gap between flat plates and cooled down. This is a free boundary value problem involving mass, momentum and energy balance equations, coupled with complicated constitutive relations for modeling phase changes and a generalized viscosity of the polymer. Furthermore, three phases are involved; polymer, air and solid flow obstacles, see Section 11.2. Writing a stand-alone piece of software that solves such a system of PDEs may easily turn into a programmer's nightmare, due to the complexity of the numerical model and the danger of generating erroneous code. From this pessimistic (and realistic) concern, it seems reasonable that the reliability of the final solver should be based on a series of step-wise refinements that can be verified individually. Furthermore, if the final solver is an assembly of modules that are known to be safe, it should be possible to pull the modules apart again and repeat independent verifications. A general rule of thumb is to avoid the copying of source code and to make extensions without editing files containing already debugged code. Although the quest for reliability has been the driving force for this strategy, practical results show a nice side-effect in a dramatic decrease of the time spent on writing and debugging the software. The proposed strategy is not only applicable to this particular simulator. It can also be successfully applied in other physical problems, which we will point out later.

Modularity and code reuse can be achieved by using traditional implementation techniques in FORTRAN. However, this requires very careful

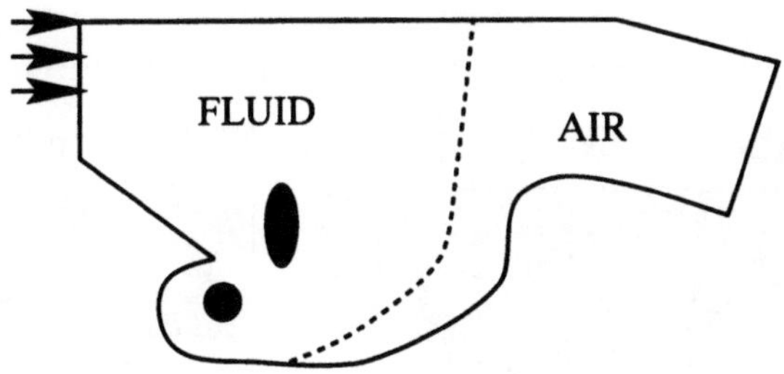

FIGURE 11.1. Sketch of the flow domain for a plastic forming process as seen from above. The dashed line is the polymer-air interface, whereas the solid obstacles are displayed as black spots.

and complicated considerations. Object-oriented design and programming techniques offer a much easier and more efficient methodology for obtaining the goals. The implementation can still be expressed in FORTRAN 77 or C, but it is obvious that this process is much more efficient when using a language that provides genuine support for object-oriented techniques. FORTRAN 90 [13] contains many useful constructs that appear as building blocks in object-oriented designs, but unfortunately lacks tools for very important principles such as polymorphism and dynamic binding [17]. Thus, C++ seems to be a reasonable alternative that offers the most important object-oriented constructs along with satisfactory computational efficiency [1]. In the present chapter we will use the C++ terminology and show some C++ code segments. Readers not familiar with C++ should consult standard textbooks [3,17].

11.2 A Plastic Forming Process

Slim plastic products, like computer keyboards and plastic bags, are usually formed by injecting hot polymer between two piecewise flat, cold plates, see Figure 11.1. The fluid is non-Newtonian, that is, the effective viscosity depends on the fluid motion. In the present problem we will use a generalized Newtonian viscosity model, where the dependency of the viscosity on the motion is given by an explicit formula [11]. The polymer displaces air, which results in a free-boundary problem since the fluid-air interface is unknown. Various solid obstacles can be installed between the plates to achieve the desired geometry of the final plastic product. The flow problem is treated as two-dimensional in the xy-plane. Velocities and other quantities are averaged in the z-direction. The normal velocity vanishes on the boundaries, while inflow and outflow are modeled by point injectors. The heat transfer problem is, however, three-dimensional with convection dominating in the xy-directions and conduction dominating in the z-direction. Note that con-

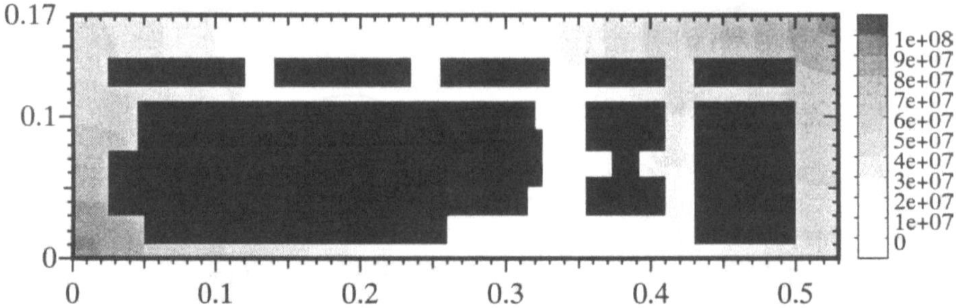

FIGURE 11.2. A snapshot of the pressure in a simulation of the injection molding of a computer keyboard. The polymer front can be detected as the isoline $p = 0$.

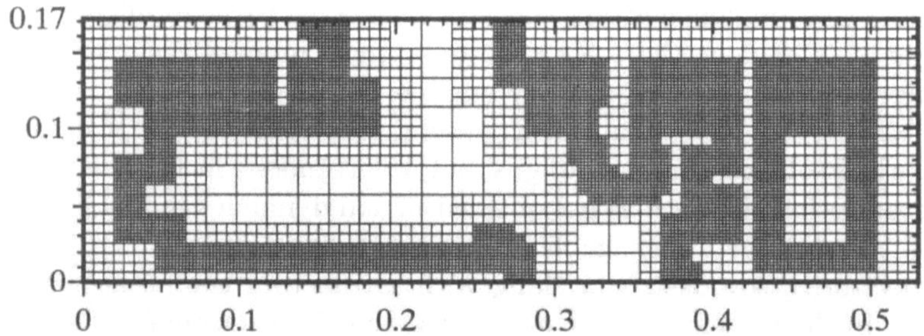

FIGURE 11.3. A snapshot of an adaptive grid used in the injection molding computations. The geometry is identical to the one shown in Figure 11.2.

duction also takes place in the solid obstacles. Since the fluid can solidify, the present application also involves phase changes. Figure 11.2 shows one snapshot of a particular simulation for a quite complicated problem.

To simplify the implementation, one can have a single grid and perform flow and heat computations in the polymer, the air and the solid obstacles. If the size of the gap between the two plates is $h(\mathbf{x})$, the obstacles are then modeled as by a very small gap $h(\mathbf{x}) = \epsilon \ll 1$, provided that h is scaled properly. The amount of mass transport through the obstacles is proportional to ϵ and can hence be controlled [15]. This approach to generating a simplified computational geometry by introducing certain modifications to the PDEs (or to the coefficients contained therein) is often referred to as *domain imbedding* or *the method of fictitious domains.* As shown in [6], the ill-conditioned linear systems arising from this procedure can be efficiently treated by certain preconditioning techniques. In order to increase the accuracy close to the advancing fluid front we have used adaptive grid refinement techniques available through the Diffpack class described in [10]. Due to the object oriented implementation of this adaptive grid class it was included in the simulator by adding just a few statements to the original code. A snapshot of a 2-irregular grid can be seen in Figure 11.3. As far as the 3D heat transfer problem is concerned, it can under certain circumstances be split into a set of 2D equations, using spectral methods to discretize the equation in the z-direction.

An appropriate mathematical model that allows computations in a single two-dimensional grid, covering polymer, air and obstacles, can be expressed as

$$\nabla \cdot [S(T,p,c,h)\nabla p] = q, \tag{11.1}$$

$$\frac{\partial c}{\partial t} + \mathbf{v} \cdot \nabla c = 0, \tag{11.2}$$

$$\rho(c,h)\frac{\partial H^{(i)}(T)}{\partial t} + \rho(c,h)\mathbf{v} \cdot \nabla H^{(i)}(T) = \nabla \cdot \left(\lambda(c,h)\nabla T(H^{(i)}(T))\right) + f(\mathbf{v}) \tag{11.3}$$

for $i = 1, \ldots, m_z$, see [11]. In this system, $p(\mathbf{x},t)$ is the pressure in the polymer or the air, where $\mathbf{x}$ is a 2D spatial point, $c(\mathbf{x},t)$ is an indicator function such that $c > 0$ implies polymer and $c < 0$ implies air. The modeling of the complex shaped front by solving (11.2) and thereby updating c is called a level set method. The function $h(\mathbf{x})$ reflects the gap between the plates (note that $h = \epsilon \ll 1$ indicates solid obstacles). Moreover, $q(\mathbf{x},t)$ models fluid injection or extraction points (normally through Dirac delta functions), T is the temperature, and $H^{(i)}(\mathbf{x},t)$ is the ith degree of freedom in a spectral expansion of the enthalpy in the third space direction [11]. The number of degrees of freedom, m_z, can usually be small, say 3–5. Furthermore, $\rho(c,h)$ represents the product of the density and heat capacity of polymer. This product varies with the medium (air or solid obstacles). Furthermore, $\lambda(c,h)$ is a heat conduction coefficient. The reason for using the enthalpy as the primary unknown in the energy equation is that phase changes (e.g. the Stefan problem) are then easier to deal with. Note that both c and h are used to identify whether the medium is polymer, air or solid obstacles. The fluid velocity is given by $\mathbf{v} = -\frac{1}{h}S(T,p,c,h)\nabla p$, where $S(T,p,c,h)$ is a known function that models the effective viscosity of the fluids. It is constant in the air ($c < 0$) and a nonlinear function of T, h, and ∇p in the polymer ($c > 0$). The source term $f(\mathbf{v})$ models internal heat generation due to friction (dissipation). The primary unknowns in the PDE system (11.1)-(11.3) are the two-dimensional quantities p, c and T. The rest of the quantities are either known or can be directly derived from the primary unknowns.

The core part of the software development techniques to be outlined does not depend on details of the discretization methods. It is therefore sufficient to assume that the initial-boundary value problem is discretized in time by some technique, such that the primary unknowns can be obtained at the same discrete time levels. The discretization of (11.1)-(11.3) in space is based on finite elements in our particular implementation.

11.3 The Basic Ideas

Most PDE systems that arise in solid and fluid mechanics involve only a few differential operators, typically $\partial/\partial t$, or $\partial/\partial t + \mathbf{v}\cdot\nabla$, and ∇^2 or $\nabla\cdot K\nabla$. If each operator could exist as an abstraction, it would be user-friendly to write the code for a system of PDEs as an assembly of various operators. However, the use of operators directly as C++ classes tend to decrease the computational efficiency dramatically. In order to achieve high efficiency, several terms in the equations must be treated simultaneously in the code. This contradicts to some extent our initial requirement of modularity. Furthermore, one might already have an existing high-quality FORTRAN code capable of solving parts of the PDE system. It is difficult to incorporate such software on the operator level.

In our experience, the *solver* for a single PDE (or a system of PDEs) has proved to be a useful abstraction in object-oriented implementations. We will therefore restrict the software design to numerical methods that are based on a particular type of so called *operator splitting* for solving systems of PDEs. Let our system of PDEs be written compactly in the form

$$\mathcal{L}_M(p;T,c) = 0, \tag{11.4}$$

$$\mathcal{L}_E(T;p,c) = 0, \tag{11.5}$$

$$\mathcal{L}_F(c;p,T) = 0. \tag{11.6}$$

These three equations correspond to (11.1), (11.2) and (11.3), respectively. The simultaneous action of $\mathcal{L}_M$, $\mathcal{L}_E$ and $\mathcal{L}_F$ is now the total differential operator in the PDE system. The idea is to split this operator and treat each of its components in sequence. For each equation, two of the primary unknowns are treated as "known" such that (11.4) becomes an equation for p, (11.6) an equation for finding T and (11.5) is used to update c. We use the most recent updates for the "known" variables in an equation, that is, the solution algorithm follows a typical Gauss-Seidel strategy. Our solution procedure can be expressed more precisely as follows.

> Use values from the previous time level as start values $T^{(0)}$, $c^{(0)}$ and $p^{(0)}$.
>
> For $k = 1, 2, \ldots$ until convergence:
> solve $\mathcal{L}_M(p^{(k)}; T^{(k-1)}, c^{(k-1)}) = 0$ with respect to $p^{(k)}$
> solve $\mathcal{L}_E(T^{(k)}; p^{(k)}, c^{(k-1)}) = 0$ with respect to $T^{(k)}$
> solve $\mathcal{L}_F(c^{(k)}; p^{(k)}, T^{(k)}) = 0$ with respect to $c^{(k)}$

When, and if, the loop is terminated, $T^{(k)}$, $c^{(k)}$ and $p^{(k)}$ are the values of T, c and p at the new time level.

Many systems of PDEs have strong couplings and nonlinearities such that this Gauss-Seidel approach may lead to convergence problems. Sometimes improved convergence can be achieved by solving a subset of the equations simultaneously. For example, we could think of replacing the $\mathcal{L}_M = 0$ and $\mathcal{L}_E = 0$ equations by a coupled subsystem $\mathcal{L}_C(p, T; c) = 0$, where p and T are solved for by an approach of implicit nature, e.g., Newton's method. The Gauss-Seidel technique can then be applied to the system of $\mathcal{L}_C(p, T; c) = 0$ and $\mathcal{L}_F(c; p, T) = 0$. If it is required to solve the whole system by a Newton type of algorithm, our basic design idea actually fails. However, it appears that only a few modifications are necessary in order to treat both Gauss-Seidel and Newton solution strategies. These modifications will be presented in Section 11.6.

The basic idea is now to develop a C++ class for the solution of each PDE (or subsystem of PDEs). The data type used to represent the coefficients should be quite flexible in order to maximize reuse of the PDE solver. Each type of equation will be modeled by a generic base class, whereas a specific equation can be implemented as a subclass. In other words, the common features of the solver are collected in a base class. A subclass inherits all the code and functionality of a base class (cf. [17]), such that the specialization of the subclass consists in programming only the *differences* between the generic and the specialized problem. In C++ this is done by including new data members plus redefining some virtual functions. The associated source code is normally very small.

A solver for the system involving p, T and c can then consist of three specialized C++ objects for the three PDEs, while the implementation of the Gauss-Seidel algorithm is placed on top of these abstractions. In each pass of the Gauss-Seidel algorithm, each PDE object is asked to solve its subproblem at the present time level. Such a design can reuse existing PDE solvers and, by a minimum of coding, tailor generic solvers to specific equations. This is obviously a desired and natural software development strategy. What is new here, is that an object-oriented implementation in C++ makes it very easy, at any time in the development process, to pull the objects apart and verify each PDE solver *without touching the code of these solvers.* This is an important feature that increases the reliability of the development process, and has clearly practical advantages during debugging and verification.

Since each PDE solver is a unit in this set-up, it is easy to use e.g. FORTRAN codes for a particular solver. To ease the integration of such a solver with the suggested C++ design, it may be a good idea to "wrap" a C++ class around the FORTRAN code. In this way the FORTRAN code will appear in the design as if it was a native C++ class. Mixing finite difference and finite element methods for the various equations is also easy when each PDE solver is a stand-alone class. The more C++ oriented details of the design will be presented in Section 11.5 using typical constructions from the Diffpack libraries.

11.4 Diffpack

Diffpack is a software system for rapid development of solvers for partial differential equations [5, 7]. Both finite difference and finite element methods can be programmed in Diffpack simulators, but at present only the latter type of methods takes full advantage of the most sophisticated software abstractions. The Diffpack libraries are coded in C++ and are based on object-oriented design and programming techniques. The object-oriented philosophy makes it easy to develop a simulator for a PDE by simply combining building blocks for vectors, matrices, linear systems, linear solvers and preconditioners, nonlinear solvers, finite element/difference grids with corresponding fields etc. The design of Diffpack is layered in the sense that high-level objects are built on top of objects from lower levels. At the most primitive level, the most important data structure is that of efficient C-style arrays, while CPU-intensive code segments employ simple FORTRAN-like constructs that are easily recognized by the optimization modules present in modern compilers. The layered design enables a high level of flexibility that allows the developer to extend and optimize Diffpack objects. An overview of the functionality in Diffpack is provided in [5].

A typical Diffpack simulator for solving a single PDE, e.g. using the finite element method [8], is implemented as a C++ class. Using the equation

$$-\nabla \cdot (K\nabla p) = 0$$

as an example, the minimalistic outline of such a class is given below. For our purpose the base class information is irrelevant. Thus, in the examples the specification of the base class(es) is omitted (`public ...`) in order to focus on the main topic. Usually, a simulator like `MyPDE` below inherits some Diffpack utilities to make, e.g., finite element programming easier, see Figure 11.4.

```
class MyPDE : public ... {
 protected:
  GridFE*    grid;     // finite element grid
  FieldFE*   p;        // finite element field over the grid
  LinEqAdm*  lineq;    // interface to linear systems and solvers
 public:
  // definition of integrands in the weak form:
  virtual void integrands (ElmMatVec& elmat, FiniteElement& fe);
  // definition of coeff. in PDE:
  virtual real K (FiniteElement& fe);
  virtual void init ();            // allocate & init grid, p, etc.
  virtual void solveProblem ();    // main driving function
};
```

In an actual Diffpack implementation one applies smart pointers with reference counting (called *handles* in Diffpack terms) instead of primitive C pointers like `GridFE*`. The classes `GridFE`, `FieldFE`, `LinEqAdm`, `ElmMatVec`

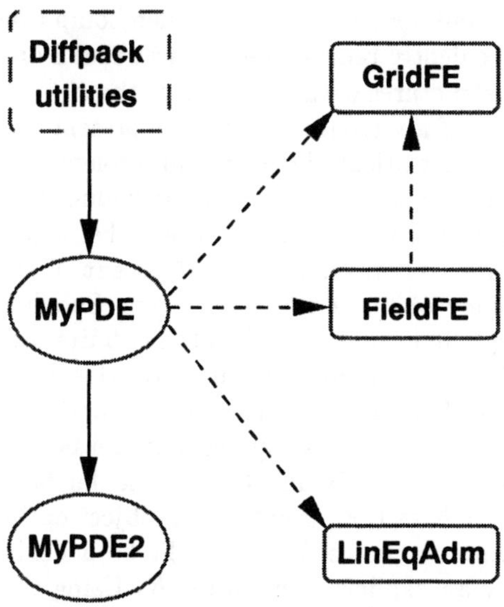

FIGURE 11.4. The simulator class `MyPDE` and its main members (finite element grid, associated field and linear system toolkit). Solid arrows indicate class derivation ("is-a" relationship) while dashed arrows represent pointers or references ("has-a" relationship). As indicated, `MyPDE` can itself serve as basis for new simulators, e.g., `MyPDE2`.

and `FiniteElement` are available in the Diffpack libraries [5]. Initialization and input to such classes are conveniently handled by the Diffpack menu system. For example, the user can through the `lineq` object get a graphical user interface for choosing matrix formats, the linear solver and associated parameters, as well as preconditioners.

The `GridFE` class is a standard finite element grid containing nodal coordinates, element connectivity and boundary information. The `FieldFE` class represents a continuous scalar finite element field over a grid and contains a smart pointer to a `GridFE` object, a vector of nodal values and information on finite elements. The `FieldFE` object offers of course interpolation of a finite element field at an arbitrary spatial point. For efficiency reasons `FieldFE`'s interpolation function can make use of precomputed basis functions. It is the responsibility of the `FiniteElement` class to compute and store the basis functions at a point in an element, their derivatives, the Jacobian of the mapping between local and global coordinates, as well as numerical integration points and weights. The `real` type represents the `float` or `double` type in C/C++. Finally, the `ElmMatVec` class represents the elemental matrix and vector in addition to a mapping from elemental to global degrees of freedom.

It is also worth mentioning that any Diffpack simulator can easily be extended with automatic report generation and other useful features for experimental scientific computing.

Since we use the finite element method when computing p, the p field in class MyPDE is conveniently represented by a FieldFE object. We have not indicated the data structures used to represent the K field, only the computation of K in terms of a virtual function was presented. The field concept naturally leads to object-oriented design and implementation in terms of a class hierarchy. A base class Field can be introduced with (pure) virtual interpolation functions of two types; one that takes a general, global point as argument and one that takes a FiniteElement object as argument to increase the efficiency of finite element solvers. Subclasses can represent constant fields, explicit formulas for functions, finite difference fields over uniform lattice grids, finite element fields, and fields over subdomains in finite element meshes. With the field hierarchy in mind, we can easily suggest a data structure[4] K_ for K in class MyPDE that can be represented in terms of a Field* pointer which is bound to a particular field subclass at run-time. For example, if $K \equiv 1$, K_ points to an object optimized for constant fields. Another common choice for K_ might be a field object where the values are given by an explicit formula for K. Using the field abstraction Field* K_ for K in the protected part of class MyPDE, the virtual function K is conveniently implemented as

```
virtual real K (FiniteElement& fe) { return K_->valueFEM(fe); }
```

The valueFEM function performs evaluation of a field inside a finite element, represented by the fe argument. This implementation works regardless of whether K is constant, a function or a precomputed finite element field.

The integrands procedure for evaluating the integrand of the weak form, in our example $K\nabla N_i \cdot \nabla N_j$, at a numerical integration point defined through the current status of the FiniteElement object, needs to sample the K coefficient. This is done by a call to the virtual function K. Hence, the integrands function is completely general for all types of variable coefficients, but the K function restricts the evaluation of K to interpolation of a Diffpack field object. Other more complicated forms of K, e.g. formulas that involve variables from other PDEs, can be implemented by overriding the K function in subclasses. This is a fundamental issue in our design of solvers for systems of PDEs.

Some readers will point out that having a virtual function, like K here, in the innermost loop of a computationally intensive code decreases the efficiency. In principle, this is true. However, we only need to evaluate K once for each numerical integration point, whereas several arithmetic operations are required to evaluate the contribution to the elemental matrix and vector from the integration point. The overhead in calling K is hence negligible. For physically realistic applications, the expressions for the vari-

[4]The identifier K_ uses an underscore character (_) to prevent name conflicts with the member function MyPDE::K.

able coefficients usually involve many operations, and the overhead of a virtual function call is even smaller.

The **solveProblem** function will compute p and, e.g., store the solution for later visualization. For debugging and verification purposes, it is a good idea first to use a simple K, e.g. $K \equiv 1$. This will only affect the initializing function `init`, which is responsible for allocating and initializing (large) data structures for the grid, various fields, and the linear system. The data structure for `K_` is quite general and the evaluation function `K` is even more general.

11.5 Systems of PDEs

In this section we will explain how PDE components in the equation system can be treated as individual and independent Diffpack simulators like class `MyPDE`. We will use the system (11.1)-(11.3) as a concrete example. Let us now consider the transition of the basic software design into working C++ classes.

We first make a generic implementation of the equation (11.1). Again it is natural to represent the variable coefficient S as a virtual function that can be redefined in subclasses. In the implementation we assume that the problem is generally nonlinear in p through the dependence of S on p. Therefore we also need a virtual companion function for $\partial S/\partial p$, since this quantity is needed in Newton's method for the nonlinear algebraic equations that arise from (11.1). Moreover, it will be an advantage to introduce a virtual function for the evaluation of T. Note that only the sign of c is really needed in the flow and energy equations. It is therefore convenient to introduce an indicator function **medium** that returns 1 if a point is inside the polymer, 2 if the point is inside the air, and 3 if the point is inside a solid obstacle. Our interface for handling a general variable coefficient $S(T, p, c, h)$ is then

```
class Flow1 : public ... {
 protected:
  GridFE*    grid;
  FieldFE*   p;
  LinEqAdm*  lineq;
  Field*     h;
  Fields*    v;  // v=-S*grad(p)
 public:
  virtual void integrands (ElmMatVec& elmat, FiniteElement& fe);
  virtual real S      (FiniteElement& fe);
  virtual real dSdp   (FiniteElement& fe);
  virtual real T      (FiniteElement& fe);
  virtual int  medium (FiniteElement& fe);
  virtual void init   (GridFE* g == NULL);  // init pointers
  virtual void solveProblem ();
};
```

This class is essentially a collection of scalar fields with shared grid data (one scalar entity for each component of the vector field) and a matrix system. Due to efficiency considerations, the argument passed to the functions evaluating the variable coefficients is of type `FiniteElement`, rather than just a point in space. The `init` function can either generate a grid or set `grid` to point to an external grid object `g`. The latter possibility will be important later. The `medium` function needs to access h and c, but in this base class we assume that we have polymer in all points. Hence, it is not necessary for the base class version of `medium` to have access to the c field. The `S` and `dSdp` functions need to access `medium` and, if the problem is nonlinear, also the `p` field. In addition, we have the velocity field **v** that is represented in terms of a `Fields` object. The interface shown above is minimalistic; a real-world example will often equip the class with additional data and functions that are not relevant to the design issues covered in this chapter.

In the base class `Flow1` the `integrands` and most other functions can handle the general case of a nonlinear equation $\nabla \cdot S(T, p, c, h)\nabla p = q$, but in the virtual functions above we implement just a simple choice of S for testing purposes. This could be $S = 1$ in the polymer and $S = \epsilon \ll 1$ (corresponding to $h \ll 1$) in the obstacles. The `T` function has no meaning and should return a constant reference temperature for the isothermal case.

From class `Flow1` we can derive a slightly more advanced class `Flow2`. For testing purposes we can assume, e.g., that there is a prescribed c field, but no temperature coupling. The S function could, e.g., be constant in each of the three media. Hence we could have

```
class Flow2 : public Flow1 {
 protected:
  Field* c_;  // pointer to external or internal front field
 public:
  virtual real S      (FiniteElement& fe); // checks medium(fe)
  virtual real dSdp   (FiniteElement& fe)  { return 0; }
  virtual int  medium (FiniteElement& fe); // checks c_ and h
  virtual void init   (GridFE* g == NULL); // call Flow1::init(g),init c_
};
```

Note that class `Flow2` can, e.g., solve a non-physical test problem for the purpose of making a small step towards a more complicated problem. The next natural extension is to couple the flow solver with a simulator that tracks the polymer/air front by solving equation (11.2). This is easily accomplished by letting `c_` point to an external field in a solver class for c. We can extend `Flow2` to handle both an internal and an external c field. This is assumed in the following.

We will refer to any numerical solution method for equation (11.2) as a *front tracker.* A key point is that class `Flow2` does not know anything about the existence of a front tracker object. It only uses general field information through its pointer `c_`. The front tracker is conveniently implemented as a base class `Front1` for solving a generic version of equation (11.2), using a virtual function for evaluating the velocity field $\mathbf{v}$. This function could simply return a constant value in order to enable easy debugging of class `Front1`. In a derived class, `Front2`, we can add a pointer `v_` to a vector field, and for example let `v_` point to the $-S\nabla p$ vector field in `Flow1`. The virtual function for the velocity evaluation must then be redefined in class `Front2` and call an interpolation function in the vector field class. The main ideas from the flow solver design can be applied directly to the front solver classes, so we omit showing details of class `Front1` and `Front2`.

Since the flow and front solvers communicate only in terms of pointers to fields, these solvers are actually completely independent of each other. We need a manager class to administer the field pointers and the solution procedure. This class will be called `HeleShawFill` (flow between two flat plates is usually referred to as Hele-Shaw flow and here we also have a filling process). For reasons of convenience the `Flow2` and `Front2` classes can have references to `HeleShawFill`. This will enable the solvers to access data in other solvers, through the manager. However, the coupling of `Flow2` and `Front2` to `HeleShawFill` should only be visible in these classes, not in their respective base classes. The manager class can then look like this:

```
class HeleShawFill : public ... {
  Flow2*   flow;
  Front2*  front;
  GridFE*  grid;                  // common grid
  TimePrm* tip;                   // time integration parameters
  void timeLoop();                // time integration algorithm
 public:
  void solveProblem ();           // main driving routine
  void init ();                   // init couplings between flow and front
};
```

Class `HeleShawFill` is in charge of making a common grid and of performing the common time stepping. It is not necessary to use the same time step size in different solvers, but it is convenient that the ratios of the step sizes are integers. The `grid` is transferred to the `init` functions of the `flow` and `front` solvers. If it is not transferred, the `init` functions in `flow` and `front` can make their own grids. The `init` function in `HeleShawFill` must create the common grid and time integration parameters, call the `init` functions in `front` and `flow`, and perform the pointer connections between the simulators. A sketch of the classes and the relations we have discussed so far appears in Figure 11.5.

The ideas that have been described so far can easily be applied to incorporate a heat transfer simulator as well. Consider for simplicity the equation (11.3) with $m_z = 1$. This equation has general variable coefficients ρ, $\mathbf{v}$, λ and f. We represent these by virtual functions. As in class `Flow1`, the coefficients will depend on the medium (polymer, air, obstacle) and it is convenient to have the same indicator function `medium`. The base class `Heat1` for solving (11.3), for the case $m_z = 1$, implements the equation in

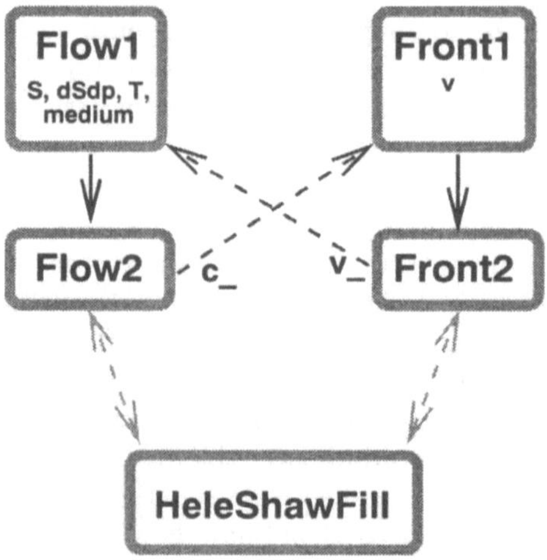

FIGURE 11.5. Sketch of the coupling of a flow solver and a front tracker for isothermal Hele-Shaw flow. Solid arrows indicate class derivation ("is-a" relationship) while dashed arrows represent pointers or references ("has-a" relationship). In the base classes **Flow1** and **Front1** we have listed some of the most important virtual functions.

a generic way in the **integrands** function, but provides trivial versions of the virtual functions for the variable coefficients. For example, we can let $\mathbf{v} = 0$, $\rho = \lambda = 1$ and assume only one medium for easy verification of the implementation.

A subclass **Heat2** can be derived from **Heat1** where we assume a varying velocity field, represented by a pointer **Fields* v**. We establish the communication with the manager **HeleShawFill** and use this communication line such that the **medium** function in **Heat2** simply calls the **medium** function in **Flow2**. An alternative is to build a local **Heat2::medium** function that uses pointers to c and h in **Flow1** and **Front1**, respectively, but in our opinion this is a less elegant (and less robust) design, since it actually implies a copy of existing (debugged) code in class **Flow1**.

The flow solver must also be extended to handle temperature effects. This is easily accomplished by deriving a class **Flow3** from **Flow2**, where we have a **Field*** pointer **T_** to some temperature field. This pointer will be set to the correct address in **Heat1** by the manager. The virtual **T** function is redefined in **Flow3** and makes use of the **T_** pointer. The manager must then be extended to have three solvers: **Flow3**, **Front2** and **Heat2**. A sketch of the new class structure is displayed in Figure 11.6.

The description of the classes and their relations is perhaps rather technical and C++ oriented. Nevertheless, the technical information will help readers with C++ experience to get a deeper insight into our ideas.

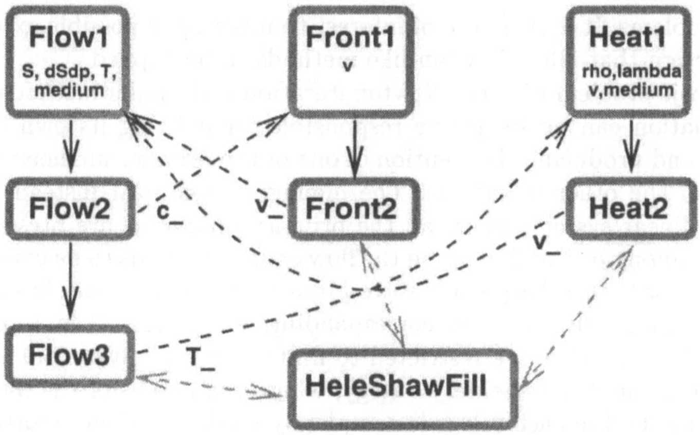

FIGURE 11.6. Sketch of the coupling of a flow solver, a front tracker and a heat transfer solver. Solid arrows indicate class derivation ("is-a" relationship) while dashed arrows represent pointers or references ("has-a" relationship). In the base classes we have listed some of the most important virtual functions.

Remarks

The proposed design of a solver for a complicated system of PDEs allows the programmer to develop the code in mathematically natural steps, and verify the implementation of each step. At any time in the process, the state of any previous step can easily be recovered for reliability tests. Inheritance is the key needed to avoid any editing of already debugged code; extensions and modifications will always appear in subclasses. Another key point is that the generic base class solvers, `Flow1`, `Front1` and `Heat1`, have no knowledge of each other, or of any other solvers. Hence, they can be reused in a wide range of applications. The base class solvers presented above are tightly connected to the equations (11.1)-(11.3), but one can think of more general Poisson, advection and energy equation solvers with a greater potential for reuse. The development steps could then easily be more refined. For example, there could be additional layers, (`Flow3`, `Flow4` and `Flow5`) before a full coupling to other equations and the manager is performed. Based on these ideas, we see the possibilities of creating very flexible solvers, within a specific application area, that can be easily combined. This will be one of the future directions of the Diffpack Project.

11.6 Extensions of the Concept

A drawback of the previously proposed design of solvers for systems of PDEs is that the equations must be solved in sequence at each time level. As already pointed out, this Gauss-Seidel type of approach may face convergence problems. It is therefore of interest to investigate possible extensions of the design that allow Newton-like methods to be applied.

The basic problem with the Newton iteration and similar methods is that each equation can no longer be responsible for defining its own discrete problem and producing the solution of one of the primary unknowns, given a value of the other unknowns. The manager class must instead build a common linear system, where all the primary unknowns are present. The coupling among p, T and c, e.g. in the flow equation, needs to be considered. In other words, the base class solvers must see more of the interface to the other equations and their corresponding weak forms. The approaches that we discuss below are restricted to finite element solvers. Some of the flexibility of the Gauss-Seidel strategy, where the details of the numerical solution method in each solver is completely hidden, is then unfortunately lost.

The simplest approach will be to consider solution methods for nonlinear systems of equations that only need to evaluate the residual of the equations, not the Jacobian or other matrices. As an alternative to the `integrands` function we can simply provide a similar function that evaluates the integrands of the residual vector of the PDE at an integration point in an element, using the available values from the previous iteration for the unknowns. The elemental vectors from each PDE solver can then be appended to each other to form the composite elemental vector for the whole system in the manager class. The manager must assemble all these composite vectors into the global residual vector for the complete nonlinear system.

Considering full Newton methods, the approach in the previous paragraph can be extended. Besides the vector containing the residuals of the PDEs, we need the Jacobian. The flow solver (`Flow1`) must hence have a function similar to `integrands` where the contributions from (11.1) to the residual and the Jacobi matrix at the elemental level are computed. This function must compute the coupling between p and itself, p and T, as well as between p and c. Assuming n_e unknowns for p, T and c in an element, the elemental contribution from `Flow1` to the Jacobi matrix for the full system is a rectangular $3n_e \times n_e$ matrix. The `Flow1` solver can still be made quite generic, although the coupling to other equations is now evident even in this base class. For example, derivatives of the PDE with respect to other primary unknowns can be accessed via virtual functions like `dSdp`. That is, we must provide $\partial S/\partial p$, $\partial S/\partial T$ and $\partial S/\partial c$. The latter can in the present physical application be complicated to evaluate. Hence, it may be natural to treat only the flow and heat equations as an implicit system, solved by Newton's method, and use the Gauss-Seidel approach for coupling the heat/flow solver and the front tracker.

The `Flow1` class can of course still be used as a stand-alone solver since the functionality we describe here is only an extension (and no modification) of the previously presented version of the class. A clean way of implementing the functionality for Newton-like methods is to derive a subclass which then can act as base class for what we previously have referred to as class `Flow2`. Another approach would be to use inexact Newton methods, where the partial derivatives needed for the computation of the Jacobian are approximated by finite differences. In this case, the requirements posed on the solver for each PDE would be similar to what we described for non-linear solvers that only utilize the residual vector of the nonlinear system.

11.7 Other Applications

The design approach in the previous sections has been tailored to a particular physical application. Nevertheless, it is obvious that this approach and the underlying ideas are of a general nature. Here we will outline some examples of other applications where the methodology can be or has been used.

Many porous media flow problems fit into the framework of equations (11.1)-(11.3). The simplest single-phase problems have a linear version of (11.1) together with (11.2) for tracking the concentration of, e.g., contaminants. Temperature effects are often of less importance in groundwater flow or hydrocarbon recovery. Seepage flow with a free surface can be solved by a linear version of (11.1) combined with (11.2) for tracking the water-air interface. Two-phase porous media flow can be formulated as (11.1) and a nonlinear version of (11.2), but without any coupling to a T equation. Finally, more general multi-phase, multi-component porous media flow models usually consist of an equation of type (11.1), possibly with a time derivative, and a set of nonlinear (almost) hyperbolic equations that are generalizations of equation (11.2). If temperature effects are important, the equations must be coupled with an energy equation, like (11.3) with $m_z = 1$.

Within computational fluid dynamics, there are obvious advantages of a flexible simulation environment of the type proposed herein. However, the equations have then a slightly different structure from (11.1)-(11.3). If we consider the fairly general problem of free thermal convection in fluids, the equations read

$$\nabla \cdot \mathbf{v} = 0, \qquad (11.7)$$

$$\varrho \left(\frac{\partial \mathbf{v}}{\partial t} + \mathbf{v} \cdot \nabla \mathbf{v} \right) = -\nabla p + \mu \nabla^2 \mathbf{v} + \alpha (T - T_0) \mathbf{g}, \qquad (11.8)$$

$$\varrho C_p \left(\frac{\partial T}{\partial t} + \mathbf{v} \cdot \nabla T \right) = \kappa \nabla^2 T. \qquad (11.9)$$

The quantity $\mathbf{v}(\mathbf{x}, t)$ is the fluid velocity, $T(\mathbf{x}, t)$ is the temperature, ϱ is the fluid density, α is a prescribed coefficient related to density changes due to temperature perturbations $T - T_0$ from a reference temperature T_0, $\mathbf{g}$ denotes the acceleration due to gravity, C_p reflects the heat capacity, and κ denotes the heat conduction coefficient in the fluid. In these equations, the Boussinesq approximation is utilized, that is, the density ϱ is treated as a constant, except in the buoyancy term in the momentum equation (11.8). It is not possible to eliminate the velocities and thereby obtain a combined mass-momentum equation like (11.1). Instead, equations (11.7) and (11.8) must be solved simultaneously. When formulating a Gauss-Seidel approach for (11.7)-(11.9), one should treat (11.7)-(11.8) as a single unit. That is, there will typically be one flow class hierarchy for (11.7)-(11.8) and one heat class hierarchy for (11.9). Two-phase problems with sharp interfaces can be treated by level set methods, and then the interface tracking can be based on an auxiliary equation identical to (11.2). This will add a front solver hierarchy as explained for the system (11.1)-(11.3).

From the discussion above, a natural direction of development is to create *application environments*. This term refers to having available a pool of abstractions that can be employed in a certain application regime. These abstractions would take the form of classes representing different solvers, mathematical models, etc., and would adopt a standard for inter-object communication based on the concepts presented in this chapter. In this environment, the user could instantiate a large number of simulators for the coupled problem, just by trying different combinations of solvers and models for the subproblems. The application areas of porous media flow and computational fluid dynamics are attractive candidates for software development along these lines, since they both can take advantage of an experimental approach to the present diversity of computational methods and model formulations.

11.8 Another Application of the Flexible Design

The representation of a system of PDEs in terms of a class hierarchy for each PDE is obviously advantageous when implementing flexible solution methods. However, the modular design has also other advantages that we will outline in the present section.

Consider a numerical solution method for a single PDE, like equation (11.1), which is based on the following local-global approach. Initially the domain is partitioned into m coarse grid elements. This coarse grid is denoted by Ω_Δ. We then define a series of locally refined grids, Ω_e, where Ω_e includes Ω_Δ, where coarse mesh element no. e and its adjacent neighbors are refined. Figure 11.7 shows Ω_e for $e = 1, 2, 3$, in a particular example. To obtain an improved solution p of (11.1), one can combine computations done on Ω_Δ and Ω_e, $e = 1, \ldots, m$. Note that Ω_e covers the whole domain (contrary to ordinary domain decomposition methods where local refinements are also local problems – the idea of Ω_e is that the problem with local refinements inherits the proper physical boundary conditions). Various methods are possible for obtaining the fine grid solution, and we will focus on a strategy [10] where one essentially solves

$$a(p, \phi) = b(\phi) + q(\phi; p_1, \dots, p_m). \tag{11.10}$$

Here, $a(\cdot, \cdot)$ is the bilinear form of the problem, and $b(\cdot)$ and $q(\cdot; \cdots)$ are contributions to the right hand side, ϕ is some test function, and p_e is the solution corresponding to a grid Ω_e.

The solution procedure will now consist of an ordinary finite element assembly process for (11.10), but where a global problem over Ω_e must be solved when visiting each element in the element-by-element assembly. In other words, there will be m global finite element computations inside the ordinary assembly process. Basically, this is easily accomplished by deriving a solver subclass, say `Flow3` from `Flow1` and including a `Flow1` solver as data member. The assembly process is a virtual function in Diffpack PDE solvers, so `Flow3` will need to redefine the default version of that function. In the redefined function, one can for each element e in Ω_Δ call the `Flow1` solver to compute a solution over the refined Ω_e grid, utilize the solution in the assembly process over Ω_Δ, throw away the solution and then proceed with the next coarse grid element. In practice, the data member in class `Flow3` will not be a pure `Flow1` object, but rather a new subclass of `Flow1`, say `Flow4`, where we can make convenient specializations for the problem over Ω_e.

When implementing the `Flow3` and `Flow4` solvers, it turned out that the associated extra code was very small, thus resulting in short development time and high reliability due to maximal reuse of code. The implementation of this numerical algorithm in a FORTRAN 77 program would require substantial modifications and associated debugging. Moreover, the readability of the source code related to the object-oriented C++ approach is much better. We believe that building PDE solvers as modular C++ objects in the way we have sketched above, opens up new possibilities for safe and fast combination of such objects to create novel software for testing new ideas regarding numerical algorithms or physical models. In other words, this modern software approach might play an important role in experimental scientific computing.

11.9 Concluding Remarks

Today, object-oriented design and implementation are beginning to demonstrate an increase of human efficiency and software reliability also in scientific computing. Examples of numerical libraries for PDEs using this programming paradigm are Cogito [18], ELEMD [14], Kaskade [4], PETSc [2,12], and Diffpack [5,7]. All these packages apply the object-oriented concept on

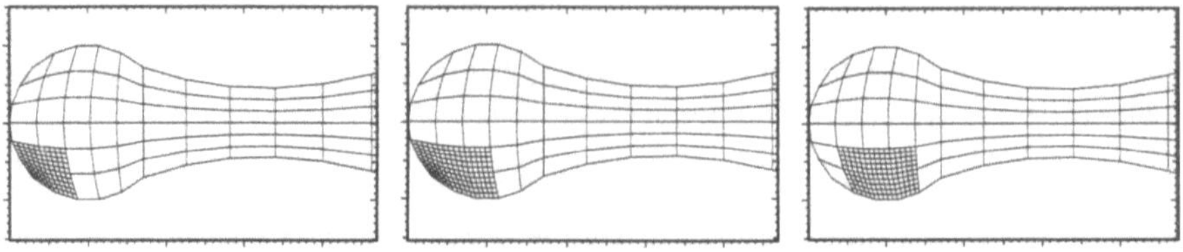

FIGURE 11.7. Example on locally refined grids Ω_e in three different cases. Removing the refinements yields a coarse grid Ω_Δ.

rather low level mathematical abstractions, like arrays, linear systems, linear solvers, preconditioners, finite elements etc. A solver can then be built by combining objects from the libraries.

The basic ideas of an object-oriented numerical library can be extended to a higher level where the objects reflect partial differential equations. This has been the topic of the present chapter. We have shown that this is a useful and powerful strategy for solving systems of PDEs. Furthermore, it opens up the possibilities of building repositories of solvers for single PDEs that can be combined with each other in a flexible way. This will dramatically reduce human efforts when developing software for advanced applications. A more detailed implementation in C++ of the design has been suggested to clarify the fundamental ideas.

Many research projects, especially in academia where students frequently enter and leave the activities, have suffered from the lack of programming standards and conventions. The approach suggested in this presentation can successfully be adopted by scientists working within a specific PDE-related application area, thus providing them with a mechanism for better organization of software contributions from different individuals. However, it should be stressed that the success of code reuse and coupling of stand-alone modules to form a complicated software environment relies heavily on a well-documented standard for building interfaces between the involved objects. In particular, there must be clear rules that dictate the behavior of the manager classes, and how the individual modules should be initialized. Such strict standards are needed if different researchers shall be able to develop new modules and reuse existing, already debugged components side-by-side in an explorative software environment.

Acknowledgments: The authors wish to thank the Research Council of Norway for the financial support through the strategic research programs STP.29643 (Toolkits in Industrial Mathematics), 110673/420 (Numerical Computations in Applied Mathematics) and the grant 100640/410. The authors also thank Xing Cai for his many useful suggestions that have spawned several significant improvements in the manuscript.

11.10 References

[1] E. Arge, A. M. Bruaset, P. B. Calvin, J. F. Kanney, H. P. Langtangen, and C. T. Miller. On the efficiency of C++ in scientific computing. In M. Dæhlen and A. Tveito, editors, *Numerical Methods and Software Tools in Industrial Mathematics*, pages 93–119. Birkhäuser, 1997.

[2] S. Balay, W. Gropp, L. C. McInnes, and B. F. Smith. Efficient management of parallelism in object-oriented numerical software libraries. In E. Arge, A. M. Bruaset, and H. P. Langtangen, editors, *Modern Software Tools for Scientific Computing*, pages 163–202. Birkhäuser, 1997.

[3] J. J. Barton and L. R. Nackman. *Scientific and Engineering C++. An Introduction with Advanced Techniques and Examples.* Addison-Wesley, 1994.

[4] R. Beck, B. Erdman, and R. Roitzsch. An object-oriented adaptive finite element code: Design issues and applications in hyperthermia treatment planning. In E. Arge, A. M. Bruaset, and H. P. Langtangen, editors, *Modern Software Tools for Scientific Computing*, pages 105–124. Birkhäuser, 1997.

[5] A. M. Bruaset and H. P. Langtangen. A comprehensive set of tools for solving partial differential equations; Diffpack. In M. Dæhlen and A. Tveito, editors, *Numerical Methods and Software Tools in Industrial Mathematics*, pages 63–92. Birkhäuser, 1997.

[6] X. Cai, B. F. Nielsen, and A. Tveito. An analysis of a preconditioner for the discretized pressure equation arising in reservoir simulation. Preprint 1995-4, Department of Informatics, University of Oslo, 1995. (Submitted for publication).

[7] Diffpack World Wide Web home page.
URL: *http://www.oslo.sintef.no/diffpack.*

[8] K. Eriksson, D. Estep, P. Hansbo and C. Johnson. *Computational Differential Equations.* Studentlitteratur (Sweden) and Cambridge University Press (UK), 1996.

[9] FEMLAB World Wide Web home page.
URL: *http://www.math.chalmers.se/Research/Femlab.*

[10] E. J. Holm and H. P. Langtangen. A unified mesh refinement technique with applications to porous media flow, 1997. (Submitted for publication).

[11] E. J. Holm and H. P. Langtangen. A unified model for injection molding, 1997. (Journal paper in preparation).

[12] PETSc World Wide Web home page.
URL: *http://www.mcs.anl.gov/petsc/petsc.html.*

[13] M. Metcalf and J. Reid. *Fortran 90 Explained.* Oxford Science Publications, 1992.

[14] G. Nelissen and P. F. Vankeirsbilck. Electrochemical modelling and software genericity. In E. Arge, A. M. Bruaset, and H. P. Langtangen, editors, *Modern Software Tools for Scientific Computing*, pages 81–104. Birkhäuser, 1997.

[15] B. F. Nielsen and A. Tveito. On the approximation of the solution of the pressure equation by changing the domain. *SIAM J. Appl. Math.*, 57:15–33, 1997.

[16] J. R. Rice and R. F. Boisvert. From scientific software libraries to problem-solving environments. *IEEE Comp. Sci. & Engrg.*, 3:44–53, 1996.

[17] B. Stroustrup. *The C++ Programming Language.* Addison-Wesley, 2nd edition, 1992.

[18] M. Thuné, E. Mossberg, P. Olsson, J. Rantakokko, K. Åhlander, and K. Otto. Object-oriented construction of parallel PDE solvers. In E. Arge, A. M. Bruaset, and H. P. Langtangen, editors, *Modern Software Tools for Scientific Computing*, pages 203–226. Birkhäuser, 1997.

Part II

Computational Geometry

12

Object Oriented Surface Design

Richard Bartels[1]

ABSTRACT

This chapter will review some work done at the Computer Graphics Laboratory of the University of Waterloo to build software abstractions with which ideas for new techniques in spline surface design can be supported. Our goal has been to facilitate the construction of prototype surface editors through the use of these abstractions. These editors are concerned with the definition of surfaces for industrial and artistic purposes.

The work built upon the most fundamental of mathematical abstractions: affine geometry; involving one of the commonest definitions of manifolds: parametric splines. Graph and tree abstractions that are linked to geometry are employed to organize the parts of the models that are edited.

12.1 Overview

In Section 12.2 we will give a sampling of basic geometry data types. There are more mathematical abstractions in this area that one could possibly use, and the main assignment has to be choosing the few that one can afford to develop and use as software. We have experimented with classes for affine spaces, affine frames, and points and vectors that live in those spaces and frames.

Section 12.3 reviews data structures that can be used to join different geometric frames of reference into composites. Typically, each node in the structure represents an affine frame, and the edge connections represent the immediate associations between two such frames, as in a jointed body, the blending of two surface elements, or the Boolean operations in a solid geometry model. Geometric data structures enhance the standard traversal from one node to the next with the corresponding coordinate transformations from one frame to the next, as is done in a standard graphics display structure.

Section 12.4 reviews a software framework for splines. These functions are regarded as the members of a hierarchy that provides general basis functions, general basic spline functions, and finally basic spline functions with particular knot structures and evaluations schemes.

[1]Computer Science Department, Computer Graphics Laboratory, University of Waterloo, Waterloo, Ontario N2L 3G1, Canada. Email: rhbartel@uwaterloo.ca.

Section 12.5 covers surfaces. These involve a parallel hierarchy to splines that provide general surfaces with basis-generated surfaces as derived classes. Surfaces distinguish themselves from general functions by their restricted domain and ranges and by certain special quantities they provide; such as normal vectors and curvature.

Section 12.6 provides a brief mention of classes that provide for subdivision, specifically for spline surfaces through knot insertion and additionally for more general surfaces through subdivision masks.

Section 12.7 will sample two of the applications that we have built using these abstractions or variants of them. We sample an editor that builds surfaces through a layering process and an editor for subdivision surfaces.

Our software development has proceeded in a very sporadic way. The material we report upon constituted the research projects of a number of graduate students covering half a decade. Periods of activity were spaced between periods of stasis. The C++ language itself mutated over that time. There was a singular need to preserve code and code knowledge from one activity and student to the next. Section 12.8 surveys the main features of a code development environment that has been set up within our laboratory to enforce a minimum of consistency between projects so that legacy code survives from each.

The environment provides a small amount of infrastructure that all further development is expected to use; of particular interest here: a class for templated, extendable, reference-counted storage (`Storage<...>`) and a variety of access iterators to such storage (for i-array-indexing, i, j-matrix-indexing, i, j, k-triangular-indexing, etc.). Furthermore, certain elements from the Standard Template Library [10] such as `list` appear in some of the listings in these sections.

The presentation of each class in the remaining sections shows only essential outlines. The destructor, other constructors, and other operators and member functions that would normally be included are not covered. Additionally, many decorations and further details have been removed or edited for the sake of brevity. To cut down on complexity, templated classes and procedures are shown in non-template form (by substituting appropriate class types for template arguments), and some class and data names have been modified as well, for presentational consistency.

12.2 Geometrical Abstractions

The fundamental abstraction supporting geometric activity is the *affine space* [7]. Briefly, this is a linear vector space augmented with a set of *points* and an operation of addition defined between points and vectors. The sum of a point and a vector is another point located at the position obtained by translating the original point along the displacement given by the vector. By extension, the difference between two points is defined as the vector displacement between them. A further extension [6] leads to the rule that combinations of the form $\sum_{i=1}^{n} c_i P_i$ are legal for points P_i provided either that $\sum_{i=1}^{n} c_i = 0$, in which case the result is a vector, or that $\sum_{i=1}^{n} c_i = 1$, in which case the result is a point. The latter case is called an *affine combination* of the points, while the former case may be called a *vector combination.*

Two basic representations for points are common. The first is in terms of a *coordinate frame*, $\{\mathcal{O}, \mathbf{v}_1, \ldots, \mathbf{v}_n\}$, $P = \mathcal{O} + \sum_{i=1}^{n} d_i \mathbf{v}_i$, where $\{\mathbf{v}_1, \ldots, \mathbf{v}_n\}$ is a set of basis vectors for the underlying vector space, $\mathcal{O}$ is a fixed reference point (the *origin*), and the coefficients $\{d_1, \ldots, d_n\}$ are arbitrary. The second is in terms of an *affine frame*, $\{P_0, \ldots, P_n\}$ $P = \sum_{i=0}^{n} a_i P_i$, where the set of points $\{P_i\}$ are in *general position* (i.e. they form a nondegenerate simplex, or equivalently, the vectors $\{\mathbf{v}_i = P_i - P_0\}$ form a basis for the underlying vector space), and the coefficients $\{a_1, \ldots, a_n\}$ must add to 1. Using P_0 as the origin, the conversion between the two representational forms involves an obvious and simple linear transformation. For a software class, the representation chosen for the private implementation is that which is most convenient for the majority of work to be done, and the other representation is provided by conversion. We have experimented with both forms of representation to define software geometry abstractions. The `Blossom` classes described in [9] work exclusively with affine combinations of points and use affine frames. The classes supporting the `PasteMaker` application described in [2] and in Section 12.7 work with displacement mappings and find it more convenient to work using coordinate frames.

Corresponding to linear transformations on a vector space are *affine transformations* on an affine space; namely, transformations that preserve vector and affine combinations $\mathrm{T}\left(\sum_{i=1}^{n} c_i P_i\right) = \sum_{i=1}^{n} c_i \mathrm{T}\left(P_i\right)$. The simplest example of an affine transformation is a translation. Restricted to the underlying vector space, an affine transformation is simply a linear transformation. A translation, for example, is the identity transformation when restricted to the vector space. Conversely, any affine transformation can be represented by the composition of a linear transformation and a translation. For geometric modeling, the affine transformations of main interest are rotations, reflections, scalings, shears, and translations. It is a common convention to store vectors as an $n+1$-tuple of real numbers $(0, d_1, \ldots, d_n)$, representing $P = 0P_0 + \sum_{i=1}^{n} d_i \mathbf{v}_i$, and to store points in coordinate-frame representations similarly as $(1, c_1, \ldots, c_n)$, representing $P = 1P_0 + \sum_{i=1}^{n} c_i \mathbf{v}_i$. For affine-frame representation this is adjusted to $P = a_0 P_0 + \sum_{i=1}^{n} a_i P_i$ with $\sum_{i=0}^{n} a_i \equiv 1$.

In the coordinate-frame convention, affine transformations can be represented by $n+1 \times n+1$ matrices whose upper-left $n \times n$ block is the matrix representation of the corresponding linear transformation on the underlying vector space. The $n+1$st row (column) provide the data for the translation that expands the linear transformation to an affine transformation. The choice between row or column is dictated by the convention of whether tuples are regarded as $1\times n+1$ or $n+1\times 1$ matrices. Changes from one affine space to another of equal dimension (or changes of frame within the space) are provided by nonsingular, $n+1\times n+1$ matrices. The same $n+1\times n+1 \rightarrow (n|1)\times(n|1)$ structure works for affine-frame representation by incorporating the matrix of the $a_i \rightarrow c_i$ transformation and its inverse into the matrix of the affine transformation.

Software abstractions that reflect this material are presented in [9]. What we present here is close to that, but adjusted to provide uniformity with the classes described in the remaining sections.

A `Space` class provides a constructor to create a space and set its dimension. Every space can report its dimension. Every space has a distinct ID to distinguish itself from all other spaces. Nothing else is needed. Spaces simply provide a way of registering a dimension and keeping points and vectors that belong to incompatible affine systems, distinguished by different IDs, from being combined in algebraic expressions.

```
class Space
{
  public:
    Space( int dim );
    int getDimension( void ) const;
    int getID( void ) const;
  private:
    int myDimension;
    int myID;
};
```

A point or vector in a space is represented by an array (of size `myDimension` or one greater, depending on frame representation) of coefficients of type `Scalar` (typically a *typedef* for `double`), and operators are defined to provide affine and vector combinations. For example,

```
class Point
{
 public:
   Point( const Storage<Scalar>& coef, Frame* frame );

   friend Point  operator+( const Point& p,  const Vector& v );
   friend Vector operator-( const Point& p1, const Point& p2 );
   // similarly other arithmetic operators

   Space* getSpace( void ) const;
   bool inSpace( const Space& s ) const;
   Frame* getFrame( void ) const;
   void setFrame( Frame* frame );
   void convertToFrame( Frame* frame );

   Storage<Scalar> getAffineFrameCoefficients( void ) const;
   Storage<Scalar> getAffineFrameCoefficients( Frame* frame ) const;
```

```
    Storage<Scalar> getAffineFrameCoefficients( const Point& p ) const;
    // similarly getCoordinateFrameCoefficients

    Storage<Scalar> getAffineWorldCoefficients( void ) const;
    Storage<Scalar> getCoordinateWorldCoefficients( void ) const;

    void setAffineFrameCoefficients( const Storage<Scalar> coef );
    void setCoordinateFrameCoefficients( const Storage<Scalar> coef );

  protected:
    Frame* myFramePtr;
    Storage<Scalar> myCoeffs;
};
```

The `Vector` class is similar.

Some of the arithmetic operators that are defined for the `Point` class may lead to geometrically meaningless results if applied without proper care. Normally, applications would not use them directly but would use a combination function.

```
Point combination( ScalarIterator coeffssbegin,
                   ScalarIterator coeffsend,
                   PointIterator  points );
```

Most of the member functions deal with registering a point to a frame, the conversion of a point into another frame or the extraction of a point's coefficients (in coordinate or affine terms) with respect to its own frame, another frame, or the frame of another point. The `inSpace` function makes use of th `Space`'s ID to test for compatibility.

The coefficients of a `Point` or `Vector` are meaningful only in the light of a `Frame`'s transformation to the `Space`'s version of those coefficients. The `Space`'s frame is taken as the standard Cartesian coordinate frame (or the standard-simplex affine frame), and naive implementation rules apply in that frame; e.g., $(1,4,-2)+(-2,1,3)=(-1,5,1)$. Similarly, naive rules apply in operations among points and vectors that reference the same `Frame` object. Thus, the arithmetic operations are implemented naively, but operand coefficients are converted to the common `Space` frame beforehand, if necessary. This is done automatically if the operand points or vectors are referenced to different frames in a compatible space. If they are in incompatible spaces, a run-time exception occurs.

The underlying engines that drive the conversions of coefficients are provided by the `Frame` class.

```
class Frame
{
  public:
    Frame( Space* s );

    Space* getSpace( void ) const;
    int    getDimension( void ) const;
    bool   inSpace( const Space& s ) const;
```

```
        Transformation*  getTransformation( void ) const;
        void setTransformation( Transformation* newtrans );
        void transformFrame( Transformation* transfToApply );

        void convertToSpace( Storage<double>* outcoeffs,
                             const Storage<double>& incoeffs ) const;
        void convertFromSpace( Storage<double>* outcoeffs,
                             const Storage<double>& incoeffs ) const;
        // similarly conversion of representation from
        // coordinate form to affine form or conversely

        Point getFrameOrigin( void ) const;
        // similarly getFrameBasis, getFrameSimplex

    private:
        Space* mySpacePtr;
        Transformation* myToSpaceTransfPtr;
};
```

The `transformFrame` function provides for an affine transformation of the entire frame system, as when the `Frame` represents the local coordinates on a rigid physical component that is moved as part of a larger system.

The `Transformation` class is merely a wrapper for matrices and matrix operations (including inversion, as implemented by LINPACK in the form of LU decomposition and back-solving), and won't be shown. The conversion routines provide for the implementation of such functions as `Point`'s `getAffineFrameCoefficients`, as well as the implicit frame conversion required by the arithmetic operators. The `getFrameOrigin` function and its cohorts provide points and vectors in the `Space`'s standard frame. These are generated through the services of the `Transformation` and are not actually stored in the `Frame`. That being the case, a mechanism is needed to create the `Transformation` from an origin point and basis of vectors, or from a simplex. Extra procedures of the following style are provided:

```
Frame makeFrameFromSimplex( PointIterator points, Space* s );
```

In graphical modeling systems, whenever possible, composite objects are defined in geometrically linked, hierarchical fashion. Each component part is represented most conveniently according to a local frame. Each linkage is associated with the transformation that represents one local frame in terms of the frame of one of its neighbors. Progressing from from one component to the next requires the conversion of one local frame to the next. The `Frame` class delivers that basic service between the members of a single pair of affine frames. The next section samples a few data structures whose nodes contain affine frames and whose traversal algorithms compose the transformations necessary to convert points from one end of an entire node path to the other.

12.3 Data Structures

Many of the abstract data structures of computer science are graph-like in nature; that is, they may be regarded as consisting of nodes and edges. The major distinction between any two static data structures lies in the *topology* of the linkage and the *traversal* used to visit the nodes. (A further important distinction, not relevant for our purposes, lies in the *updating* algorithms used upon the structure, if it is dynamic.) We survey two abstract base classes in this section, FrameNode and FrameStructure, that are used to augment data structures. They provide to the nodes of the structure a way to contain a Frame object that may be used by associated Point and Vector objects, and they provide to the structure itself the means to maintain "ToRoot" and "toParent" transformations efficiently during traversal.

The essence of the FrameNode class is the following:

```
class FrameNode
{
  friend class FrameStructure;

  public:
    FrameNode( Frame* myFrame )

    virtual void convertToParent( Point*  p ) const;
    virtual void convertToParent( Vector* v ) const;
    // similarly convertToSpace, convertToRoot,
    // convertFromParent, convertFromSpace, convertFromRoot

  protected:
    void setStructure( FrameStructure* structure );
    FrameStructure* getStructure( void ) const;

  private:
    bool ImDirty;
    Frame* myFramePtr;
    Transformation* myToParentTransfPtr;
    Transformation* myToRootTransfPtr;
    FrameStructure* myStructure;
};
```

And the essence of the FrameStructure class is the following:

```
class FrameStructure
{
  public:
    FrameStructure( void );
    virtual FrameNode *getParent( const FrameNode* frameNode );
    virtual void frameDirty( const FrameNode* frameNode );
    virtual void updateHierarchy( const FrameNode* rootNode );

  protected:
    bool structureDirty;
    list<FrameNode*> structureList;
    void includeFrameInStructure( FrameNode* frameNode );
```

```
        void removeFrameFromStructure( FrameNode* frameNode );
        void setFrameDirty( FrameNode* frameNode );
};
```

These two classes are designed to work with data structures constructed in the classical manner, with one base class to provide the update and traverse mechanisms, and the other base class providing the node objects. Augmenting such a data structure with geometric frames involves using each of the above classes as a *mixin*. For example, our `FrameNTree` data structure for n-way geometric trees is formed from our `NTree` data structure for ordinary n-way trees in the following manner:

```
class FrameNTreeNode : public FrameNode, public NTreeNode
{...};
class FrameNTree : public FrameStructure, public NTree
{...};
```

We have provided such "frame aware" structures for lists, general graphs, balanced binary trees, and directed acyclic graphs as well as n-way trees.

The `FrameStructure` mixin class is responsible for maintaining a separate list to record the node's `FrameNode` objects in parent-child (traversal) order from a given root. The `FrameStructure` and `FrameNode` classes together maintain composed `Transformation` objects (which are, in turn, responsible for maintaining inverses) to important target frames: the parent's, the space's, the root's. The maintenance is designed to be "lazy;" that is, the inverses (that is, the LU matrix factors) of composed transformations are brought up to date only when they are required to transform a point, vector, or frame, hence the presence of "`Dirty`" flags. A `FrameNode` becomes dirty, for example, when its `Frame`'s `setTransformation` or `transformFrame` method is executed. The `FrameStructure` becomes dirty when any of the `FrameNode` objects in its list becomes dirty.

12.4 Splines

The coefficients for affine combinations of points must sum to 1. If these coefficients are functions of a single variable, they may serve to generate curves, and if they are functions of two variables, they may serve to generate surfaces. An important class of functions that satisfy this summation to 1 property are polynomial *B-splines* (short for *basis splines*). They are further agreeable in having compact support, being nonnegative, and providing simple and numerically stable evaluation algorithms. This section reviews a software design for such functions, reflecting the material in [1].

Splines lie at the leaves of a `FuncBasis` hierarchy. The essence of the classes in this hierarchy are that they provide for the evaluation of a basis of functions through virtual `evaluate` member functions. Derived from `FuncBasis` is the `BBasis` class, which implements evaluation for B-splines.

At the bottom of the hierarchy, are varieties of spline representations, such as nonuniform B-splines and Bézier splines, together with the knot structure that defines them, which may override the `BBasis` evaluation with one that is more efficient. The following is an outline:

```
class FuncBasis
{
  public:
    virtual int number( void ) const=0;
    Storage<Scalar> evaluate( Scalar x ) const;
    Storage<Scalar> evaluate( Scalar x, int derivative ) const;
    virtual void    evaluate( double x,
                              Storage<Scalar>* results ) const=0;
    virtual void    evaluate( double x, int derivative,
                              Storage<Scalar>* results ) const=0;
};

class BBasis: public FuncBasis
{
  public:
    int number( void ) const;
    int order(  void ) const;
    int degree( void ) const;

    void evaluate( double x,
                   Storage<Scalar>* results ) const;
    void evaluate( double x, int derivative,
                   Storage<Scalar>* results ) const;

    virtual Storage<Scalar> getKnots( void ) const=0;
    virtual void setKnots( const Storage<Scalar>& knots ) const=0;
};

class NUBBasis: public BBasis
{
  public:
    void evaluate( double x,
                   Storage<Scalar>* results ) const;
    void evaluate( double x, int derivative,
                   Storage<Scalar>* results ) const;

    Storage<Scalar> getKnots( void ) const;
    void setKnots( const Storage<Scalar>& knots ) const;

  private:
    Storage<Scalar> myKnots;
};
```

Derived also from `BBasis`, for example, is the `BezBasis` class that stores breakpoints instead of knots, redefines the `getKnots` and `setKnots` members to convert to and from breakpoints, and adds `getBreakPoints` and `setBreakPoints` members. The `evaluate` routines are correspondingly overridden for efficiency.

12.5 Surfaces

A substantial amount of work on the design of an object oriented surface system has been carried out by Slusallek et. al. at the University of Erlangen [5,8,13]. Their work is worth reading.

We have been far less ambitious, providing only for tensor product spline surfaces; more precisely, surfaces of the form $S(u,v) = \sum_i \sum_j b_i(u) b_j(v) P_{ij}$, where the function values $b_i(u)$ and $b_j(v)$ are supplied as the contents of `Storage<Scalar>` by the `BBasis` class or one of its derived classes, and the P_{ij} are to be provided as the contents of `Storage<Point>`. The `evaluate` functions of the `BSurf` class and its subclasses merely need to carry out the summation using the procedure `combination` mentioned in Section 12.2 using iterators that provide i, j-indexed storage mapping.

At the base of the surface hierarchy is a generic `TPSurf` class responsible merely for defining the public services every surface should provide:

```
class TPSurf
{
  public:
    virtual Point  evaluate(  Scalar u, Scalar v )  const=0;
    virtual Vector evaluate(  Scalar u, Scalar v,
                              int du, int dv )      const=0;
    virtual Vector normal(    Scalar u, Scalar v )  const=0;
    virtual Scalar curvature( Scalar u, Scalar v )  const=0;
    // also gridEvaluate routines to provide meshes
};
```

Below the abstract `TPSurf` class is the class `LCSurf` class that specifically accounts for surfaces defined by `FuncBasis` objects.

```
class LCSurf : public TPSurf
{
   public:
     // evaluate functions defined
     // in terms of FuncBasis evaluations
   protected:
     FuncBasis* myUBasisPtr;
     FuncBasis* myVBasisPtr;
     Storage<Point> myCoeffs;
};
```

Below `LCSurf` comes the class `BSurf` that redefines `myUBasisPtr` and `myVBasisPtr` to be of type `BBasis*`. Below `BSurf` come a number of classes, such as `BezSurf` and `NUBSurf`, that further redefine `myUBasisPtr` and `myVBasisPtr` in terms of `BezBasis*` and `NUBBasis*` respectively. Thus, generic code can be written for `TPSurf` objects and executed for objects of type `BezSurf`, for example, with evaluation being supplied by the specifically tuned Bézier evaluation routines supplied by `BezBasis`.

12.6 Refiners

Certain extra functionality appears in the surface classes at the `BSurf` level in the hierarchy. B-spline surfaces with a certain knot structure can be re-represented as an identical surface with a finer knot structure by a process of basis refinement (knot insertion) [3]. Abstractly, given an original knot set and its coefficient points together with a finer knot set, a set of coefficient points corresponding to the finer set is obtainable. Each coefficient point of the new set is an affine combination of some local collection of the original coefficient points. The weights defining the affine combinations are determined by the original knots and the difference set of the knots. Thus, to convert from one `BSurf` object (or a derived surface type) to one of finer knot structure requires the replacement of a pair of `BBasis` (or derived) objects by objects with the finer knot structure and the creation of new coefficient points from old ones. A `Refiner` base class and two derived base classes provide such a service.

A `Refiner` object represents a transforming engine that converts old coefficient points to new ones. At this level of abstraction, knots are not involved, because the `Refiner` class is intended to define the interface for any mechanism that converts meshes of coefficient points.

```
class Refiner
{
  public:
    virtual Storage<Points>* newCoeffs(
                                 const Storage<Points>& old
                               ) const=0;
    virtual void newCoeffs( const Storage<Points>& old,
                            Storage<Points>* new ) const=0;
};
```

Derived from this class is a `KnotInserter` class that defines the `newCoeffs` procedure at construction time, given the old and new knots. A `KnotInserter` object is a more specific transforming engine that converts old coefficient points to new ones for given sets of old and new knots. As such, its use is restricted to meshes that represent B-spline surfaces. The cost of computing the weights is paid when the `Refiner` object is constructed, and that cost is amortized over the number of times the object performs a conversion.

```
class KnotInserter : public Refiner
{
  public:
    KnotInserter( const Storage<Scalar>& oldKnots,
                  const Storage<Scalar>& newKnots );
    // newCoeffs method defined
};
```

More specific subclasses, such as `NUBKnotInserter`, provide more specific definitions of the `newCoeffs` methods. An additional member, not present in `LCSurf`, is provided in `BSurf` and its descendants to make use of this functionality.

```
class BSurf : public LCSurf
{
  public:
    void refine ( const Refiner& rU, const Refiner& rV );
    void refineU( const Refiner& rU );
    void refineV( const Refiner& rV );
    // Change coefficients to reflect
    // the knot refinement process.
};
```

Also derived from the `Refiner` class is a `SurfaceRefiner` class whose method of defining the `newCoeffs` method is though the specification, at construction, of a mesh refinement *mask* that works directly on the coefficients without any assumption that the surface has an underlying B-spline basis. This provides a means of treating more general kinds of *subdivision surfaces.*

12.7 Applications

These classes have been developed by a number of students over the last half decade in the support of research thesis projects that include a spline surface fitter, a spline volume fitter, a general blossoming library, two surface-pasting editors, and a refinement surface editor. Other classes not mentioned, developed in the same style and environment, have supported a full-spectrum ray tracer, image processing tools, and a system to construct reflectance distribution functions from models of surface micro-structure. We will briefly mention one of the surface pasting editors and the refinement editor.

We have been experimenting with ways to design composite surfaces that generalize the hierarchical B-splines introduced in [4]. The essential idea in this body of work, which appears in [2], consists of designing compound surfaces by beginning with a base spline surface

$$S^0(u,v) = \sum_{i^0}\sum_{j^0} b_{i^0}(u) b_{j^0}(v) P^0_{i^0 j^0}$$

and "pasting" sequence of spline *offsets* upon it

$$\begin{aligned} S^k(u,v) &= S^{k-1}(u,v) \\ &\quad + \sum_{i^k}\sum_{j^k} b_{i^k}(r(u,v)) b_{j^k}(s(u,v)) \left(P^k_{i^k j^k} - S^{k-1}(u^k_{i^k}, v^k_{j^k}) \right) . \end{aligned}$$

The composite surface exists over the domain $(u,v) \in \mathcal{D}^0$ of the initial surface. Each offset is defined from some spline, $\sum_{i^k}\sum_{j^k} b_{i^k}(r) b_{j^k}(s) P^k_{i^k j^k}$, on some domain $(r, s) \in \mathcal{D}^k$. The pasting process requires selecting a mapping of this domain *into* $\mathcal{D}^0$, for which the functions $r(u, v)$, $s(u, v)$ represent the inverse mapping. The pasting also requires that a selection of sample points be made: $(u^k_{i^k}, v^k_{j^k}) \in \mathcal{D}^0$.

The mapping of domains is the mechanism by which the offset is positioned on the "base" surface $S^{k-1}(u, v)$. The selection of sample points is made so that the offset blends smoothly with the base around the "margins;" that is, for r or s fixed at either their maximum or minimum domain values. At these margins the offset should have minimal effect on the base surface or its normal:

$$0 \approx \sum_{i^k} \sum_{j^k} b_{i^k}(r(u,v)) b_{j^k}(s(u,v)) \left(P^k_{i^k j^k} - S^{k-1}(u^k_{i^k}, v^k_{j^k}) \right)$$

$$\vec{\mathbf{N}} \left[S^{k-1}(u,v) \right] \approx \vec{\mathbf{N}} \Bigg[S^{k-1}(u,v) + \sum_{i^k} \sum_{j^k} b_{i^k}(r(u,v)) b_{j^k}(s(u,v)) \left(P^k_{i^k j^k} - S^{k-1}(u^k_{i^k}, v^k_{j^k}) \right) \Bigg] .$$

The sample-point selection is best made using the tools of *quasi-interpolation* ([12], page 358).

From the perspective of the software abstractions being surveyed, the utility of the affine, spline, and surface classes is clear. The geometric data structures, however, are also useful. The mappings $\mathcal{D}^k \rightarrow \mathcal{D}^0$ implicitly build up a directed, acyclic graph of domain frames that must be traversed each time the composite surface is to be rendered and whose links must be adjusted each time an offset is moved on its base surface.

A second project that has benefitted from the software abstractions presented here has been the exploration of an editor capable of handling subdivision surfaces [11]. These surfaces have come into prominence in recent years. They combine the arbitrary topology of polyhedral meshes with the smoothness of splines. In abstract terms: one begins with a course mesh and, by applying a refinement process iteratively, the mesh is transformed into a smooth surface. Typically, a refinement process is described in terms of one or more *masks*, each of which is a pattern of weights whereby a vertex in the refined mesh is defined as the affine combination of a cluster of vertices in the unrefined mesh. A mask constitutes the essential data for defining a replacement rule. In the most regular situations, all mesh vertices are connected to their neighbors in an identical way, and as few as one mask may be all that is needed. In more general situations, there may be several categories of vertices (e.g., on the boundary or of different connectivity) with a different mask that applies to each.

The `Point` and `SurfaceRefiner` classes, as well as a variety of iterators for `Storage` that access points and neighbors in a regularly connected mesh, have provided the infrastructure for the subdivision editor. At an abstract level, the refinement process acts as a black box that takes a control mesh as input and generates a refined control mesh. This refinement process can be applied repeatedly to the resultant control mesh, and the process can be stopped after the required degree of accuracy has been achieved. A good approximation is usually achieved after only a few refinement steps. The editor was written entirely in terms of the `SurfaceRefiner` base class and in terms of a `LatticeIndex` base iterator class. The original mesh is manipulable, through an interface built upon *OpenInventor*, and an n-level refinement of the mesh (n selectable) provides an approximate rendering of the limit surface. The editor's `LatticeIndex` iterator class currently supports only 4-connected meshes, but subdivision rules on such surfaces can produce any box-spline surface or tensor-product B-spline surface, and the editor manages several such surfaces by menu selection. This enables the user to adjust the points of a mesh and then observe what the resulting effect would be if the mesh is taken as the defining coefficients of a variety of B- and box-spline surfaces.

As examples of the two systems just surveyed, we offer Figure 12.1, which is a screen shot of the `PasteMaker` editor [2], and Figure 12.2, a screen shot of the `RefEdit` editor [11]. Figure 12.2 displays a four-connected mesh and a box-spline surface resulting from refining the surface several times with an appropriate mesh. Figure 12.1 shows, in the lower window, the domains of five spline surfaces, a base surface and four feature surfaces, and in the upper window the surfaces themselves in pasted configuration.

12.8 Development Environment

All code described has been produced in an environment that is the collective design of the students and myself. It has managed to preserve a substantial portion of each student's software to serve as infrastructure for the next student's work. The environment, initially called "the Splines Project" but now more neutrally named "the GP Environment," has the following features:

1. all development is ultimately installed in a project directory tree that enforces standardized `include` file and library locations and conventions as well as a standardized organization and location for class files and documentation;

2. all development code is set up in a local environment that is completely consistent in include and library usage with the installed code;

3. development and installation versions of code may coexist, with the development version taking precedence for the developer and the installed code being provided to everyone else;

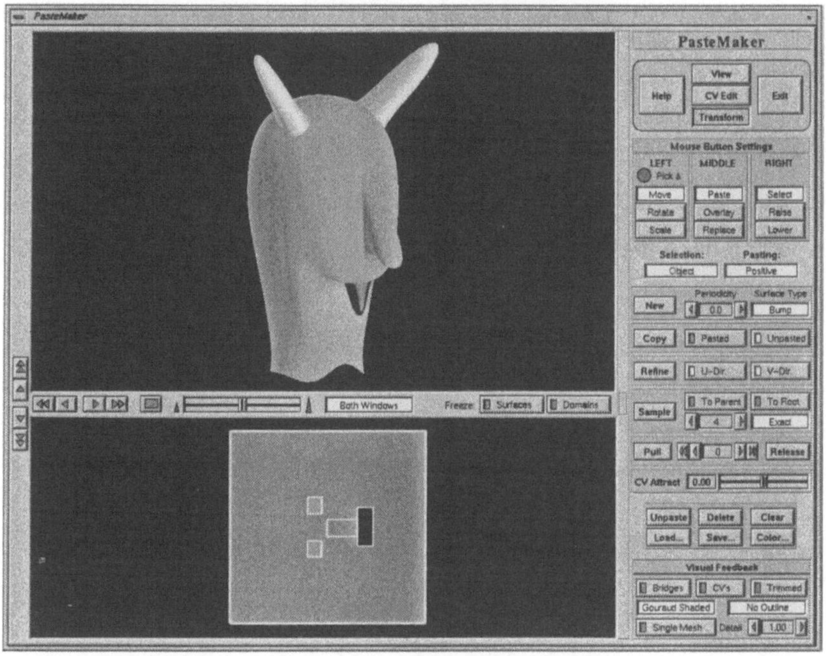

FIGURE 12.1. Surface pasting editor; the bottom window shows the domains.

4. all **Makefiles** are provided through **Imakefiles** that are built from a small number of macros designed for ease of use and for enforcement of common development practices;

5. development of every class is begun by executing a "newclass" command that provides both skeleton code and an **Imakefile** conforming to a common style and content;

6. all skeleton code enforces a common set of decisions on a class' default constructor, copy constructor, destructor, equals operator, and input and output operators;

7. the interface (.h) file for each class, as provided by the "newclass" command, outlines a form whereby a *man* page can be generated for the class automatically from the file;

8. common version control and version numbering through *rcs* is provided;

9. a common discipline of test examples is enforced using a standardized development directory structure and a "newexample" command that provides a skeleton testing harness for any class;

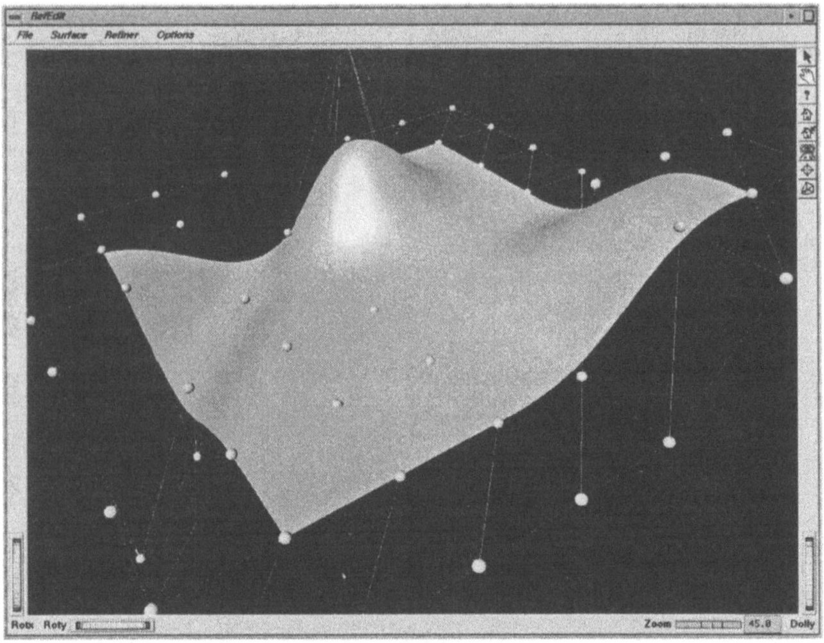

FIGURE 12.2. Refinement editor; a mesh and its $M_{(2,2,1,1)}$ box-spline surface.

10. all common actions such as: compile, create man page, create library, install, archive, de-install, de-archive, set version number, execute example, and verify output are provided by simple *make* command targets.

The environment was designed to use the C++ compiler together with only stock UNIX facilities: the Bourne shell, *man*, *nroff*, *nawk*, *rcs*, *imake*, and *make*. It can be installed on any system with a minimum of tuning (typically only the editing of two files is needed, one file to set paths to the development project's root directory, to *imake*, and to C++, and one file to set default compiler and load flags). The environment has run successfully on *DEC Ultrix* and *DEC OSF1* systems, *SGI IRIX* systems, *SunOS* systems (with *Sun* C++ and g++), and *Linux* systems (with g++). The "reference version" of the environment, however, is the one for the *SGI Irix* platform. A full manual describing the latest version is available as

`ftp://ftp.cgl.uwaterloo.ca/pub/users/rhbartel/Manual.ps.gz`

The latest version of the environment itself is available as

`ftp://ftp.cgl.uwaterloo.ca/pub/users/rhbartel/GP.tar.gz`

Currently the project environment and software is undergoing a thorough cleaning and conformity check. At this writing, only the basic infrastructure routines are provided, although this will improve as time progresses. For a sample of the environment bundled together with a student project, the reader might prefer the *Blossom* classes (partly described in Section 12.2) and its corresponding version of the environment. These are available in the directory

`ftp://ftp.cgl.uwaterloo.ca/pub/software/blossom/`

and they are more fully described in [9].

12.9 References

[1] Richard Bartels. Object oriented spline software. In P-J. Laurent, A. Le Méhauté, and L. L. Schumaker, editors, *Curves and Surfaces in Geometric Design*, pages 27–34. A K Peters, Ltd., 289 Linden Street, Wellesley, MA 02181 USA, June 1994.

[2] Cristin Barghiel, Richard Bartels, and David Forsey. Pasting spline surfaces. In T. Lyche and L. L. Schumaker, editors, *Mathematical Methods in Computer Aided Geometric Design III*, pages 31–40. Vanderbilt University Press, Nashville, Tenn. 37235 USA, 1995.

[3] Gerald E. Farin. *Curves and Surfaces for Computer Aided Geometric Design: A Practical Guide.* Academic Press, San Diego, California, third edition, 1993.

[4] David R. Forsey and Richard H. Bartels. Hierarchical B-spline refinement. *Computer Graphics*, 22(4):205–212, August 1988. SIGGRAPH '88 conference proceedings.

[5] Günther Greiner, Andreas Kolb, Ronald Pfeifle, Hans-Peter Seidel, Philipp Slusallek, Miguel Encarnação, and Reinhard Klein. A platform for visualizing curves and surfaces. *Computer Aided Design (CAD)*, 27(7):559–566, July 1995.

[6] Ronald N. Goldman. Illicit expressions in vector algebra. *ACM Transactions on Graphics*, 4(3):223–243, July 1985.

[7] Alexei Kostrikin and Yu Manin. *Linear Algebra and Geometry.* Gordon and Breach Science Publishers, PO Box 786, Cooper Station, New York, NY 10276 USA, 1989.

[8] Reinhard Klein and Philipp Slusallek. An object-oriented framework for curves and surfaces. In J. Warren, editor, *Curves and Surfaces in Computer Vision and Graphics III (Proceedings SPIE 1830)*, pages 284–295. SPIE, 1992.

[9] Wayne Liu and Stephen Mann. Programming support for blossoming. In *Proceedings of Graphics Interface '96*, pages 95–106. Morgan Kaufmann Publishers, 340 Pine Street, San Francisco, CA 94104 USA, May 1996.

[10] D. R. Musser and A. Saini. *STL Tutorial and Reference Guide.* Addison-Wesley Publishing Company, One Jacob Way, Reading, Mass. 01867 USA, 1996.

[11] Haroon Sheikh and Richard Bartels. Towards a generic editor for subdivision surfaces. In A. Pasko, editor, *Shape Modeling International '97.* IEEE Computer Society Press, 10662 Los Vaqueros Circle, Los Alamitos, CA 90720 USA, March 1997. to appear.

[12] Larry L. Schumaker. *Spline Functions: Basic Theory.* John Wiley & Sons, New York, New York, 1981.

[13] Philipp Slusallek, Reinhard Klein, and Hans-Peter Seidel. An object–oriented framework for curves and surfaces with applications. In Pierre-Jean Laurent, Alain Le Méhauté, and Larry L. Schumaker, editors, *Curves and Surfaces in Geometric Design*, pages 457–466. A K Peters, 289 Linden Street, Wellesley, MA 02181 USA, 1994.

13

Object-Oriented Scattered Data Modelling with Siscat

Erlend Arge
Øyvind Hjelle[1]

ABSTRACT
The construction of geometry from scattered data brings together the problem of data approximation and the problem of geometric design. Most specific algorithms for scattered data approximation have limited flexibility for incorporating design issues from an engineer's point of view. Such considerations have to be offered the engineer via flexible software systems. Siscat (The SINTEF Scattered Data Library) is an object-oriented toolkit with the aim of providing the engineer with tools for building geometries that approximate data and satisfy his intuition as well. The present paper discusses some basic design issues of Siscat, and illustrates these concepts on a model problem taken from cartography.

13.1 Introduction

Scattered data approximation deals with the problem of constructing geometry from unevenly distributed spatial points. Industrial applications of scattered data approximation are found in almost every area dealing with geometry, including Computer Aided Design (CAD) applications such as aeroplane-, automobile- and shipbuilding-industries, oil and gas industries, medicine and cartography. The underlying data is often collected with some kind of measurement equipment, such as laser scanners, light pens, ultrasound probes and seismic shots, but scattered data might well be derivates of already processed data, such as contour curves digitised from paper maps.

However, independent of the application and measurement equipment, a central issue is that we are given a finite number of points from which a continuous geometry is to be constructed. Even if we assume that the data is exact, and the geometry is required to *interpolate* the data, we face the problem of choosing "the correct" version among an uncountable

[1]SINTEF Applied Mathematics, Box 124, 0316 Oslo, Norway.
Email: {Erlend.Arge, Oyvind.Hjelle}@math.sintef.no

number of interpolants. If we do not have any information about the underlying geometry, except the measurements, a mathematician might argue that linear interpolation between neighbouring points would be the best we could do. This procedure is indeed often applied, leading to a triangulated geometry description, which is one of the most common ways of constructing geometries from scattered data. Again, from a mathematician's point of view, one could argue that this procedure is good if we know that the underlying geometry is differentiable and that the data is uniformly dense. Then one could obtain error estimates, and in this way control the quality of the approximation.

However, this idealised problem is only of marginal interest in connection to most real life applications. There are a number of reasons for this:

- The data is always subject to noise and errors, usually not quantified.
- The data distribution is often very irregular.
- Data might come from different sources, possibly with mismatch.
- The engineer might possess additional information about the data, not related directly to spatial position, that used correctly would improve the geometry.
- The engineer will have strong opinions about what kind of properties the geometry should possess. This type of information relates directly to the application area and the engineer's experience and knowledge of the application.

These issues enlighten the true nature of scattered data approximation. The spatial measurements are most often only one of several information carriers for the underlying geometry. A good result can only be achieved if the engineer is also allowed to draw upon other types of information in the construction process. Such information would be different for e.g. a CAD engineer and a geophysicist. While the CAD engineer would be extremely preoccupied with surface smoothness, light reflection and other aesthetic measures [12], the geophysicist would be interested in global geological trends and perhaps correctness of geological faults [8].

Scattered data approximation is therefore not alone "pure approximation". In addition, application dependent *design* issues play an extremely important role. In the literature [1, 7, 15] numerous methods have been proposed for the basic scattered data approximation problem. However, it should be clear from the above that no single one of these methods can act as a de facto solution for a range of situations. In fact, from an engineer's point of view, each single method will have its strong and weak sides in light of a given problem. What the engineer would need is to be able to exploit the aspects of a given method which in a given situation appears to be strong. Thus, he will need to mix different methods, and in such a way use the characteristics of the basic methods as design parameters in the construction process.

The SINTEF Scattered Data Library (Siscat) is an object-oriented C++ software library which represents an effort towards filling the gap between what the engineer requires and what is offered by basic scattered data approximation methods. Thus the design goals of Siscat are to provide a range of methods, ways to combine these methods, and ways to interpret data in order to offer the engineer a flexible set of tools to build geometry satisfying specific design constraints. A number of these issues are discussed in [2]. See also the Siscat WWW Home Page [17] which contains reports, manuals and examples of Siscat applications.

In the present paper we will discuss some of these topics in light of a model problem taken from cartography. This model problem exemplifies application-dependent constraints to be met in the geometry construction process. Although these constraints come from a specific application area, they represent typical surface design issues of a more general nature. We illustrate how the characteristics and requirements of the model problem are handled by the Siscat toolkit, and how the design elements in Siscat for explicit surfaces, i.e. surfaces of the form $z = f(x, y)$, fit into this particular problem.

13.2 A Cartographic Model Problem

Terrain models, as the carrier of topographic information, are an extremely important entity in cartography. They form the basis for a wide range of specific applications where examples are the production of contour maps, road planning, simulation of hydrology and navigation of airborne vehicles, to mention a few. For such diverse applications, the focus on what to be treated as important aspects will of course be very different.

The problem we shall consider here is to construct terrain models based on the data shown in Figure 13.1, where the basic idea is to make models that respect certain physical properties inherent in a terrain.

13.2.1 The Data

The data set represents a hilly region of Norway covering about 1.5 x 2.5 square kilometers with a total difference in altitude of 230 meters. The data set originally consisted of three basic entities:

- A set of terrain measurements collected by means of digital photogrammetry devices on the basis of stereo photographies taken by airplane. This data set is quite dense, although some parts of the data are missing.

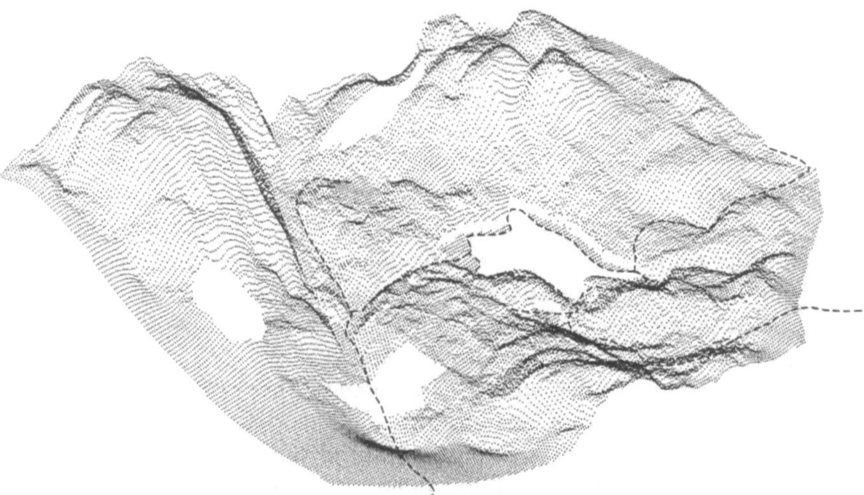

FIGURE 13.1. A data set representing a terrain. The data includes photogrammetric 3D data and polygons describing rivers and the shore of a lake.

- A closed 2D polygon representing the shore of a lake, digitised from a topographic map on scale 1:5000.
- A set of open 2D polygons representing rivers, also digitised from a topographic map on scale 1:5000.

The only 3D information in the data set is thus contained in the photogrammetric data material. Although assigned height values in the figure, the polygonal 2D information only represents qualitative aspects of the terrain. In the geometry construction process this information is used to make the terrain model consistent with the requirements of the cartographer.

13.2.2 Cartographic Requirements

As mentioned, our aim is to construct terrain models that are consistent with inherent physical properties of a terrain. By consistent we here mean fulfilling the following requirements:

- The elevation of the terrain model inside the lake should be constant.
- The shore of the lake should be sharp and follow the closed polygon digitised from the topographic map.
- The terrain model must respect the fact that rivers run downstream (this is in fact a common problem with commercial paper maps). Moreover, we here assume that rivers lie in the bottom of V-valleys.

- The holes in the data should be filled in a way that will be conceived as a natural extrapolation of the surrounding terrain.

In the context of this model problem, these considerations represent what we may regard as the design issues of the problem. Of course, other aspects of a terrain model are also of interest to the cartographer, e.g. adjusting the level of detail to a pre-described map scale, ensuring the production of smooth and easily conceivable contours from the terrain model, and respecting the extreme points of the data. However, we will not pursue such issues here.

13.2.3 Data Preparation

In the original data set only the photogrammetric data represent 3D information, while the polygonal data are qualitative in the sense that they only indicate where the polygonal features lie. We used a simple method to assign 3D information also to the polygonal data. This was done by using Siscat to construct a Delauney triangulation of the photogrammetric data, and sample the triangulation at the vertices of the polygons.

Since the photogrammetric data might be inaccurate, this procedure does not guarantee consistency of the polygonal information in the sense of downstream rivers and constant height values at the lake shore. To obtain consistency, the river polygons were made non-increasing along the stream direction by hand modification, and the shore polygon was assigned a single value being the average of the samples.

This modified data set was then used as the basis for the terrain model constructions described below.

13.3 The Basic Surface Hierarchy in Siscat

The terrain model we seek is simply a function

$$f : \mathbb{R}^2 \to \mathbb{R}$$

which approximates the data set well, and in addition respects the cartographic constraints given in Section 13.2.1.

In Siscat, problems of this type can be addressed through the `Explicit` hierarchy (cf. [10]) illustrated in Figure 13.2. This figure depicts a C++ class hierarchy containing methods for scattered data approximation and schemes for handling the input data. The abstract base class `Explicit` represents a common interface to the methods implemented in derived classes. In the figure three derived classes are shown, these methods will form the basis for experiments in later sections.

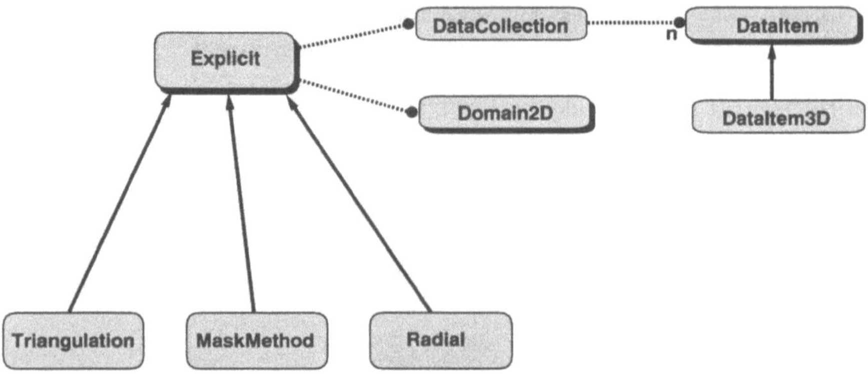

FIGURE 13.2. Class hierarchy for approximants on the form $z = f(x, y)$. Solid arrows indicate class inheritance and the dotted lines indicate containment. The 'n' indicates a list. The classes depicted as boxes with shadows are abstract.

The interface provided by **`Explicit`** basically includes data initiation, creation of the approximation and evaluation, and makes it possible to operate in a unified way with different methods.

The class `Domain2D` defines the subset of $\mathbb{R}^2$ where the function representing the surface approximant is defined. This is an abstract entity which basically only can tell if a point is inside or outside the domain. The class hierarchy derived from `Domain2D` contains rectangular, circular and polygonal domains. In addition, domains can be formed by any number of unions and intersections of other domains, which makes it easy to define domains of arbitrary complexity.

The underlying data for the surface approximants is held by the class **`DataCollection`** which is a container class for holding different types of data. Data of different types is organised in a list of objects derived from the the abstract type **`DataItem`**. This entity does not contain any data storage, it only defines the interface to the storage. Typically, the user will implement a class derived from **`DataItem`** which encapsulates his data storage and defines how to address this storage by implementing pure virtual functions specified by **`DataItem`**. This avoids copying of data, however, each point provided by the user then has to be addressed by a call to a virtual function. This may seem utterly inefficient, but experience shows that the main chunk of CPU consumption is found other places in the algorithms than in connection with data access.

In the model problem there will be one object holding the photogrammetric data, one object holding the closed polygon embedding the lake, and one object for each open polygon representing a river.

Tags

The software handling input data is equipped with mechanisms for managing data of different types. Each **`DataItem`** may be tagged with information about what kind of data it contains, and one DataItem may contain multiple tags[2] indicating multiple properties of the DataItem. The DataCollection administrates these tags, and the member functions of the DataCollection makes it easy to access and traverse data of common types for different purposes.

The Explicit methods address the input data via a DataCollection and are capable of recognising the tags. However, each Explicit method has a fixed configuration of tags that it can recognise. DataItems that are not tagged according to a method's understandable tags will simply be ignored during the course of constructing the approximation. One basic tag is the one indicating which data that is to be approximated - this tag is understandable for all methods.

In the context of the model problem, we have assigned 3D values to all data points (see Section 13.2.3), and therefore all these data may be tagged to be approximated. In addition, the data can be tagged to reflect other properties such as for a polygon, a boundary, a grid etc. Some methods are able to use this information and treat, for example, polygons in a certain way. This feature is used to treat the river and lake shore polygons in the examples below. More information about these structures can be found in [9].

13.4 Aspects of Basic Methods

In the framework of the model problem we shall take a closer look at three of the methods in the Explicit hierarchy, and discuss their performance on two different local areas taken from the model problem data set in Figure 13.1. These local data sets are shown in Figure 13.3 and 13.4 and correspond to the data segment where the river runs through a hole in the photogrammetric data and a segment around the lake.

13.4.1 Three Methods for Scattered Data Approximation

The methods we will consider are the *radial basis function method*, a *global smoothing method* and *triangulation* methods. We give a short description of these below.

Radial Basis Function Method

With this method the approximating function is of the form

$$f(\mathbf{x}) = \sum_{i=1}^{n} c_i \phi(\|\mathbf{x} - \mathbf{x}_i\|)$$

[2]Such tags are also called *item types* in the Siscat documentation.

where $\mathbf{x} = (x, y)$, the c's are coefficients, $\|\cdot\|$ is some norm on $\mathbb{R}^2$ (usually the Euclidean norm) and $\phi : \mathbb{R}_+ \to \mathbb{R}$ is the radial function.

The coefficients are determined by solving a $n \times n$ system of linear equations arising from the corresponding interpolation problem. Popular choices for ϕ are $\phi(r) = (k^2 + r^2)^{1/2}, r^3, r \log r, e^{-r^2}$ and others. Good solutions are generally obtained only for radial functions with global support. This also implies that the linear system to be solved is dense and thus prohibits the interpolation of large data sets. The practical limitation would be around 1000 points. The method is sensitive to the data distribution in the sense that the interpolation collocation matrix is singular if points coalesce. For suitable data sets the surfaces are generally pleasant looking with smoothness properties inherited from the smoothness of the basis function. More information about this can be found in [13].

The implementation of radial basis functions in Siscat allows us to attach any type of basis function. However, the method can only treat one type of data, which in the wording of Section 13.3 means that the method only understands one tag, the tag that signals which data are to be approximated.

Triangulations

A triangulation can briefly be described as a collection of planar triangular faces with vertices at the data nodes. The edges of the triangulation connect the data nodes and define the neighbouring relationship among the nodes. The neighbour relations can be defined in various ways, and the configuration will strongly influence the characteristics of the surface. Numerous methods have been proposed to construct triangulations to explicit scattered data. Such methods are designed to optimise the triangulation with respect to different features including predescribed edge configurations (constraints), data dependency, minimum variation of surface normals, contour smoothing, etc. For more information about triangulations see [5,6,14].

Siscat contains a rich menu of functionality for the construction and manipulation of triangulations. For the model problem we will utilise the notion of constraints to exactly represent the linear features described by the polygonal data. Thus we will use two tags - all data will be tagged to be approximated, and in addition the polygonal data will be tagged to be predescribed edges in the triangulation. When using only the approximation tag, the result will be a *Delauney triangulation*. Additional use of the tag for treating the polygonal data as constraints will result in a *constrained Delauney triangulation*.

Grid Generation with Smoothing

Regular grids are a common representation format for surfaces due to their simple structure. The particular method for constructing regular grids we will describe here is well suited for approaching problems with severely irregular distribution of the data points, e.g. clustered data or data with holes.

A brief description of the method is that we solve the problem

$$\left(\frac{\partial^4}{\partial x^4} + \frac{\partial^4}{\partial y^4}\right) f = 0,$$

subject to the constraint that the surface should approximate the data. This corresponds to the idea of minimising second order partial derivatives subject to the constraint of data approximation.

The grid comes into play when discretising the problem with a finite difference scheme over a regular grid. This is done in two steps, first local approximation methods are used to compute grid values at nodes "close" to the scattered data points. These values remain fixed. Next the values at the remaining nodes are computed by solving a linear equation system corresponding to a finite difference formulation of the fourth order problem, where the nodes computed in the first step act as boundary conditions. Since the approximation is defined by a regular grid, it does not interpolate the data, which is the case for the radial basis function method and the triangulation method. The grid generation method is described in detail in [3].

Moreover, this method is extended to handle feature lines that allow edges and discontinuities in the surface. Roughly speaking, this is achieved by modifying the difference stencils along polygonal lines in a way that disconnects the solution on one side of the polygon from the solution on the other side. This extension is presented in [4].

In Siscat, this grid generation strategy is part of the `Explicit` hierarchy as a derived class called `MaskMethod`[3]. As for the triangulation we will use two tags - all data will be tagged to be approximated, and in addition the polygonal data will be tagged to be discontinuity features as described above. The use of the approximation tag alone will be referred to as *grid generation with smoothing*, while use of the other tag as well will be referred to as *constrained grid generation with smoothing*.

13.4.2 Test of Approximation Properties - Part I

In our first test example we have applied the three methods outlined above to a local part of the data shown in Figure 13.1. In particular we study the data surrounding the "hole" crossed by one of the river polygons. The extracted data set is shown in the upper left frame of Figure 13.3. The characteristic of this data set is that the photogrammetric data contains a large hole. Moreover, the polygon indicating the river runs over this hole. From the cartographic requirements described in Section 13.2.2 we would

---

[3]This method of grid generation is used by the oil and gas system vendor Geomatic AS in the IRAP RMS system for constructing grid-based geological horizons with faults.

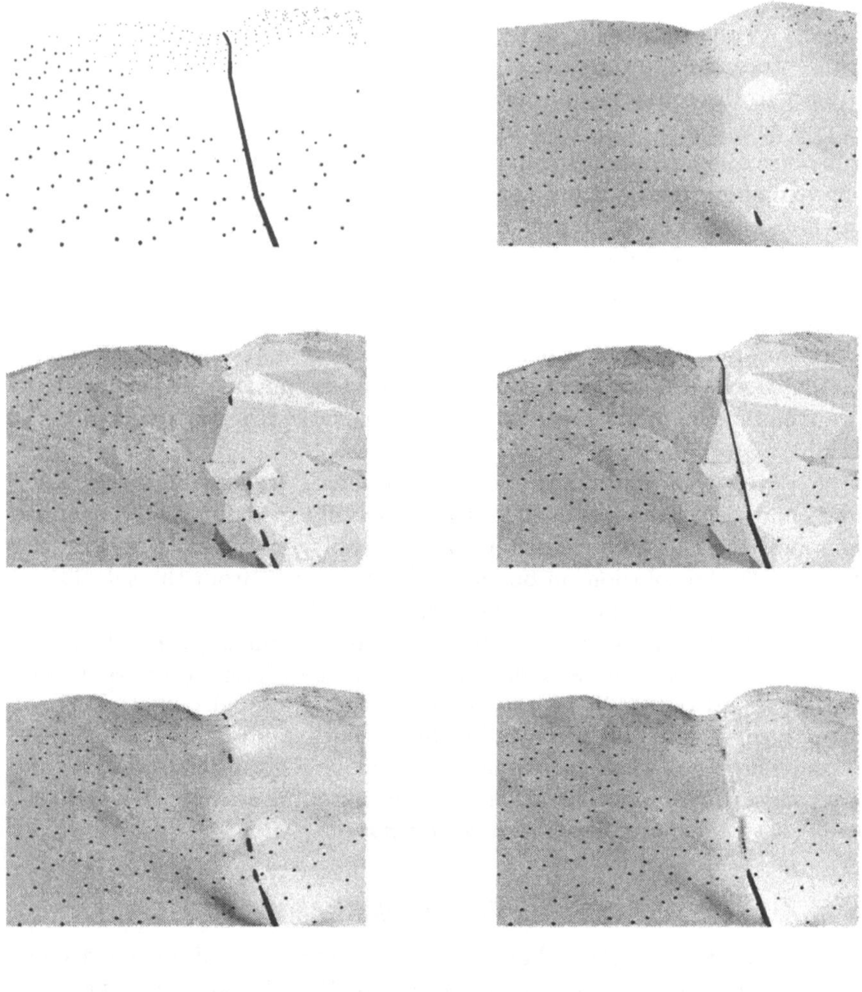

FIGURE 13.3. Results of test of approximation properties - Part I. The underlying data including photogrammetric data with a hole and a river polygon (upper left). Radial basis function interpolant (upper right). Delauney triangulation (middle left) and constrained Delauney triangulation (middle right). Grid generation with smoothing (lower left) and constrained grid generation with smoothing (lower right).

like the resulting terrain model to be a natural continuation of the photogrammetric data into the hole. In addition the river should clearly be indicated in the terrain model by forming a V-valley with the river running in the bottom.

The result for the radial basis function method with $\phi = r^3$ is shown in the upper right frame of the figure. The way the hole is filled is acceptable with respect to a natural extrapolation of the photogrammetric data. However, since the method can take the polygon into consideration only as points to be interpolated, there is no clear visible valley indicating the riverbed.

The Delauney triangulation is shown in the middle left frame. No characteristic riverbed is visible. In fact, we experience a "zig-zag" behaviour along the riverbed which is typical for a Delauney triangulation of such data. Moreover, the triangles in the hole are large and would most likely be unacceptable as a natural model of the terrain, at least presented at this scale.

The constrained Delauney triangulation shown in the middle right frame removes the zig-zag behaviour in the unconstrained version, and has a clear and visible riverbed, in particular away from the hole. However, still there is a problem with large triangle faces inside the hole.

In the lower left frame the result of applying grid generation with smoothing is shown. We observe that the surface is generally slightly more oscillating than the radial basis function. In particular, the oscillations are prominent along the riverbed, which is probably the most unpleasing feature of the approximation. However, the hole is filled in an acceptable way.

Finally, the constrained grid generation with smoothing is shown in the lower right frame. Compared to the unconstrained grid generation, the riverbed lies in a V-valley, although there are still some oscillations. In fact, the density of points on the river polygon is small compared to the grid spacing, and this explains why the polygon lies under the surface several places. Resampling the polygon would have improved the situation. However, the way the hole is filled is probably the most acceptable among the five variants illustrated.

Among the variants shown, the constrained Delauney triangulation and the constrained grid generation with smoothing appear to be the ones that best fit the cartographic requirements discussed in Section 13.2.2. However, due to the superior quality of the way the hole is filled, the cartographer would probably conceive the grid generation method to be the one that best fulfills the criteria.

13.4.3 Test of Approximation Properties - Part II

In our second test example we shall look into an area containing a part of the lake and the lake shore in Figure 13.1. The extracted data set is shown in the upper frame of Figure 13.4. From the cartographic requirements, we

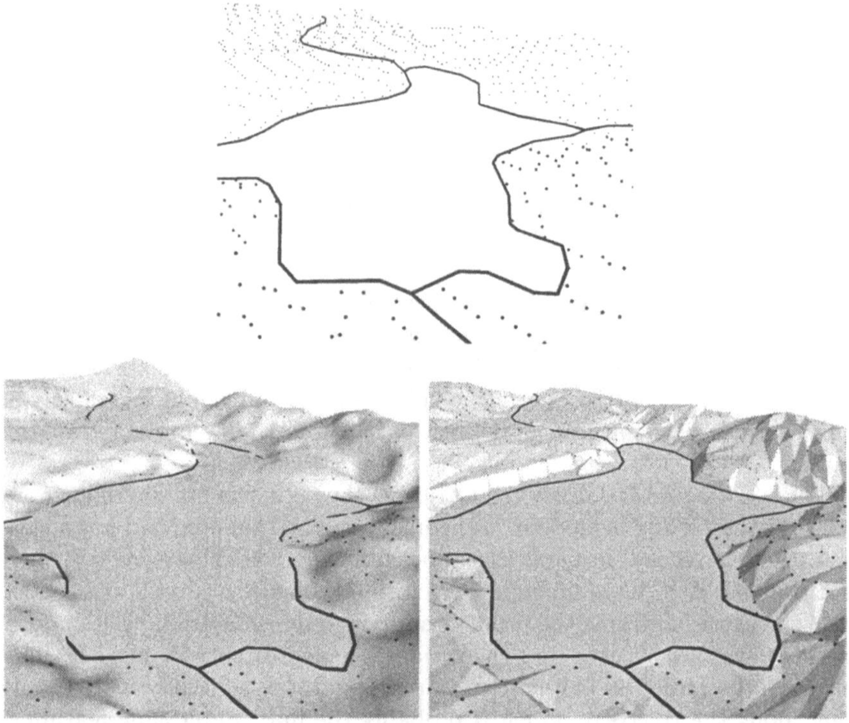

FIGURE 13.4. Results of test of approximation properties - Part II. The underlying data including photogrammetric data and the lake shore polygon (on top). Constrained grid generation with smoothing (lower left) and constrained Delauney triangulation (lower right).

would like the resulting terrain model to sharply reflect the shoreline and in addition, the value of the terrain model inside the lake should be constant. From the former example it should be clear that only the constrained Delauney triangulation and the constrained grid generation with smoothing will have a chance to meet these requirements. Therefore, we only present the results of these methods.

First, we note that the photogrammetric data and the lake shore polygon come from different data sources. Probably there are some discrepancies in these data sets, because the natural interpretation of the shore from the photogrammetric data does not fall in line with the polygon. Adjustments that would make these data sets more concurrent would probably improve the quality of the approximations, but we will not pursue this problem any further here.

The result of constrained grid generation with smoothing is shown in the lower left frame. The problem with the surface covering the polygon is even more pronounced here than in the previous example. Moreover, the lake surface is slightly oscillating and this is in contradiction to the requirement that the surface of the lake should be a constant value.

The constrained Delauney triangulation is shown in the lower right frame. The construction ensures that the surface has the desired properties. Therefore, this version would probably be the one chosen by a cartographer.

13.4.4 Test Conclusion

The two simple tests reported above illustrate the fact discussed in the introduction of this paper: different methods are superior in different situations. Moreover, for the model problem these situations apply within what should constitute a single terrain model.

This illustrates a more general problem in scattered data approximation, not present in cartographic applications. The engineer needs to be able to locally modify the geometry in order to fulfill local shape constraints. Given the limited abilities of a single method, this raises the need for being able to freely utilise a spectrum of methods within one single geometric model.

In Siscat this challenge is met through the concept of *composite methods.* In the next section we will discuss this concept and illustrate how object-oriented design techniques can be used to embed the variety of methods from the `Explicit` hierarchy in a powerful framework that resolves the problem of local design.

13.5 Composite Methods

In the concept of composite methods, the key idea is to split the approximation domain into several smaller domains and perform approximations separately on each subdomain. The results on each subdomain are then combined seamlessly by blending the approximations on each subdomain to a continuous surface approximation on the whole domain.

13.5.1 Brief Description

In order to give an idea of the construction let $D \subset \mathbb{R}^2$ be the domain of interest and let $D_i \subset \mathbb{R}^2$ be subdomains such that

$$D \subset \bigcup_i D_i.$$

Now, let $v_i : \mathbb{R}^2 \to \mathbb{R}$ be weight functions satisfying

$$v_i(x, y) \quad = \quad 0 \quad \forall (x, y) \notin D_i,$$

$$\begin{aligned} v_i(x,y) &\geq 0 \quad \forall (x,y) \in D_i, \\ \sum_i v_i(x,y) &> 0 \quad \forall (x,y) \in D. \end{aligned}$$

From the v_i's we construct normalised weight functions w_i by

$$w_i = \frac{v_i}{\sum_i v_i},$$

which gives us weight functions with the properties

$$\begin{aligned} w_i(x,y) &= 0 \quad \forall (x,y) \notin D_i, \\ w_i(x,y) &\geq 0 \quad \forall (x,y) \in D_i, \\ \sum_i w_i(x,y) &= 1 \quad \forall (x,y) \in D. \end{aligned}$$

If f_i is a local approximation to the data on the subdomain D_i we can define the global approximation f on D by

$$f(x,y) = \sum_i w_i(x,y) f_i(x,y).$$

Clearly, the value of f at any point in D is determined only by those local functions f_i for which the point lies in the domain D_i. Moreover, if the f_i's are interpolants, then the convex combination f will also interpolate.

The weight functions can be constructed in several ways. One convenient way is to set

$$v_i(x,y) = \text{dist}\{(x,y), \partial D_i\}^k$$

if (x,y) is interior to D_i and 0 elsewhere. Here the exponent $k > 0$ serves as a smoothing factor, in the sense that the normal derivatives of v_i across the boundary ∂D_i will vanish for suitable k.

In Figure 13.5 we have shown the domain of the model problem in Figure 13.1 together with a partition into circular overlapping subdomains D_i placed on a hexagonal grid. The subdomains of the approximating functions f_i that have significant influence on the approximation properties for the global approximation f in the vicinity of the holes are highlighted in the figure.

The process of constructing the approximation to the data set will then simply be done by collecting all data belonging to a subdomain D_i, and thereafter using some scattered data approximation method f_i to approximate these local data.

In addition to the advantage of local control as discussed in Section 13.4.4, we also note that the size of the problem for each subdomain can be kept at a manageable level. This property is important from the viewpoint of computational efficiency. It comes into play, for example, in the choice of grid size for the method of grid generation with smoothing, and

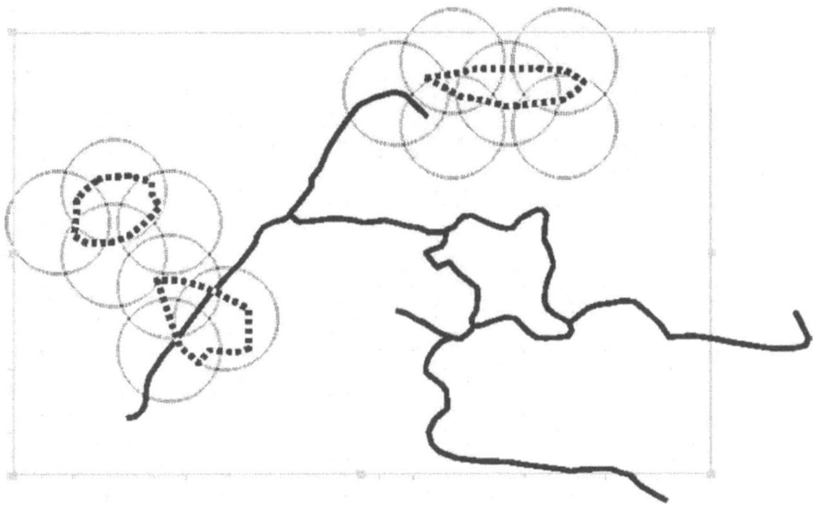

FIGURE 13.5. The domain of the model problem (Figure 13.1) partitioned into circular overlapping subdomains. The solid thick lines indicate river polygons and the lake shore polygon. The dotted thick lines illustrate where there are holes in the photogrammetric data. The highlighted circles indicate subdomains having influence on the approximating surface in the vicinity of the holes.

in the number of points when applying the radial basis function method. Using the composite approach it is easy to construct localised radial basis function interpolants to a million data points or more, while the unlocalised version would be limited to, say, one thousand points.

The method is local, and hence, we cannot take into account global trends that are not reflected in the subdomains. In addition, with a fixed configuration of subdomains, as for the example shown in Figure 13.5, we might come into situations with too little data in some of the subdomains. Thus, if the subdomains are too small, or the data is severely irregular, we might experience problems with the method.

13.5.2 Software Design

From the discussion in Section 13.5.1 it is clear that the natural entities connected to each local approximation problem are a local domain D_i, the data local to D_i, an approximating function f_i and a weight function v_i.

In Figure 13.6 these entities appear as the classes `Explicit`, `Domain2D` and `DataCollection` (which are already discussed in Section 13.3), and

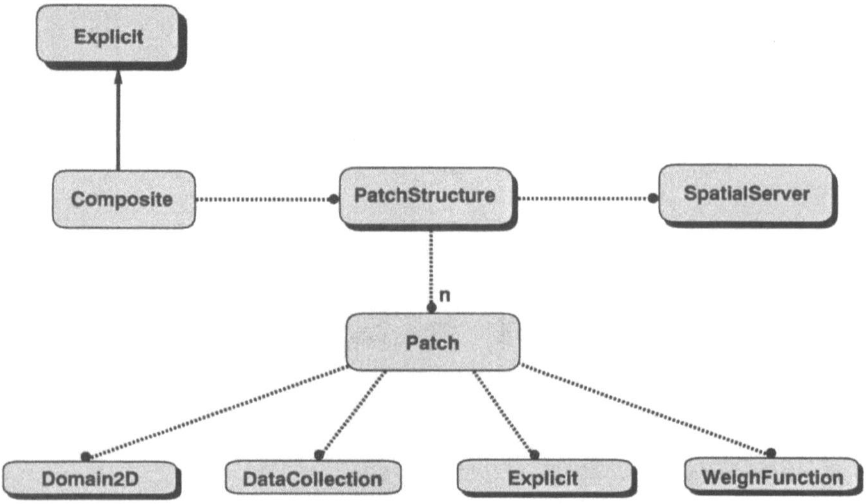

FIGURE 13.6. The class hierarchy for composite methods. Solid arrows indicate inheritance and dotted lines indicate containment.

in addition the class **WeightFunction** represents the weight function v_i. These classes are collected together in a **Patch**.

The **PatchStructure** keeps track of the collection of patches. It is able to answer queries on which patches intersect a certain point or domain in the plane by examining the patches' domains. It is also responsible for the evaluation of the surface. This is accomplished by collecting all patches overlapping the point where the evaluation should take place. Then the weight function and approximating functions are evaluated and the results are put together according to the formulae in Section 13.5.1. The composite method is made available to the **Explicit** hierarchy by a simple class **Composite** that encapsulates the **PatchStructure**. In this way the composite methods are available to service functionality that address the interface to **Explicit**. However, the flexible configuration of the method has to be done via the **PatchStructure** interface.

To facilitate the efficient collection of patches overlapping a point or a domain, the **PatchStructure** is in fact an abstract class. The actual implementation of different configurations of structures belongs to derived classes. The circular patch structures illustrated in Figure 13.5 are thus one choice. Presently, structures with rectangular subdomains are also implemented in Siscat.

As will be clear later on, more flexible domain structures such as polygonal domains would greatly improve modelling flexibility. Such structures will be implemented in the near future. It should be mentioned though, that more complex domain structures will make point-domain operations more costly. The circular and rectangular subdomains facilitate fast access via simple formulae and array operations.

The `PatchStructure` gets data from the `SpatialServer` and feeds the `DataCollection` object of the patches. The basic idea of the server is to cope with huge sets of data where it would not be possible nor efficient to store all the data in the computer memory. The server accesses data for local areas from disk on request from the `PatchStructure`. For this to be efficient, disk storage must be organised as a spatial database. Siscat provides a simple data base solution for this purpose. The report [11] gives a more in-depth description of the composite method.

13.5.3 Composite Surfaces for the Model Problem

In order to apply the composite method to the model problem we used the circular subdomains shown in Figure 13.5. The highlighted circles indicate where constrained grid generation with smoothing is applied and the non-highlighted circles indicate the domains belonging to the constrained Delauney triangulation. The basis for this particular choice is the result of tests performed in Sections 13.4.2 and 13.4.3. Thus the constrained grid generation with smoothing is used where there are holes in the photogrammetric data set, and the constrained Delauney triangulation is used elsewhere. The result of the composite approximation is shown in Figure 13.7.

Since the different approximation methods constituting the composite surface have different internal data structures, in the present case grids and triangulations, it is not practical to use these structures directly in the visualisation. The composite surface for the model problem was therefore sampled on a 400x400 grid which is shown in Figure 13.7.

We observe that in the areas dominated by the triangulation, more surface details are visible. The grid generation method smooths the data, a property which is clearly apparent in the figure. However, the basic requirements introduced in Section 13.2.2 are satisfactorily fulfilled by the surface model. The lake surface has a single value, the holes are filled naturally, and the riverbed is decreasing downstream and lies in a V-valley.

The smoothing effect of the grid generation method might be considered too prominent. Moreover, by its construction it does not interpolate the data. To overcome this, it would be possible to introduce more refined techniques.

One way would be to interpolate the error between the surfaces obtained with the grid generation method and an interpolation scheme such as Shepard's method [16]. There are tools in the `Explicit` hierarchy combining two methods in this way. It would therefore be easy to substitute the grid generation method with a "constrained grid generation with smoothing and error interpolation" method in order to accomplish this.

FIGURE 13.7. The result of approximating the model problem using the composite method with underlying subdomains as shown in Figure 13.5.

Another possibility would be to constrain the domain for the grid generation method to better fit the holes in the photogrammetric data. This would require the usage of more advanced subdomain structures than the circles used here. By the object-oriented design of the composite method, it is straight forward to replace the patch structure based on circles with a more general patch structure as soon as this has been implemented.

The two extensions outlined above are both motivated from design requirements present for this particular application. Once more this emphasises that the problem of scattered data approximation is a discipline at the crossroad of data approximation and geometric design.

Acknowledgments: This work was supported by the Research Council of Norway under the research programs STP.29643 (Toolkits in Industrial Mathematics), 110673/420 (Numerical Computations in Applied Mathematics).

13.6 References

[1] P. Alfeld. Scattered data interpolation in three or more variables. In T. Lyche and L. Schumaker, editors, *Mathematical Methods in Computer Aided Geometric Design*, pages 1–33. Academic Press, 1989.

[2] E. Arge and Ø. Hjelle. Software tools for modelling scattered data. In M. Dæhlen and A. Tveito, editors, *Numerical Methods and Software Tools in Industrial Mathematics*, pages 47–62. Birkhäuser, 1997.

[3] E. Arge, M. Dæhlen, and A. Tveito. Approximation of scattered data using smooth grid functions. *J. Comput. Appl. Math.*, 59:191–205, 1995.

[4] E. Arge and M. Floater. Approximating scattered data with discontinuities. *Numerical Algorithms*, 8:149–166, 1994.

[5] A. K. Cline and R. J. Renka. A constrained two-dimensional triangulation and the solution of closest node problems in the presence of barriers. *SIAM J. Numer. Anal.*, 27(5):1305–1321, 1990.

[6] L. De Floriani and E. Puppo. An on-line algorithm for constrained delaunay triangulation. *Graphical Models and Image Processing*, 54:290–300, 1992.

[7] R. Franke. Scattered data interpolation, tests of some methods. *Math. Comp.*, 38:181–200, 1982.

[8] N. P. Fremming, Ø. Hjelle, and C. Tarrou. Surface modelling from scattered geological data. In M. Dæhlen and A. Tveito, editors, *Numerical Methods and Software Tools in Industrial Mathematics*, pages 305–320. Birkhäuser, 1997.

[9] Ø. Hjelle. Data structures in Siscat. Report STF42 A96007, SINTEF Applied Mathematics, 1996.

[10] Ø. Hjelle. Explicit surfaces in Siscat. Report STF42 A96008, SINTEF Applied Mathematics, 1996.

[11] O. Kloster. Composite surfaces in Siscat. Report STF42 A96010, SINTEF Applied Mathematics, 1996.

[12] E. Mehlum and V. Skytt. Surface editing. In M. Dæhlen and A. Tveito, editors, *Numerical Methods and Software Tools in Industrial Mathematics*, pages 385–400. Birkhäuser, 1997.

[13] M. J. D. Powell. The theory of radial basis function approximation in 1990. In W. Light, editor, *Advances in Numerical Analysis, Vol II*, pages 105–210. Oxford Science Publications, 1992.

[14] F. P. Preparata and M. I. Shamos. *Computational Geometry, an Introduction.* Springer-Verlag, New York, 1985.

[15] L. L. Schumaker. Fitting surfaces to scattered data. In G. G. Lorntz, C. K. Chui, and L.L Schumaker, editors, *Approximation Theory II*, pages 203–268, New York, 1976. Academic Press.

[16] D. Shepard. A two-dimensional interpolation function for irregularly spaced data. In *Proc. 23rd Nat. Conf. ACM*, pages 517–524, 1968.

[17] The Siscat Home Page. World Wide Web document at `http://www.oslo.sintef.no/siscat`.

Part III

Software Development

14

Is the Quality of Numerical Subroutine Code Improving?

T. R. Hopkins[1]

ABSTRACT

We begin by using a software metric tool to generate a number of software complexity measures and we investigate how these values may be used to determine subroutines which are likely to be of substandard quality.

Following this we look at how these metric values have changed over the years. First we consider a number of freely available Fortran libraries (Eispack, Linpack and Lapack) which have been constructed by teams. In order to ensure a fair comparison we use a restructuring tool to transform original Fortran 66 code into Fortran 77.

We then consider the Fortran codes from the Collected Algorithms from the ACM (CALGO) to see whether we can detect the same trends in software written by the general numerical community.

Our measurements show that although the standard of code in the freely available libraries does appear to have improved over time these libraries still contain routines which are effectively unmaintainable and untestable.

Applied to the CALGO codes the metrics indicate a very conservative approach to software engineering and there is no evidence of improvement, during the last twenty years, in the qualities under discussion.

14.1 Introduction

The last two decades have witnessed dramatic advances in the way we view the creation of software. Since *software engineering* was born in 1969 [5] an ever growing number of techniques have been put forward in an attempt to promote software production from an art to a science. For almost as long attempts have been made to quantify the quality of software. This has led to a bewildering number of software metrics being proposed; an excellent review of many of these may be found in the book by Zuse [31]. In addition, a number of software tools have appeared which compute a variety of metrics for code written in most of the commonly used programming languages. Such tools include QAFortran [24], QAC and QAC++,

---

[1]Computing Laboratory, University of Kent, Canterbury, Kent, CT2 7NF, UK. Email: `trh@ukc.ac.uk`.

the nag_metrics tool from the NagWare f77 tool suite [22] and the LDRA Testbed [16].

In this paper we use a small number of these metrics to investigate whether we can detect an improvement in the quality of numerical software written in Fortran over the last twenty years. In Section 14.2 we briefly discuss these metrics and, in the following section, we look at how successful they are at identifying software modules that suffer from readability, maintainability and testability problems.

We use the QAFortran tool to generate these metrics for the routines in a number of public domain packages. We also note, in this section, that during the time spanned by these libraries, the Fortran language has evolved from Fortran 66 [2], through Fortran 77 [3] to Fortran 90 [15]. The additional control structures made available with each new code standard means that it is unfair to make direct comparisons of the code metrics between packages without taking into account the version of Fortran being used. In order to 'normalise' the metrics we have only considered code written in Fortran 66 and Fortran 77 and we have used an automatic code restructurer, *spag* [23], to translate from Fortran 66 to Fortran 77. Metrics from the restructured Fortran 66 codes may then be compared.

The packages used in Section 14.3 have all been produced by teams of researchers which have included wide ranges of expertise from software engineers to numerical analysts. To try to gauge how the quality of scientific software has varied among the community in general we consider the Collected Algorithms from the ACM (CALGO). These codes have been published since 1960 and are widely regarded as being state-of-the-art both in algorithmic and coding terms at the time of publication. In Section 14.4 we report on how our chosen metric values have changed with time by considering the Fortran routines that have appeared in CALGO since 1975.

Finally, in Section 14.5, we draw some conclusions from the results we have presented.

14.2 Software Metrics

In addition to the number of executable *L*ines *O*f *C*ode (LOC) and the number of explicit GOTO statements in a subprogram, we will also consider three further metrics: cyclomatic complexity [19], knot count [30] and a variant of the path count metric proposed by Nejmeh [21].

It should be noted that the metrics used in this paper have been chosen to measure qualities of Fortran code although some of them may also be applied successfully to other procedural languages. Object-oriented languages require a different approach and there appears to be, as yet, no general consensus on which metrics are most appropriate. A general discussion of object-oriented software metrics may be found in Lorenz and Kidd [18].

Cyclomatic Complexity

The control graph of a program unit is a directed graph whose nodes are the basic blocks of code and whose edges are directed arcs corresponding to the flow of control between the basic blocks. A basic block is a section of code which contains no transfer of control (for example, a sequence of assignment statements). The cyclomatic complexity, $V(G)$, is defined to be the cyclomatic number of the control graph, i.e.,

$$V(G) = \text{Number of edges} - \text{Number of nodes} + 1$$

This may be shown to be equivalent to one more than the number of predicates (decision statements) used in the code and hence, we believe, it is a good indicator of the complexity of the underlying algorithm. This metric was extended by Myers [20] who suggested the use of a cyclomatic complexity interval to take account of the additional complexity caused by compound predicates. The lower bound of Myer's interval is the cyclomatic complexity and the upper bound is defined as one more that the total number of conditions appearing in the code.

Cyclomatic complexity was originally advocated both as a measure of the testing effort required for a module and as an effective way of dividing software into subroutines. Shepperd [26] and Shepperd and Ince [27] have questioned the use of the metric for measuring testing effort as it appears to be extremely insensitive to the structure of the software. They state that this may be due to the fact that the measure is based on a lexical rather than a structural view of the code.

In the present paper the metric is used as an indicator that a piece of code would probably benefit from being broken down into a number of simpler program units. McCabe [19] suggested a maximum value of 10 for an individual program unit while Grady [10], after analysing the relationship between the cyclomatic complexity and the number of updates required to each module in 830,000 lines of Fortran code, suggests a maximum value of 14.

Knot Count

A knot occurs in a piece of code whenever the paths associated with two transfers of control intersect (see Figure 14.1 for two examples). Code with large knot counts is generally extremely difficult to read and understand. The number of knots is, therefore, a good indicator of code clarity.

Path Count

The path count metric provides an upper bound on the number of possible static paths through the code and thus gives an estimate of the amount

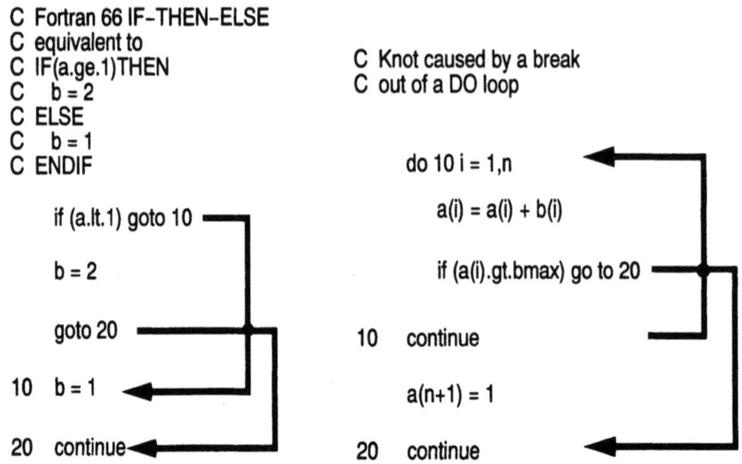

FIGURE 14.1. Examples of knots in Fortran code.

of effort required to thoroughly test a routine. It should be noted that the path count is calculated solely from the syntax of the code; some paths may be counted which are not executable. When computing path counts QAFortran will return a maximum value of 5×10^9.

Using a large library of numerical Fortran subroutines, Hatton and Hopkins [13] reported a strong link between very large path counts and the number of defects reported in routines after they had been released. Thus the path count metric may also be regarded as an indicator of the reliability of program units.

14.3 A Comparison of Freely Available Packages

Table 14.1 shows the values obtained for these metrics, along with the number of executable and explicit GOTO statements, for the single precision level 1 BLAS [17]. The routines which stand out, SNRM2 and SROTMG, both exhibit a very complicated code structure. Indeed Hanson and Krogh [14] remark that they did not translate SROTMG into assembler as part of their INTEL specific version of the level 1 BLAS due to the complexity of its specification.

It has been shown by Hopkins [11, 12] that even using Fortran 66 the knot and path counts for these routines may be substantially reduced by more careful structuring of the code. This leads in both cases to codes that are easy to read and understand. In the case of SROTMG the restructuring exercise also discovered a minor error in the code.

Routine	LOC	GOTOs	CC	Knots	Paths
isamax	22	3	8	3	8
sasum	22	3	8	4	8
saxpy	29	2	11	2	16
scopy	31	2	10	2	16
sdot	29	3	10	4	32
snrm2	44	15	13	41	1024
srot	22	1	7	1	8
srotg	22	2	5	2	16
srotm	84	7	13	17	144
srotmg	121	32	18	92	98304
sscal	22	2	8	2	8
sswap	37	2	10	2	16

TABLE 14.1. Metric values for the real level 1 BLAS.

The cyclomatic complexity intervals of these two routines changed from 13:14 to 11:16 for SNRM2 and from 18:18 to 13:17 for SROTMG, showing that, although restructuring led to fewer conditional statements being used, more of the resulting logical expressions were compound. Basically the underlying algorithms for these routines require a number of conditional tests to be made and restructuring has little or no effect on this.

The metric values obtained for the newer Fortran 77 level 2 [7] and level 3 [6] BLAS show a distinct improvement. There are no GOTO statements (and therefore no knots) in any of the routines and the highest path counts are all less than 10^4. These large path counts are almost invariably associated with routines which carry out a large number of input parameter checks and do not, therefore, reflect complexity in the numerical part of the code.

Eispack (see [9, 28]) was one of the first attempts to create a library of high quality, freely available, numerical routines. Many of the codes were Fortran 66 versions of the Algol 60 procedures presented in Wilkinson and Reinsch [29]. Figure 14.2 shows a scatter plot of knots against number of executable statements for the 75 double precision routines along with the regression line through this data. We note that there is a very strong correlation between the number of knots and the number of executable statements in a routine and this may be used to identify three routines which exhibit unusually high knot counts for their length; BISECT, TRIDIB and TSTURM.

Generating similar figures for the double precision real and complex versions from two other similar projects, Linpack [8] (91 routines) and Lapack [1] (542 routines), allow further subprograms to be identified which have unusually high knot counts for their size. Table 14.3 gives a summary

Library	Slope of Regression Line
Eispack	0.262
Linpack	0.145
Lapack	0.048

TABLE 14.2. Slopes of regression lines for knots against executable statements.

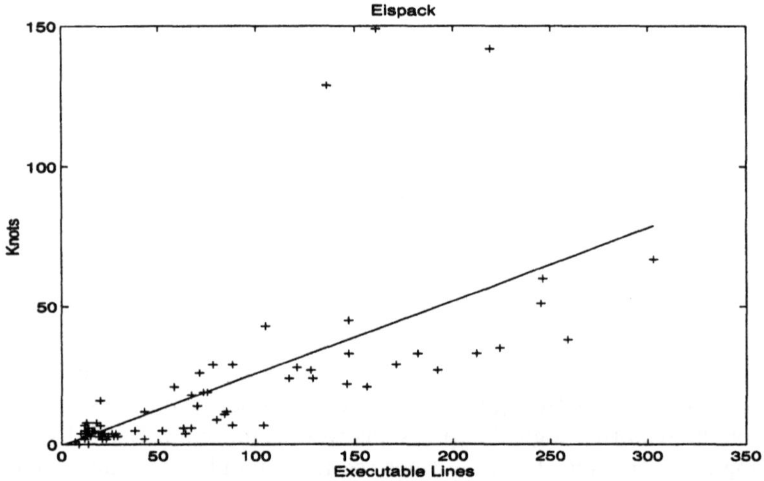

FIGURE 14.2. Scatter plot and regression line for double precision EISPACK routines.

of the routines with large standardised residuals; these are the routines which the Minitab statistical package [25] flags as being outliers and thus worthy of further investigation. In the cases given the standardised residual values are extremely high.

In Table 14.2 we present the slope of the regression line for the Eispack, Linpack and Lapack packages. This indicates a general improvement in the readability of code over time. It should be expected that the Lapack value would be the smallest since this is the only one of the three packages to have been written in Fortran 77. Although both written in Fortran 66, the Linpack routines are generally better structured and therefore more comprehensible than the Eispack routines; a fact reflected in the smaller slope value.

Fortran 77 provided one major new control structure that was missing from Fortran 66; the block IF. Using Fortran 66 to simulate a simple IF-THEN-ELSE construction required the use of a pair of GOTO statements and resulted in a knot appearing in the code (see Figure 14.1). In order to make a fair comparison between the quality of Fortran 66 and Fortran

Routine		LOC	Knots	Residual
Eispack				
	BISECT	136	129	4.43
	TRIDIB	161	149	5.09
	TSTURM	219	142	4.11
Linpack				
	DSVDC	285	54	3.18
	ZSVDC	310	58	3.35
Lapack				
	DLARFX	440	132	11.23
	ZLARFX	440	132	11.23
	ZHGEQZ	340	60	4.46

TABLE 14.3. Routines identified as difficult to read/maintain due to the large knot count relative to the number of executable statements.

77 software it is necessary to 'normalise' the codes in some way to take account of the availability of the block IF.

A number of code restructuring tools are available which translate between Fortran 66 and Fortran 77. These tools are based on the work of Baker [4] and will, amongst other changes, create block IFs from equivalent Fortran 66 code. The *spag* restructuring tool [23] was used here for this purpose.

Table 14.4 shows the effects of using *spag* on all the Eispack, Linpack and Lapack codes detailed in Section 14.2. As well as showing the aggregate values for the number of executable lines of code, GOTO statements and knots for all the routines considered from each package, we also provide the maximum value, in any routine, of each of these metrics along with the maximum value of the cyclomatic complexity (CC). Note that each of these maximum values may be associated with a different routine.

For the Eispack routines only the number of executable statements has increased during restructuring due to *spag* duplicating small blocks of code in order to obtain better code structure. Although the number of GOTO statements has been substantially reduced the number of knots has only decreased to just over a half of the original number, pointing to substantial amounts of unstructured code. Of the remaining 807 knots 268 occur in the three routines, BISECT, TRIDIB and TSTURM, identified above as possible problem routines.

Automatic translation into Fortran 77 was far more successful on the Linpack routines eliminating 86% of the GOTO statements and 77% of the knots. This indicates that the original Fortran 66 code was well structured and the majority of the GOTO's and knots were only present due to the lack of adequate control structures within the implementation language. As

	Aggregate			Routine Maxima			
	LOC	GOTO	Knots	LOC	GOTO	Knots	CC
Eispack							
f66	5774	863	1482	303	50	149	76
spag	6482	298	807	337	25	116	76
Linpack							
f66	7381	1022	1003	310	50	58	86
spag	7250	139	226	300	11	27	85
Lapack							
f77	52001	993	1972	1159	26	132	199
spag	51101	672	1469	1145	21	61	199

TABLE 14.4. Effects of using the *spag* restructurer on routines from the Eispack, Linpack and Lapack libraries.

with Eispack the two routines identified in Table 14.3 account for a good proportion (just over a third) of the remaining knots.

Perhaps the biggest surprise was that an automatic restructing tool should have such a marked effect on the Lapack software. While GOTO statements are still necessary in Fortran 77 (for example, to construct while loops) and may generate knots (for example, when breaking out of a loop) such constructions would not be removed by *spag* when generating Fortran 77. The routines DLARFX and ZLARFX have all 132 knots removed, the restructured code containing no GOTO statements at all and this accounts for almost half the total number of knots removed. The other routine, ZHGEQZ, however is largely unaffected with a reduction from 60 to 52.

Restructuring has no effect on the cyclomatic complexity, the values being largely unchanged. Perhaps the only worrying trend is that the maximum value has increased quite dramatically with time, the routine ZHBEST has a cyclomatic complexity of 199 although it contains just 653 executable statements; given the suggested maximum values for this metric quoted in Section 14.2 this routine looks ripe for modularisation.

Table 14.5 gives the density of both GOTO statements and knots in terms of the number of executable statements. The original Eispack code has an extremely high knot density (one for every four executable statements) which, even with restructuring, is only marginally better than the original Linpack code. The values obtained for the restructured Linpack code are extremely close to those of the original Lapack although the Lapack knot density requires the use of *spag* to improve on Linpack. The improvement due to restructuring the Lapack code is startling for a translation of Fortran 77 into itself.

Considering only restructured codes the percentage of routines with path counts exceeding the maximum value returnable by QAFortran has in-

		LOC per GOTO	Knots	% of routines with Path Counts of Max Path (5×10^9)
Eispack	f66	7	4	6.6%
	Spag	22	8	1.3%
Linpack	f66	7	7	8.7%
	Spag	52	32	2.2%
Lapack	f77	52	26	3.6%
	Spag	76	35	3.3%

TABLE 14.5. Density of knots and GOTO statements and the percentage of routines exhibiting high Path Counts.

creased. This, along with the occurrence of large values for the cyclomatic complexity, implies that neither time nor software engineering principles have done anything to reduce the fraction of long and, logically, extremely complex routines appearing in state-of-the-art numerical Fortran libraries.

14.4 CALGO Fortran Codes

Since 1975 all the Collected Algorithms from the ACM have appeared as fully refereed papers in ACM Transactions on Mathematical Software and the accompanying code has been thoroughly tested prior to publication. It was felt that these sources should provide evidence of how both research workers and referees, who are mostly scientists whose primary expertise is not in computer programming, have altered the way they write Fortran code. It might be expected that there would be a lag between new software engineering principles being expounded and their practical application by the scientific community.

There are few restrictions imposed on the languages that may be used to implement algorithms, nevertheless, as Figure 14.3 shows, the majority of the published code appearing over the last twenty years has been in Fortran. It is interesting to note how long it took both for Fortran 66 to die out and for Fortran 77 to become the dominant dialect (almost 10 years after the appearance of the standard). Up until the end of 1995 only a single algorithm has been published in Fortran 90.

Many of the published codes use previously published or freely available software (for example, BLAS, Linpack, etc); in generating the data presented in this section all such routines were removed along with any drivers and other testing code.

In all the following figures the asterisks signify the average of the plotted quantity over all subroutines appearing in all the algorithms implemented

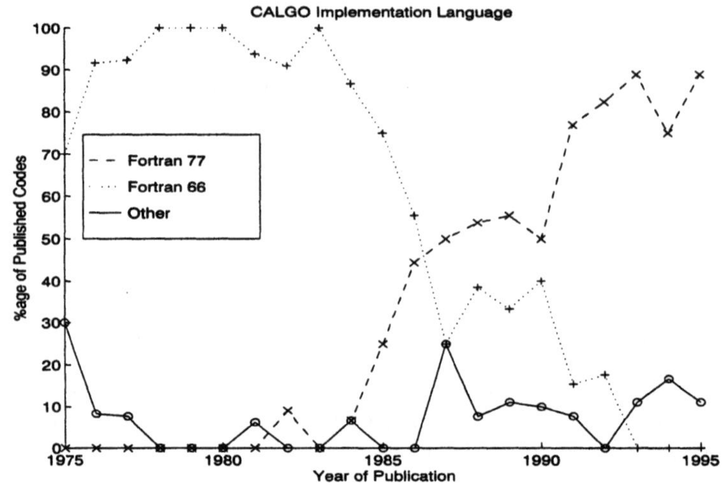

FIGURE 14.3. (1975–1995) CALGO algorithms by implementation language.

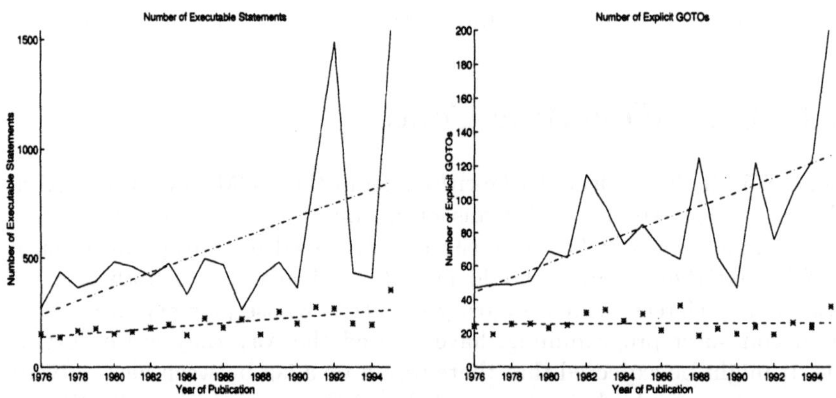

FIGURE 14.4. CALGO Algorithms: Executable Lines (left) and GOTO statements (right). The piecewise solid line shows the maxumum values encountered each year, whilst the asterisks indicate the average values over all subroutines. The straight lines signify the corresponding linear regression models.

in Fortran that were published in that year, whilst the piecewise, solid line represents the maximum values. Both also have plotted their respective regression lines; as a dashed line for average values and a dot-dash line for the maximum values. Note that none of the Fortran 66 codes have been restructured.

It may be observed from Figure 14.4 that on average the number of executable lines per subroutine has gradually increased whilst the maximum number shows a more rapid upward trend. Even with the introduction of Fortran 77, Figure 14.4 shows that the average number of explicit GOTO statements has remained almost level. This is surprising as we would have

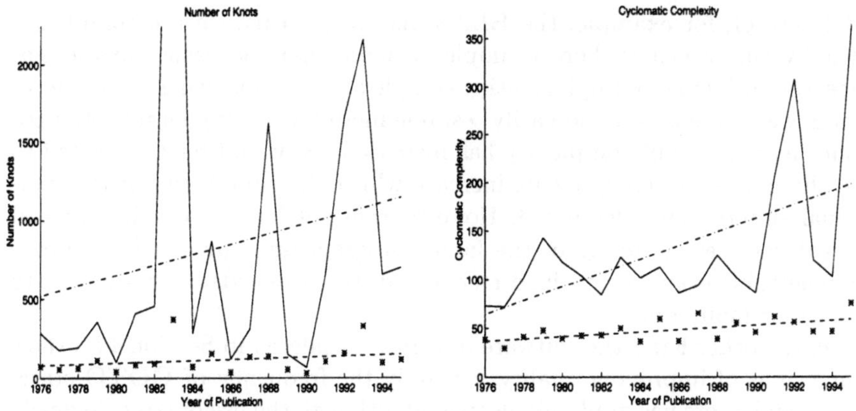

FIGURE 14.5. CALGO Algorithms: Knots (left) and Cyclomatic Complexity (right). The piecewise solid line shows the maxumum values encountered each year, whist the asterisks indicate the corresponding average values over all subroutines. The straight lines signify the corresponding linear regression models.

expected to see a sharp decrease in these values as authors switched to Fortran 77. It is certainly worrying that in 1995 we are still seeing published routines containing around 200 explicit GOTO statements even if they may have been machine generated. The effect of the growth in GOTOs, is clearly reflected in the knot counts (Figure 14.5) which exhibits a very similar trend.

Figure 14.5 also shows the maximum values for the cyclomatic complexity of each routine. Again the average values show a gradual increase with time showing a general unwillingness by authors to increase the modularity of their code. We also note that the average value of around 50 is far in excess of the value of 14 suggested by the study reported in Section 14.2.

14.5 Conclusion

The overall picture obtained from Section 14.4 is that the general practitioners of scientific software production are an extremely conservative group of people. The past twenty years have witnessed a general reticence to embrace new versions of Fortran and on average the values of the metrics used in this paper have at best remained static but are generally showing a gradual increase for the worse.

There is no doubt that there has been a dramatic increase in the underlying algorithmic complexity of computational numerical methods over the period studied and that this has resulted in an overall increase in the complexity of their software implementations.

On the positive side many authors do reuse existing software both for portability purposes and as building blocks for their own code. However, while use of, for example, the BLAS may help to reduce the complexity of individual routines there is ample evidence that the basic software engineering principle of implementing complex algorithms using a number of small, well understood and easily testable modules is being largely ignored. Thus, as the overall complexity has increased, we would have expected to see the number of components increase whilst the complexity of the individual subprograms decreased. However routines have grown in size and logical complexity; many of the larger program units are likely to prove impenetrable even to a trained reader and would be extremely difficult to change or maintain.

On the other hand the software libraries considered in Section 14.3 show some signs of improving quality, certainly the frequency of GOTO statements and knots has gradually decreased although the use of large, logically complex and virtually untestable routines has increased.

Tools to generate metric values could usefully be employed at all levels of numerical and scientific Fortran software production and reasonable restrictions on, for example, knots and cyclomatic complexity, would certainly improve the overall readability of the resultant code. Restructuring tools like *spag* have also proved to be useful not only in improving the structure (and thus decreasing the knot and path counts and the number of explicit GOTO statements) in dusty deck code but also in code written from scratch in Fortran 77.

Finally we note again the strong link, reported by Hatton and Hopkins [13], between very large path counts and the number of defects reported in released program units. On the above evidence this surely cannot bode well for the future.

Acknowledgments: My thanks to the two anonymous referees whose comments helped to improve this paper considerably.

14.6 References

[1] E. Anderson, Z. Bai, C. Bischof, J. Demmel, J. J. Dongarra, J. Du Croz, A. Greenbaum, S. Hammarling, A. McKenney, S. Ostrouchov, and D. Sorensen. *LAPACK: Users' Guide.* SIAM, Philadelphia, second edition, 1995.

[2] ANSI. *Programming Language Fortran X3.9-1966.* American National Standards Institute, New York, 1966.

[3] ANSI. *Programming Language Fortran X3.9-1978.* American National Standards Institute, New York, 1979.

[4] B. S. Baker. An algorithm for structuring flowgraphs. *JACM*, 24(1):98–120, January 1977.

[5] J. N. Buxton and B. Randell, editors. *Software Engineering Techniques; Report on a Conference Sponsored by the NATO Science Committee, Rome, Italy, 27th to 31st October 1969.*, Brussels, 1970. NATO, NATO Scientific Affairs Division.

[6] J. J. Dongarra, J. Du Croz, I. S. Duff, and S. Hammarling. Algorithm 679: A set of level 3 basic linear algebra subprograms. *ACM Trans. Math. Softw.*, 16(1):18–28, March 1990.

[7] J. J. Dongarra, J. Du Croz, S. Hammarling, and R. J. Hanson. Algorithm 656: An extended set of basic linear algebra subprograms: Model implementation and test programs. *ACM Trans. Math. Softw.*, 14(1):18–32, March 1988.

[8] J. J. Dongarra, C. B. Moler, J. R. Bunch, and G. W. Stewart. *LINPACK: Users' Guide.* SIAM, Philadelphia, 1979.

[9] B. S. Garbow, J. M. Boyle, J. J. Dongarra, and C. B. Moler. *Matrix Eigensystem Routines – EISPACK Guide Extension*, volume 51 of *Lecture Notes in Computer Science.* Springer-Verlag, New York, 1977.

[10] R. B. Grady. Successfully applying software metrics. *Computer*, 27(9):18–25, September 1994.

[11] T. R. Hopkins. Restructuring software: A case study. *Software — Practice and Experience*, 26(8):967–982, July 1996.

[12] T. R. Hopkins. Restructuring the BLAS level 1 fast Givens routines. Technical Report 13-96, Computing Laboratory, University of Kent, Canterbury, Kent, CT2 7NF, UK, July 1996.

[13] T. R. Hopkins and L. Hatton. Experiences with FLINT, a software metrication tool for fortran 77. In *The Symposium on Software Tools*, Edinburgh, 1988. Napier Polytechnic.

[14] R. J. Hanson and F. T. Krogh. Translation of Algorithm 539: PC-BLAS basic linear algebra subprograms for Fortran usage with the INTEL 8087 80287 numeric data processor. *ACM Transactions on Mathematical Software*, 13(3):311–317, September 1987.

[15] ISO/IEC. *Information Technology – Programming Languages – Fortran (ISO/IEC 1539:1991(E)).* ISO/IEC Copyright Office, Geneva, 1991.

[16] LDRA Group, Liverpool, UK. *LDRA Testbed.*

[17] C. L. Lawson, R. J. Hanson, D. R. Kincaid, and F. T. Krogh. Basic linear algebra subprograms for Fortran usage. *ACM Trans. Math. Softw.*, 5:308–323, 1979.

[18] M. Lorenz and J. Kidd. *Object-Oriented Software Metrics.* Object-Oriented Series. Prentice Hall, Englewood Cliffs, New Jersey, 1994.

[19] T. J. McCabe. A complexity measure. *IEEE Transactions on Software Engineering*, SE-2(4):308–320, 1976.

[20] G. J. Myers. An extension to the cyclomatic measure of program complexity. *Sigplan Notices*, 12(10):61–64, 1977.

[21] B. A. Nejmeh. NPATH: A measure of execution path complexity and its applications. *Commun. ACM*, 31(2):188–200, 1988.

[22] Numerical Algorithms Group Ltd., Oxford, UK. *NAGWare f77 Tools*, second edition, September 1992.

[23] Polyhedron Software, Oxford, UK. *plusFORT*, Revision B edition, 1993.

[24] Programming Research Ltd, Hersham, Surrey. *QA Fortran 6.0*, 1992.

[25] B. F. Ryan, B. L. Joiner, and T. A. Ryan Jr. *MINITAB Handbook.* Duxbury Press, Belmont, California, second edition, 1992.

[26] M. Shepperd. A critique of cyclomatic complexity as a software metric. *Software Engineering Journal*, pages 30–36, March 1988.

[27] M. Shepperd and D. C. Ince. A critique of three metrics. *J. Systems Software*, pages 197–210, 1994.

[28] B. T. Smith, J. M. Boyle, J. J. Dongarra, B. S. Garbow, Y. Ikebe, V. C. Klema, and C. B. Moler. *Matrix Eigensystem Routines – EISPACK Guide*, volume 6 of *Lecture Notes in Computer Science.* Springer-Verlag, New York, second edition, 1976.

[29] J. H. Wilkinson and C. Reinsch. *Linear Algebra*, volume 2 of *Handbook for Automatic Computation.* Springer-Verlag, New York, 1971.

[30] M. R. Woodward, M. A. Hennell, and D. Hedley. A measure to control flow complexity in program text. *IEEE Transactions on Software Engineering*, SE-5(1):45–50, 1979.

[31] H. Zuse. *Software Complexity: Measures and Methods.* W. de Gruyter, Berlin, 1991.

15

Object-Oriented Redesign of a Real-World Fortran 77 Solver

Matthias Weidmann[1]

ABSTRACT
This chapter describes the ongoing work of applying modern software development techniques to a real-world computational fluid dynamics software package in the interdisciplinary project SEMPA. First results are given on a C++ implementation of the Fortran 77 algebraic multigrid solver of this package. The resulting benefits for this first prototype are the improved readability and increased flexibility of the code, but this first design and implementation also shows a significant increase in execution time. Efficiency improvements are discussed at the end of the chapter.

15.1 Introduction

In the domain of computational fluid dynamics (CFD), fluid flow is simulated on a computer to enhance the study of physical and chemical behavior and effects of fluids on industrial geometries (like turbines, cars, planes, etc.). The results of the computer simulations contribute to the understanding of the usability of the geometric design and shorten the development cycle of a new product.

The fluid flow is modeled by partial differential equations, which are solved numerically on the respective geometry. This is a complex task with large requirements for computational resources and involves expert knowledge from the domains of computational fluid dynamics, numerical mathematics and computer sciences. This interdisciplinary combination of mathematical methods, computer science methods and technical applications form the area of scientific computing.

Traditional technical applications are mainly implemented in the programming language Fortran 77, which was developed explicitly to handle computationally intensive problems. Due to the simplicity of this language and its long use, Fortran 77 compilers of today provide a satisfactory op-

[1]Lehrstuhl für Rechnertechnik und Rechnerorganisation (LRR-TUM), Institut für Informatik, Technische Universität München, D-80290 München. Email: weidmann@informatik.tu-muenchen.de.

timization, thus producing highly efficient object code. Moreover, many numerical programming libraries have been developed in Fortran 77 over the years, which can save a lot of programming effort by reuse.

However, as more complex real-world problems are modeled, the complexity of technical applications is growing. Fortran 77 does not explicitly support software engineering methods which lead to an enhanced reliability, easier maintainability and therefore shorter development time of software. Object-oriented programming (OOP) is able to address these software design issues, but is quite uncommon in traditional scientific computing. In particular, real-world pure object-oriented technical applications are still rare.

Furthermore, the parallelization of technical applications (especially CFD applications) with message passing libraries like PVM [9] or implementations of the MPI standard [4, 5]) is being done at a low-level code view, which – in terms of convenience and productivity – can be said to correspond to assembler-level programming of sequential computers. High-level numerical parallel programming models [6, 7, 10, 15] and concurrent object-oriented environments and languages [2, 17] are areas of strong research efforts today and may provide means to overcome the low-level programming difficulties of parallel scientific computing.

In the SEMPA Project, a state-of-the-art Fortran 77 CFD application is being redesigned for an object-oriented re-implementation in C++. Here, the first step of this approach is described: the redesign and re-implementation of one part of the CFD application — the numerical solver. This solver is an algebraic multigrid method (AMG) with an additive correction scheme for unstructured grids [16]. Unlike the work in [1], the approach here does not wrap up basic Fortran 77 numerical library functions with C++ classes but is a complete redesign of the numerical heart of the CFD application (see Sections 15.2 and 15.4). There was no approach to object-oriented AMG for unstructured grids found in the literature.

This first step – though there exists a working C++ implementation which yields the same computational results as the Fortran 77 implementation – is not completed, as it could be improved (see Section 15.6).

The rest of this chapter is organized as follows: Section 15.2 gives a short overview of the SEMPA Project. Section 15.3 describes the CFD application, Section 15.4 presents the object-oriented analysis and design of the AMG, and Section 15.5 compares the original and the new codes. Section 15.6 presents the discussion of this approach and outlines further work.

15.2 The SEMPA Project

SEMPA – which started in spring 1995 and is scheduled for three years – unites researchers from university and industry (CFD engineers, numerical mathematicians and computer scientists) [13]. Three major directions of activity exist:

1. Research in *s*oftware *e*ngineering *m*ethods for the design and implementation of *p*arallel *a*pplications in scientific computing (understanding complex software systems, documentation, general approaches to the parallelization of scientific applications, standards for scientific software development, new software design paradigm),
2. the single-process, multiple data (SPMD) parallelization of the state-of-the-art, real-world CFD Fortran 77 software package *TASCflow for CAD* (*TfC*) with message-passing libraries like PVM [12], and
3. the research in the design of a distributed resource manager that executes production runs of parallel applications in batch mode on networks of workstations [14].

The emphasis in SEMPA is on *TfC*. The – still to be derived – parallel prototypes will run on parallel supercomputers as well as on networks of workstations, managed by the resource manager in the latter case.

The object-oriented redesign of *TfC* is placed in the new software design paradigm area in SEMPA (see item 1 above). The goal of this redesign is

1. to show a better software quality with object-oriented software design methods in a real-world technical application, and
2. to enhance the applicability of high-level object-oriented parallelization tools in a real-world object-oriented technical application.

Starting the redesign with the AMG module was motivated by the following points:

1. Understanding the AMG method is a key issue in redesigning the whole *TfC*, as it comprises the most complex algorithm in *TfC*.
2. The inside of the AMG method at code level could benefit from modern software engineering methods like object-orientation to make it more readable and maintainable. This solution method in *TfC* is advanced and performs well, but still needs to be extensible to meet future requirements of CFD problems.
3. The logical entities on which the AMG method operates are not restricted to the solution method but are basic to the whole program. For example, the global equation matrix solved by the AMG is built up in the discretization phase.

4. The AMG algorithm shows a natural possibility for parallelization during the building of the coarse grid blocks. This parallelization could be exploited by high-level, automatic parallelization concepts, if the present data representation allowed this (which it does not). The redesign can take this into account, leading to a new algorithm-based parallelization approach.

5. A manageable size of one module is important if one starts to redesign a part of a big code. Even more, a good modularity is important. In spite of its complexity, both are the case with the AMG method module.

Through this choice, we decided to go in the – for us new – field of object-oriented numerics. The choice of C++ as the object-oriented programming language was motivated by the fact that an implementation of this language can be found on nearly every computer system, even on supercomputers (for example, Cray delivers the CRAY T3D MPP with the Cray C++ 1.0 Compiling Systems and Class Libraries for MPP). Moreover, in the field of concurrent object-oriented programming, most today's research efforts are based on C++.

15.3 The CFD Program

TfC is developed and marketed by Advanced Scientific Computing GmbH and runs on most common UNIX platforms. It is a widely applicable CFD program package that is used to predict laminar and turbulent viscous flows in complex, 3D geometries with unstructured grids. Flows can either be steady or transient, isothermal or convective, incompressible or compressible (subsonic, transonic and supersonic). Application fields include pump design, turbo machines, heat exchangers, automotive aerodynamics and so on.

The CFD simulation part of *TfC* consists of an unstructured grid element-based finite volume discretization of the 3D Navier-Stokes and scalar transport equations and the AMG solution. This is an implicit approach using a coupled iterative solution algorithm.

For a more detailed view of *TfC*'s implementation, see Figure 15.1: Both for the coupled mass-momentum equation and the scalar equations, the equation system set up by the discretization is solved iteratively in space and time. Depending on the convergence rate, the next iteration is performed or the flow field distribution is written to a file.

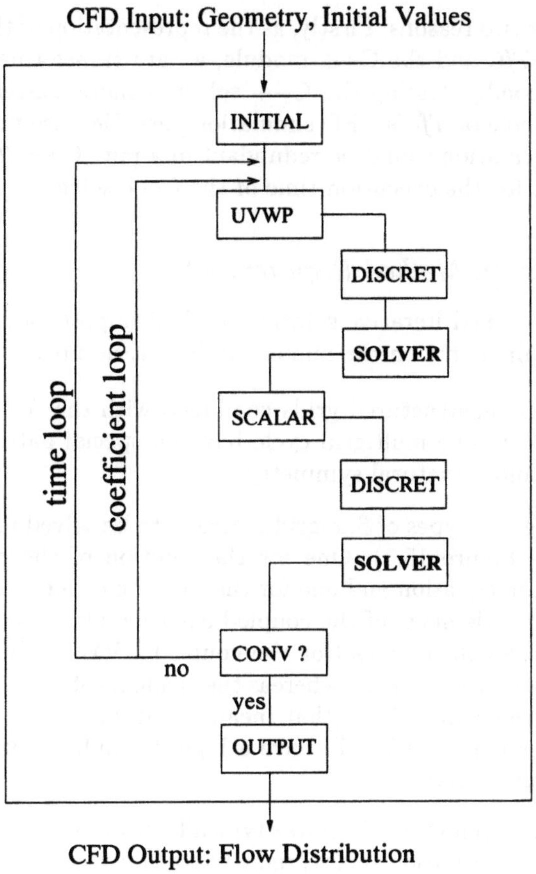

FIGURE 15.1. Overview of *TfC*'s CFD simulation part.

15.4 Analysis and Redesign of the Solver

The strategy of the approach is to replace the Fortran 77 SOLVER module (Figure 15.1) with a pure C++ solver implementation. *TfC* proceeds as before, passing the equation set to the solver and receiving the solution. So, the most important requirement for the redesign has been to have the numerical results of the C++ solver identical with the Fortran 77 solver. As the AMG is not trivial, all further requirements concerning an efficient implementation should be addressed with a first, working version.

The achievement of the same numerical behavior was based on a pseudocode formulation of the existing AMG implementation, according to which the C++ algorithm has been realized.

For this first version, a data file interface serves for the data transfer between *TfC* and the C++ solver. *TfC* writes the contents of the interface data structures to this file which the C++ solver reads to fill its objects.

This is done for two reasons: Firstly, as the representation of the data structures differ in *TfC* and the C++ module, a data transformation needs to take place. Secondly, testing the C++ solver is more convenient without invoking the whole of *TfC*'s CFD simulation part. Because this data transfer and transformation would be redundant in a pure C++ *TfC* version, it does not count for the execution time of the C++ solver.

15.4.1 Solution Method Requirements

The implicit, coupled iterative solution method imposes some general development requirements on the numerical data structures:

1. Due to the unstructured grids combined with the AMG, the matrices solved in one multigrid cycle have no special pattern except for sparsity and structural symmetry.

2. There are two types of fine-grid matrices to be solved in one iteration step (see Figure 15.1): One for the solution of the coupled mass-momentum equation and one for the solution of the scalar equations. The matrix elements of the coupled equation matrix consist of dense sub-matrices themselves (for the coupled UVWP solution, the sub-matrix dimension is 4x4), whereas the elements of the scalar equations matrix are single values (that means, of dimension 1x1). Moreover, future versions of *TfC* will have a higher coupling degree which the solver must handle.

3. The same applies for the vectors typically involved in an iterative solution process (solution vector, right-hand-side-vector, residual-vector): Each vector element is either a sub-vector (with dimension 1x4 for the coupled mass-momentum equation) or a single value (again, that means dimension 1x1).

The solver must therefore be able to adjust to the dimensions of the matrix elements. So, a general sparse matrix structure is required where the elements of the sparse matrix can either be dense sub-matrices or single values. A vector data structure must exist of which elements can either be sub-vectors or single values.

15.4.2 Object-Oriented Solver Design

To meet these general method requirements in the redesign, ideas of design concepts found in the object-oriented sparse matrix library SparseLib++ [8] have been adopted and extended to a sparse matrix and vector structure with adjustable element dimensions.

From an object-oriented viewpoint, it would have been natural to reuse a complete sparse matrix library where the data structures of the matrix elements could be specified, for example, via the *template* mechanism. The introduction of these deep level changes in the existing library did not seem reasonable. Today, SparseLib++ version 1.5c incorporates these demands.

The AMG method consists of two building blocks: an ILU0-smoother and the algebraic grid management. Starting from a fine (the initial) grid, each grid in the multigrid cycle is algebraically built and smoothed. So, both building blocks operate on grid level representations, whereas the smoother operates independently on each grid and the algebraic grid management on all grids. Thus, the smoother part has been redesigned first for later integration in the total AMG cycle.

This analysis led to the definition of the objects in Figure 15.4.2 (given as Booch diagram [3]): On the left side the objects for the smoother are shown, on the right side the objects for the algebraic multigrid management.

The Smoother Objects

The central object is the *Sparse Matrix* object, which relies on four other objects: the *Matrix Element* object together with the *Matrix Element Values* object, the *Column Row Access* object and the *Original Matrix Access Pattern* object. The latter object is not really a part of the design and algorithm, as it serves only for initializing the *Sparse Matrix* with the Fortran 77 matrix data from the interface file.

The *Column Row Access* object serves for the access on the sparse matrix elements, which are stored in a column-compressed order.

The objects *Matrix Element* and *Matrix Element Values* represent the required adjustable matrix structure.

The *Main Diagonal* object is required by the smoother algorithm and represents a copy of the sparse matrix main diagonal.

The *Vector*, *Vector Element* and *Vector Element Values* objects represent the required adjustable vector structure.

The AMG Objects

The *Equation* object is the logical interface between the discretization (not treated here) and the solution in *TfC*. *Equation* contains the equations to be solved and the solution method chosen. It relies on the objects *Algebraic Multigrid* and *Algebraic Grid*.

Algebraic Multigrid represents and controls the AMG cycle, *Algebraic Grid* represents one (to be smoothed) grid level.

The object *Coarse Block List* supports the formation of the coarser grids (see Section 15.6.2).

Implementational Aspects

Concerning the 1-dimensional case, these matrix and vector representations introduce objects at a very fine-grained level. Using the same representation for the 1-dimensional matrix and vector elements as for the

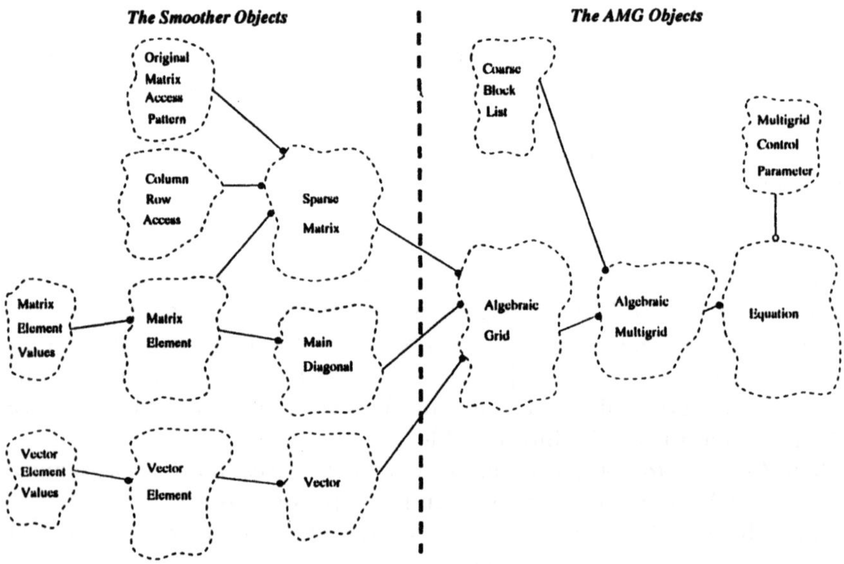

FIGURE 15.2. The class diagram of the solver.

higher-dimensional elements has reduced the development effort, since the same code is dynamically applicable to the different element structures. For an efficient solver implementation in C++, this special case has to be reconsidered in the design (Section 15.6.1).

In the C++ implementation, operator overloading is used for numerical and arithmetical operations to increase the readability of the numerical expressions in the code: The objects *Matrix Element Values* and *Vector Element Values* both overload the arithmetical operators (+,-,*,/) and assignment expressions (+=, -=). The matrix element access operator *()* is overloaded in the following objects: *Sparse Matrix*, *Main Diagonal*, *Vector*, *Matrix Element*, *Vector Element*, *Matrix Element Values* and *Vector Element Values*.

Again, using operator overloading in C++ introduces a source of inefficiency and should be reconsidered for an efficient implementation (Section 15.6.1).

Design Aspects

Main Diagonal consists of the same logical entities as *Sparse Matrix*, but has a simpler element storage and access logic. The different access logics are modeled in *Matrix Element*. This reduces the coding effort in *Sparse Matrix* and *Main Diagonal*, but introduces an additional indirection to be resolved during run-time for the access on the actual elements of both objects (see Section 15.6.2).

Vector is designed the same way as *Sparse Matrix* and *Main Diagonal* to facilitate the implementation logic. Again, this introduces an additional indirection resolving during run-time (Section 15.6.2).

High-Level Parallelization Support

The actual design supports high-level parallelization by introducing basic simple data structures as in *Vector*, which are not hidden and distributed in self-managed memory areas as in the Fortran 77 code. No parallel execution has been implemented yet, as this requires further investigation (see Section 15.6.3).

15.5 Original and New Module Comparison

The Fortran 77 and C++ solver modules are compared regarding the quality of the source code and the execution time.

15.5.1 Source Code Quality

The number of code lines of the Fortran 77 solver module is 3966 and of the C++ module is 9617. This increase is due to the higher coding effort for the abstractions additionally to the AMG method. The object *Equation* has 1962 code lines. *Equation* relies on the abstractions and contains the formulation of the multigrid algorithm.

The AMG C++ code is nearly identical to the AMG pseudo-code formulation, which the AMG Fortran 77 code is not. This is because of using the abstractions in, and the re-movement of memory management logic from, the AMG formulation at code level.

Furthermore, the pseudo-code like C++ formulation is robust and can be kept in case of changing code requirements due to the encapsulation of the abstractions.

Because of the reduced number of code lines for the AMG algorithm and its abstract formulation, the C++ solver is superior to the Fortran 77 solver regarding its readability (and thereby the ease of maintainability).

15.5.2 Execution Times

The test environment was a SUN SparcStation 10 with the Fortran 77 SUN SC 2.0.1 and GNU C++ 2.7.1 compilers. Both compilers were driven with -O3 optimization level.

Three geometries and flow conditions have been computed with the solvers:

1. A laminar flow in a straight duct with two different sized grids. On the first grid (225 nodes, 1345 matrix elements) the C++ solver shows an execution time increase rate of 3.6. For the second grid (1377 nodes, 8865 matrix elements) this rate is 4.4.

2. A laminar cross flow in a cylinder. For this grid (3150 nodes, 20450 matrix elements) the increase rate is 7.4.

3. A turbulent flow in a curved duct (3627 nodes, 23791 matrix elements). The overall increase rate is 15.8. Looking at the different equation sets involved in this test case, gives more detailed increase rates: 9.7 for the UVWP equations, 24.6 for the turbulent kinetic energy equation and 23.9 for the dissipation equation.

The C++ solver shows significant increase rates in execution time. For the coupled equations, these increase rates rise linearly with the complexity of the test cases and can be associated with increased dynamic object management at run-time. The two scalar equations involved in the turbulent flow show an even more dramatic increase. These increase rates are strongly related to the current C++ solver code implementation, which has been done according to design and implementation decisions inevitably leading to run-time inefficiency. Design and implementation improvements are discussed in the next section.

15.6 Discussion of the Approach

Usually, execution time is considered as the most important criterion to evaluate a task like the present. This attitude can have following consequences in the redesign phase: In order to beat the existing Fortran 77 implementation, new abstract concepts are often not really developed because the thinking is too close to the existing code and data representation. Instead of applying a new way of programming, the old way undergoes some sort of polishing (for example, implementing a statically allocated list dynamically, or removing GOTO's from a loop formulation by inserting the appropriate loop syntax construct of the implementation language), which enhances the programmer's task, but cannot show the same software quality benefits as a re-consideration of the representation of the logical entities an algorithm operates on.

The execution time increase rates shown here can by no means be an argument for not using object-orientation or C++, as this would be the wrong conclusion. There was deliberately no attention paid to implementation efficiency aspects to avoid these redesign pitfalls.

Of course, execution time is important for the evaluation of a new software design paradigm and therefore important for this work. As the sources of inefficiency are identified in the current C++ solver design and implementation, they can now be tackled. The good thing is that the elimination of the inefficiency does not contradict the source code quality improvements achieved so far.

15.6.1 Towards an Efficient C++ Implementation

Access on the matrix elements and arithmetical operations with the matrix elements (+,-,*,/) take most of the execution time of the C++ solver. These operations have been modeled by operator overloading – the worst thing one can do regarding efficient large arithmetical expressions in C++: Operator overloading is resolved by the C++ compiler with temporarily allocated copies of the objects involved in the computation. For a code which relies on numerical operations, this is intolerable. A solution is to have the operations realized in dedicated functions, and functional instead of operator overloading with the expression of operations' semantics through the function interfaces. This solution nearly maintains the readability achieved by operator overloading.

In the case of the solution of the scalar equations, each matrix element (one floating point value) is an object itself in the current design. The increase in execution time for the solution of the turbulent kinetic energy and the dissipation shows the disastrous effects of doing this in C++. Managing a built-in data type (like a floating point value) as an object introduces too much overhead and should not be done. The correction of the current design will yield two different solver object designs: one for the coupled equations solution and one for the scalar equation solution.

15.6.2 Improving the Design Efficiency

In the current design, the objects *Matrix Element* and *Vector Element* are a source of inefficiency, because they incorporate an additional execution of the overloaded operator (). These actions should be directly integrated in the according objects *Sparse Matrix*, *Main Diagonal* and *Vector* to avoid this additional resolving.

The improvements proposed so far are relatively straight forward. Profiling of the C++ code has revealed another, more hidden, consumer of execution time: During the algebraic formation of the coarse grids, the coarse grid blocks are formed in a redundant and disordered way. *Coarse Block List* handles this by keeping a list with unique entries, which is ordered before the coarse grid matrix is filled. This is circumstantial, as the coarse grid matrix should be filled immediately with a direct in-place insertion mechanism. Providing this may mean adjusting *Sparse Matrix* or the AMG algorithm. This requires further investigation.

15.6.3 Future Work

The C++ solver implementation should be improved according to the deficiencies identified to receive an efficient code.

With an optimal sequential version, the prepared parallelization support in the design should be exploited.

Presently, the discretization in *TfC* is the subject of an object-oriented redesign. Combining this part with the object-oriented solver should show further significant improvements in code quality.

Acknowledgments: For this work, I have received a lot of input from all SEMPA team members, which I appreciate very much. Special thanks to Werner Krötz, a computer science student at the TUM, who has done the implementation as part of his diploma master thesis [11].

This work has been funded by the German Federal Department of Education, Science, Research and Technology, **BMBF** (**B***undes***m***inisterium für* **B***ildung, Wissenschaft,* **F***orschung und Technologie*).

15.7 References

[1] Ian G. Angus and W. Dan Curtis. From Fortran to Object Orientation: Experiences with a Production Flutter Analysis Code. In *Proceedings of the Second Annual Object-Oriented Numerics Conference (OON-SKI'94)*, pages 174–180, April 24–27, 1994.

[2] Peter Beckman, Dennis Gannon, and Elizabeth Johnson. Portable Parallel Programming in HPC++. In *WWW*, http://www.extreme.indiana.edu/hpc++/docs/ppphpc++/icpp.html.

[3] Grady Booch. *Object Oriented Design With Applications.* Addison-Wesley, 2. edition, 1994.

[4] Patrick Bridges, Nathan Doss, William Gropp, Edward Karrels, Ewing Lusk, and Anthony Skjellum. *Users' Guide to* `mpich`*, a Portable Implementation of MPI.* Argonne National Laboratory, 9700 South Cass Avenue, Argonne, IL 60439-4801, July 1995.

[5] Gregory D. Burns, Raja B. Daoud, and James R. Vaigl. LAM: An Open Cluster Environment for MPI. In *Proceedings of the Supercomputing Symposium '94*, Toronto, Canada, June 1994.

[6] Lois Curfman, McInnes, and Barry F. Smith. PETSc 2.0: A case study of using MPI to develop numerical software libraries. In *MPI Developers Conference*, June, 1995.

[7] Jack Dongarra, Ed D'Azevedo, and Jim Demmel. ScaLAPACK Project. In *WWW*, http://www.netlib.org/scalapack/index.html.

[8] Jack Dongarra, Andrew Lumsdaine, Xinhui Niu, Roldan Pozo, and Karin Remington. Sparse Matrix Libraries in C++ for High Performance Architectures. In *Proceedings of the Second Annual Object-Oriented Numerics Conference (OON-SKI'94)*, pages 122–138, April 24–27, 1994.

[9] Al Geist, Adam Beguelin, Jack Dongarra, Weicheng Jiang, Robert Manchek, and Vaidy Sunderam. *PVM: Parallel Virtual Machine — A Users' Guide and Tutorial for Networked Parallel Computing.* Scientific and Engineering Computation. The MIT Press, Cambridge, MA, 1994.

[10] F. Guidec and J.-M. Jézéquel. Design of a Parallel Object-Oriented Linear Algebra Library. In K. M. Decker and R. M. Rehmann, editors, *Programming Environments for Massively Parallel Distributed Systems*, pages 359 - 364. Birkhäuser Verlag, 1994.

[11] Werner Krötz. Entwurf und Implementierung einer objektorientierten Gleichungslösung in C++ zum Einsatz im industriellen Strömungssimulationsprogramm TASCflow. Diplomarbeit, Technische Universität München, August 1996.

[12] P. Luksch, U. Maier, S. Rathmayer, F. Unger, and M. Weidmann. Parallelization of a State-of-the-Art Industrial CFD Package for Execution on Networks of Workstations and Massively Parallel Processors. In *Third European PVM Users' Group Meeting, EuroPVM 96*, München, October 1996.

[13] Peter Luksch, Ursula Maier, Sabine Rathmayer, and Matthias Weidmann. Software Engineering Methods for Parallel and Distributed Scientific Computing. In *HPCN Europe 1996*, Lecture Notes in Computer Science. Springer-Verlag, April 1996.

[14] Ursula Maier. Verteilte Ressourcenverwaltung in einem Netzwerk von Workstations. Technical report, Technische Universität München, Institut für Informatik, Lehrstuhl für Rechnertechnik und Rechnerorganisation, August 1996.

[15] Rebecca Parsons and Daniel Quinlan. A++/P++ Array Classes for Architecture Independent Finite Difference Computations. In *Proceedings of the Second Annual Object-Oriented Numerics Conference (OON-SKI'94)*, pages 408–418, April 24–27, 1994.

[16] Michael J. Raw. A Coupled Algebraic Multigrid Method for the 3D Navier-Stokes Equations. In *Fast Solvers for Flow Problems*, Proceedings of the 10th GAMM-Seminar, Notes on Numerical Fluid Mechanics Vol. 49. Vieweg-Verlag, Braunschweig, Wiesbaden, 1995.

[17] Gregory V. Wilson and Paul Lu, editors. *Parallel Programming Using C++*. Scientific and Engineering Computation. The MIT Press, http://www-mitpress.mit.edu:80/mitp/recent-books/comp/wilrp.html, July 1996.

16

Automating the Debugging of Large Numerical Codes

Fredrik Manne
Svein Olav Andersen[1]

ABSTRACT
The development of large numerical codes is usually carried out in an incremental fashion and over a long period of time. In this process, a segment of code might, for reasons of speed or accuracy, be replaced with a new code segment. If the resulting program contains a fault or generates a completely different result than the old one, it is crucial to find the place where the discrepancy first occurs. Finding this place might be a very tedious and time-consuming operation.

We present a framework called *comparative debugging* specifically designed to aid the programmer in debugging such situations. In this setting, the programmer can run both the old and the new program simultaneously, while comparing the execution flow and the values of certain variables in the two codes. If any discrepancy is discovered between the two programs, this is automatically reported back to the user who can then take appropriate action. The presented ideas have been implemented in a debugging tool.

We believe this to be a simple yet powerful way of automating the debugging process that can easily be incorporated into existing debugging programs.

16.1 Introduction

It has been estimated that 25-50 % of the total cost and time in system development may be spent on software testing and debugging [6, 21], and as much as 95 % of the time spent on debugging may again be spent on fault-location [13]. Therefore it is important to automate and speed up the process of locating where an error occurs in the code.

For this purpose programmers apply debugging tools such as the generic Unix debugger *dbx*, the GNU debugger *gdb* [17], and other vendor-specific debuggers. These tools let the programmer keep track of the state of the

[1]Department of Informatics, University of Bergen, N-5020 Bergen, Norway. Email: {fredrikm,sveinoa}@ii.uib.no.

program and the values of certain variables while stepping through the execution. Even with these tools the debugging process can be time consuming since the programmer must inspect values manually to determine where an error occurs.

In order to automate the process most debuggers can set conditional breakpoints (watch-points). In this way, the debugger stops only when some condition is evaluated to be true. For example, one might know the expected range of a variable. However, if the computation is "almost" correct such information will be of little help in determining where the error occurs.

Program slicing [3,15,19] is a way of determining which lines of code affect a certain variable. Thus if one knows that a certain variable is incorrect this method gives an overview of the other variables that affect it.

One must, however, know in *which* variables the error occurs and also *when* it first occurs. An incorrect end result might be the result of a long chain of events in the program. For this reason execution backtracking [3] has been suggested. This method lets the user backtrack through the execution of the program recreating the different states of the execution. For memory intensive programs with long execution times this is not a practical procedure as it would require storing vast amounts of data. A limited time window, in which one can backtrack, will save memory but might not be sufficient to backtrack to where the error originated. Moreover, this process still requires the user to inspect the program manually.

The motivation for the current work comes from debugging large timestepped simulation codes. Examples of such codes are fluid dynamics computations, oil reservoir modeling, weather forecasting, etc. These codes have in common that they try to simulate the behavior of a complex system over time. The simulation is broken down into shorter time steps on which the model is able to predict the outcome within reasonable accuracy. Such simulations tend to be computationally intensive and may run for hours and even for several days. These codes are also often very large involving up to several hundred thousands of code lines. If an error should occur in such a program, the task of locating it might be very difficult and time-consuming.

The development of large numerical codes is usually carried out in an incremental fashion and over a long period of time. Hence it is not unlikely that if parts of the code are changed and an error occurs one would have two versions of the program, one giving the correct answer and a new version giving the wrong one. In this case, it could be argued that the programmer has sufficient knowledge of what has been changed in the program in order to locate the error. But adding new features and capabilities to a program might initiate an error in the old parts of the code that have already been tested. Moreover, if the error only occurs for specific data sets the code might have continued developing for some time before one realizes that it contains an error.

We present a framework called *comparative debugging* that automates this particular debugging situation. In this framework, the programmer runs the old and the new versions of the program simultaneously, while monitoring the values of a set of selected variables. If the values of any variable differs between the two programs the programmer is automatically notified. This will direct the programmer to the first instance where an error occurs. It is also possible to trace the execution and ensure that the same statements are executed in the two programs.

We believe this to be a novel approach to automating and thus speeding up the debugging process that can easily be incorporated into existing debugging tools. In order to illustrate this we present a prototype tool called the *Wizard* that realizes the presented methods. This tool is developed using standard software components thus making it easily portable.

Apart from speeding up the debugging process the Wizard can also be used to verify that two programs that reach the same answer do so through the same set of computations.

The rest of the paper is organized as follows. In Section 16.2 we set up the theoretical framework for comparative debugging. In Section 16.3 we point to some situations where this might be of use. The Wizard itself is presented in Section 16.4 before we conclude in Section 16.5.

16.2 Comparative Debugging

Based on the observation that most of the work in debugging is in locating where the error first occurs we suggest the following approach to automate the debugging process.

It is assumed that the user has two programs available: An old program A giving the correct answer and a new program B giving the incorrect answer. The task we wish to solve is to determine where the error first occurs in the new program. It is also conceivable that instead of an old program one has file access to the correct values of certain variables at different stages of the computation.

16.2.1 Comparative Breakpoints

We start by defining a comparative breakpoint $(E_A, l_A), (E_B, l_B)$. Here E_A is an expression involving variables from program A and l_A is a line number in A defining where E_A is to be evaluated. The variables used in E_A must be in scope in line l_A. The expression E_B and line number l_B is defined similarly for program B.

Each time program A executes line l_A the value of E_A is evaluated using the current values of the variables in A. The value of E_A is then stored in a queue. Similarly when program B executes line l_B the value of E_B

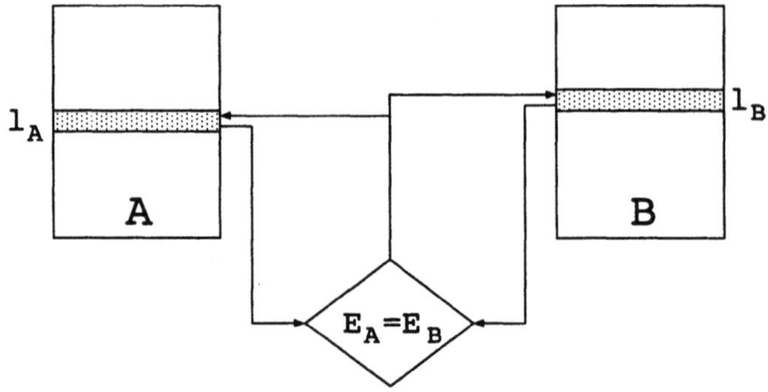

FIGURE 16.1. A blocking breakpoint.

is evaluated and stored in a separate queue. Whenever both queues are non-empty the first elements of the queues are compared for equality and deleted. Thus the value of E_A and E_B at the k'th time lines l_A and l_B are executed will be compared. This is true even if the programs should execute asynchronously.

We define two types of comparative breakpoints: blocking and non-blocking. When a program encounters a blocking breakpoint the program halts after storing its value to the queue. When both programs have stored their associated values and the comparison has been performed the programs are notified of the outcome. If $E_A = E_B$ both programs continue execution, but if $E_A \neq E_B$ they halt. Figure 16.1 illustrates the data-flow for a blocking breakpoint.

The rationale for this behavior is that if A and B are running inside two separate debuggers, control is passed back to the debuggers. So that in the event of a halt message the programmer can perform traditional debugging on the two programs before deciding on whether to terminate or continue the execution.

When a non-blocking breakpoint is encountered the program does not wait for the result of the comparison, but continues execution as soon as the associated value has been stored in the queue. If a mismatch is found when comparing the values in the queues an error message is output to the user. Consequently non-blocking breakpoints are mainly for detection of where an error occurs.

The main reason for including non-blocking breakpoints is speed of execution. A blocking breakpoint will cause the programs to synchronize at each breakpoint, slowing down the overall execution. Also if E_A and E_B are matrix values, or if breakpoints are encountered frequently, it may take a significant amount of time to transmit the necessary values and perform the comparison.

Apart from speeding up the execution, non-blocking breakpoints are slightly more flexible than the blocking ones. If A encounters a blocking breakpoint fewer times than B, this would cause B to hang indefinitely. Also as shown in Figure 16.2, if two blocking breakpoints have been de-

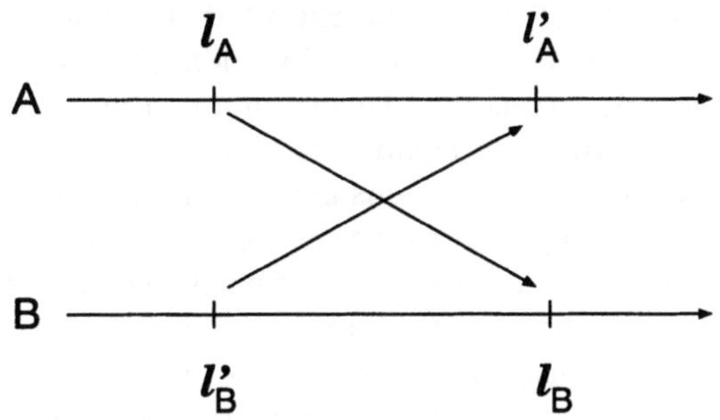

FIGURE 16.2. Blocking breakpoints may cause a deadlock.

fined such that $l_A < l'_A$ and $l'_B < l_B$, a deadlock might occur with program A waiting for input at l_A and program B waiting for input at l'_B. Both of these situations can be avoided by using non-blocking breakpoints.

Comparative breakpoints should preferably be used in such a manner that deadlocks are avoided. It is, however, the responsibility of the user to ensure this.

In Section 16.3 we present some examples to motivate where comparative breakpoints can be of use.

16.2.2 Execution Tracing

The idea of comparative debugging can be extended to other areas in addition to comparing variables. We define a special type of comparative breakpoint called a *logical breakpoint* that can be used to trace and compare the execution of two programs.

A logical breakpoint is basically defined in the same way as the blocking and non-blocking breakpoints. The main difference is that when a program encounters a logical breakpoint only the identity of the breakpoint is used for comparison. So instead of associating an expression with each breakpoint we use an identifier i that is the same in both program A and B but is unique for each breakpoint. The formal definition of a logical breakpoint then becomes $(i, l_A), (i, l_B)$.

The only thing being compared during the execution of the two programs is the order in which the logical breakpoints are encountered. If program A encounters logical breakpoint i before breakpoint i' while program B encounters i' before i, an error will be reported.

To see how this might be used consider the following code segment:

```
while (sum < limit) {
        sum = sum + a[i];
        i = i + 1;
}
return;
```

Assume that this particular segment of code is in both programs and that the user wants to know if it is being executed the same number of times in the two programs. This can be done by adding two logical breakpoints to the codes, the first attaching the two summation lines to each other and the second attaching the `return` statements to each other. Then an error will be reported if the programs execute the loop a different number of times. If a blocking logical breakpoint has been used, the program that executes the loop the fewest times will halt on the `return` statement, while the other program will halt on the summation statement. If non-blocking logical breakpoints are used an error will be reported but the programs will continue their execution. Note that in the above example, it is not sufficient to add a comparative breakpoint that compares the values i in the two programs. This is because the initial values of i might differ between the two programs.

Logical breakpoints should only be used where the execution flow of a program can branch. The obvious place being inside a loop or in an `if` statement. Other settings might be a subroutine with multiple `return` statements or a `case` statement.

To ensure that any discrepancy is discovered, each possible branch of execution that one wants to trace must be tagged by a logical breakpoint. This means that in a loop one should tag both a statement inside the loop as well as the first line following the loop. Similarly in an `if` statement, both the `then` and the `else` statement should be tagged. If there is no `else` statement the first line of code following the `if` statement should be tagged.

If one avoids tagging a certain branch of execution one of the programs can escape through this and later return to execute the necessary logical breakpoint, thus avoiding the detection of an error. A basic block of code is a sequence of source lines such that if the first line is executed the program cannot branch until the last source line has been executed. Thus it is only necessary to add logical breakpoints to the first statement of a basic block in each program.

We note that in some cases it is possible to simulate a logical breakpoint by the use of comparative breakpoints. This can be done by adding comparative breakpoints that compares the logical expressions that govern the branching of the execution flow. This is, however, more cumbersome and not as flexible as using a logical breakpoint. In particular, this is true if one is comparing two code segments where the structure of the governing logical expressions are different.

16.2.3 Floating Point Comparisons

The exact value of a floating point value might vary depending on issues such as the order of the operands, the machine precision, the implementation of standard numerical functions, etc. For these reasons it is seldom meaningful to compare two floating point values to determine if their values are exactly equal. To circumvent this problem we introduce a user specified value ϵ which is the largest tolerated difference when comparing two floating point values before an error is reported.

16.3 Examples of Use

In this section, we give a few examples where comparative breakpoints might aid the programmer both in verification of code and in the debugging process. In general, it is possible to use comparative debugging in any setting where a secondary program can generate the expected answer of a code segment.

16.3.1 Matrix Multiplication

The core of many block algorithms used in linear systems solvers, is a matrix multiplication routine. The time complexity of calculating the matrix product $C = AB$, where A, B, and C are $n \times n$ matrices, using traditional matrix multiplication, is $O(n^3)$. Strassen's matrix multiplication algorithm is able to perform this calculation in time $O(n^{2.807})$ [10,18]. Thus one might be able to speed up the code by using Strassen's algorithm instead of the traditional algorithm. Both IBM and Cray support routines for fast matrix multiplications using Strassen's algorithm. See also [4,5] for examples of implementations of Strassen's algorithm. However, the error bound given by Strassen's algorithm is weaker than that of the traditional algorithm [7,11]. Therefore the two approaches might give different answers. If one suspects that this is the case one can use comparative breakpoints to determine if and when the result of Strassen's algorithm differs significantly from that of the ordinary algorithm.

16.3.2 Order of Execution (Sequential - Parallel)

For our second example we consider parallel computing. It is well known that the order of the operands in a computation can have significant impact upon the end result. See [8] for a discussion of floating-point summations. A particularly computationally intensive segment of code might be rewritten to run in parallel. If the end results of the sequential and parallel codes differ, one might wonder if this is the result of a coding error or of performing the computations in a different order. In this case, comparative debugging can be used to determine the magnitude and the first instance where the codes produce different results.

16.3.3 Different Input

If the input data to a program has been slightly perturbed the program might perform other actions then what it did previously. The user might then want to catch the first instance where the program behaves differently on the two data sets. This can easily be accomplished with the use of logical breakpoints.

16.3.4 Cross-platform Debugging

Finally, we note that if an error occurs only on a particular computer platform it is possible to use comparative debugging while running the same program simultaneously on two different computers and performing the communication between the two programs through the network.

16.4 The Wizard

To illustrate that the concept of comparative debugging is feasible and useful, we have implemented the features described in Section 16.2 in a debugging tool called the *Wizard.* The Wizard lets the user run two programs simultaneously, each running inside a debugger extended with comparative breakpoints.

16.4.1 Implementation

The Wizard is comprised of a monitor program and two debuggers. These are again merged inside a graphical user interface. The two debuggers are slightly modified version of the GNU debugger *gdb* [17]. The *gdb* debugger has been extended with comparative breakpoints as described in Section 16.2. These breakpoints are constructed in a similar way as the ordinary breakpoints only that the required expression has to be evaluated and transmitted to the monitor program. The comparative breakpoints use *gdb*'s own internal representation of variables. With this implementation it is possible to use all the ordinary debugging features included in *gdb*. It is also possible to make the evaluation of a comparative breakpoint conditional.

The monitor program is implemented in C and performs the actual comparisons of data transmitted from the two debuggers. The communication between the two debuggers and the monitor program is done using the PVM message-passing system [9]. The programs are assembled inside a user interface implemented using Tcl/Tk [14,20] and Expect [12]. By using software that is available on most Unix platforms there is no major problem in porting the tool to new systems. Figure 16.3 illustrates the design of the Wizard.

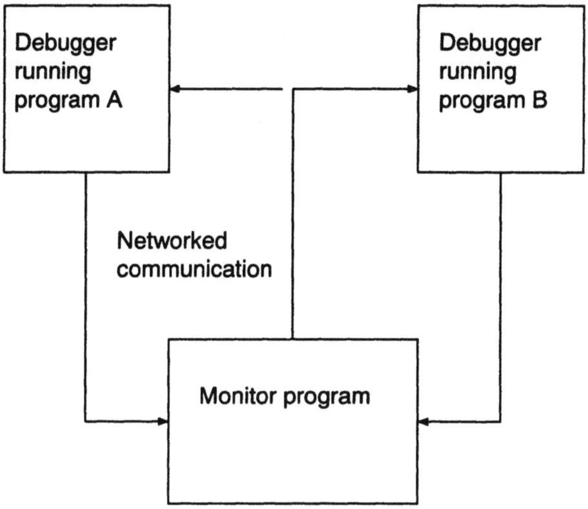

FIGURE 16.3. The design of the Wizard.

16.4.2 The User Interface

When the Wizard is started, the user first specifies on which computers the programs are to run. The two programs to be executed are then loaded into the debuggers. Figure 16.4 shows the main view of the program.

There are four menus at the top of the screen. Their functions are as follows:

- `File`, This menu is used to specify the executable programs to run in the debuggers.
- `Pvm`, This menu lets the user specify on which computers the programs are to be run. It is also possible to configure PVM through this menu.
- `Views`, This allows for pop-up windows containing graphical interfaces to a number of *gdb* commands.
- `Help`, Displays various help messages.

There are two windows listing the source codes of the two programs. Below these are two interactive windows used for input to the programs and for displaying messages from *gdb*. Below these windows are five push-buttons. The effect of these are as follows:

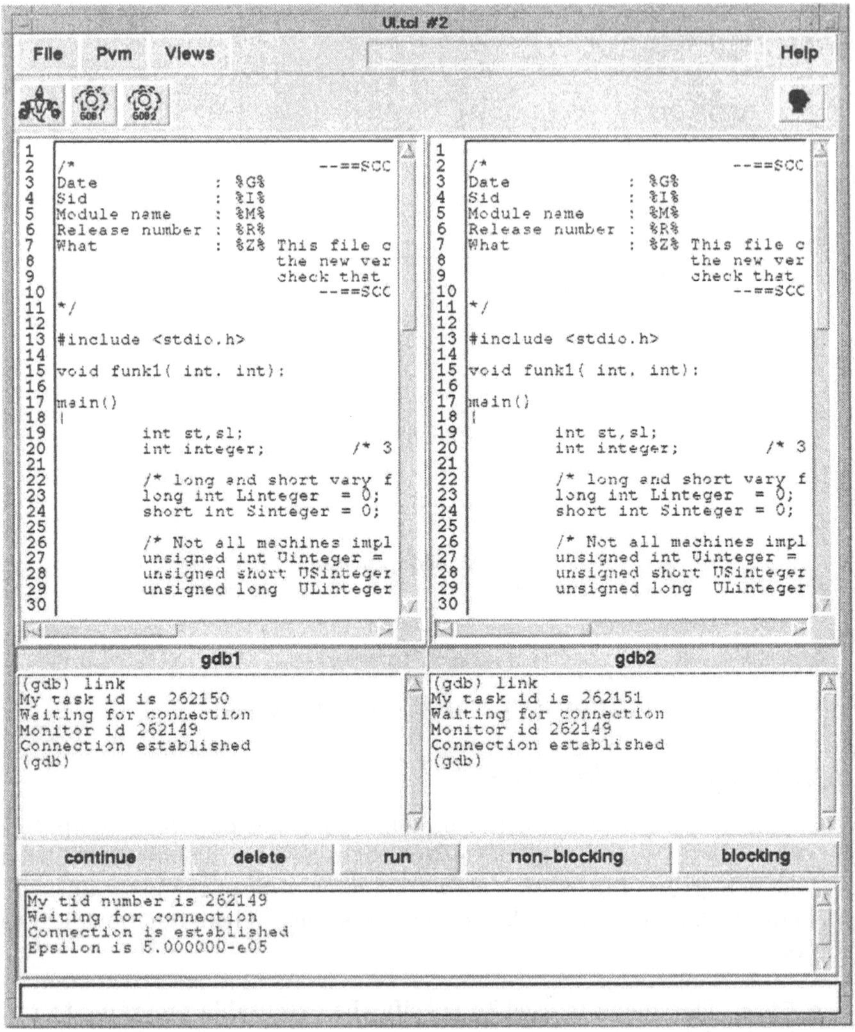

FIGURE 16.4. Main view of the Wizard.

- **`continue`**, After a mismatch has been found at a blocking breakpoint both *gdb* processes halt. With this button the user can continue execution of the two programs.

- **`delete`**, This button deletes a comparative breakpoint (from both programs). One may thus be assured that both programs contain the same number of comparative breakpoints.

- **`run`**, This button starts the execution of the two programs.

- **`blocking`**, This button lets the user specify a blocking comparative breakpoint. The user gives the line numbers of the breakpoint and specifies if this is a logical breakpoint or not. If not the user gives the expressions to be evaluated in each of the two programs.

- `non-blocking`, This button is similar to the "blocking" button only that a non-blocking comparative breakpoint is specified.

At the bottom of the user interface is a window displaying the output of the monitor program. Whenever a comparison of the values from a comparative breakpoint results in a discrepancy an error message is printed in the bottom-most window specifying the line-numbers of the breakpoint and the values that caused the mismatch.

Below the top-most menus there are three push-buttons, the left-most containing an image of a wizard and the two others displaying "GDB1" and "GDB2". The "GDB1" button is used only to view the first of the two programs. The user interface then changes so that the second program is no longer shown and a number of new push-buttons appear. These buttons display standard debugging commands for *gdb*. In this setting, the Wizard functions like a standard *gdb* debugger. Similarly "GDB2" displays only the second program, while the wizard button resets the display to that of Figure 16.4.

Since the Wizard has been implemented using explicit message passing it is possible to execute the two programs to be debugged on different computers. As mentioned in Section 16.3 this is advantageous if the error only occurs on a particular computer platform or if the memory requirements of the program are such that two versions cannot be executed simultaneously on the same computer.

16.5 Conclusion

We have introduced the notion of comparative debugging for automating the debugging process applied to large numerical codes. Comparative breakpoints were implemented in the Wizard thus showing the feasibility of this idea.

Tests performed so far indicate that the Wizard can be used to detect a number of errors. It can also be used to compare and verify the results of a segment of code continuously through a computation. Although this work is motivated from debugging numerical codes, the described methods can be applied to any debugging situation where it is possible to generate the expected answer to be produced by a segment of code.

The main obstacle to using comparative debugging is that one must have access to two very similar programs. Although this might not be a common situation for most programmers, we believe that when this does happen, using a tool like the Wizard will be worthwhile.

One of the main objectives of this study has been to show that comparative debugging can be incorporated in existing debugging tools. We believe that this has been achieved. Several debuggers have the possibility to control multiple threads or programs from one debugging session. Thus it should be fairly easy to extend these with comparative breakpoints.

We have been made aware of a similar effort to the one described here presented by Abramson et. al [2,16]. They describe a software system that can be attached to and control two existing debuggers. In this way they supply the user with comparative breakpoints similar to those described in Section 16.2.1, but not with logical breakpoints. Their approach to comparative debugging requires the use of an software system in addition to the debuggers. Thus to run their system on a new platform requires porting the software system, while the approach presented here uses a debugger that already exists on several platforms.

There are several extensions to the current work that would be useful in a future, more complete tool. We mention some of these here:

- The current version of the program can only be used to debug codes written in C. We are currently working on extending the program to also handle Fortran programs.
- As of now the program is restricted to only comparing scalars. One should allow for sending of whole arrays and array segments.
- A comparative breakpoint inside a tight loop will generate a large number of data packages to be sent to the monitor program. This overhead will slow down the Wizard. One way this could be remedied is by specifying that data should only to be sent every n'th time a breakpoint is executed. The extra data could then either be packed into larger messages or discarded.
- Currently we do not allow a program to be run against data stored in a file.
- The tolerance ϵ used in comparisons is the same for every breakpoint. It might be useful to have the option to specify an individual tolerance for each breakpoint.

It is our intention to make the Wizard available to other users. There is currently an extended web presentation available on the Internet [1].

16.6 References

[1] Automating the debugging of large numerical codes.
http://www.ii.uib.no fredrikm/fredrik/debug/.

[2] D. Abramson, I. Foster, J. Michalakes, and R. Sosic Relative debugging and its application to the development of large numerical models. In *Proceedings of IEEE Supercomputing 1995*, December 1995.

[3] H. Agrawal. *Towards Automatic Debugging of Computer Programs.* PhD thesis, Purdue University, West Lafayette, IN, 1991.

[4] D. H. Bailey. Extra high speed matrix multiplication on the Cray-2. *SIAM J. Sci. Statist. Comput.*, (9):603–607, 1988.

[5] P. Bjørstad, F. Manne, T. Sørevik, and M. Vajteršic. Efficient matrix multiplication on SIMD computers. *SIAM J. Matrix Anal. Appl.*, 13(1):386–401, 1992.

[6] B. W. Boehm. *Software Engineering Economics.* Prentice Hall, Englewood Cliffs, NJ, 1981.

[7] R. P. Brent. Algorithms for matrix multiplication. Technical Report CS 157, Computer Science Department, Stanford University, Stanford, CA, 1960.

[8] T. Espelid. On floating-point summation. *SIAM Rev.*, 37(4):603–607, 1995.

[9] G. A. Geist, A. Beguelin, J. Dongarra, W. Jiang, R. Manchek and V. S. Sundaram. *PVM 3 User's Guide and Reference Manual.* Oak Ridge National Laboratory, 1993.

[10] G. H. Golub and C. F. Van Loan. *Matrix Computations.* North Oxford Academic, 2nd edition, 1989.

[11] N. J. Higham. Exploiting fast matrix multiplication within the level 3 BLAS. *ACM Trans. Math. Software*, 16:352–368, 1990.

[12] D. Libes. *Exploring Expect.* O'Reilly, 1994.

[13] G. J. Myer. *The Art of Software Testing.* Wiley-Inter-Science, New York, 1979.

[14] J. K. Ousterhout. *Tcl and the Tk Toolkit.* Addison-Wesley, 1993.

[15] H. Pan. *Software Debugging with Dynamic Instrumentation and Test-based Knowledge.* PhD thesis, Purdue University, West Lafayette, IN, 1993.

[16] R. Sosic and D. A. Abramson. Guard: A relative debugger. To appear in Software Practice and Experience.

[17] R. M. Stallman and R. H. Pesch. *Debugging with GDB, The GNU source-level debugger*, 4.12 edition, 1994.

[18] V. Strassen. Gaussian elimination is not optimal. *Numer. Math*, 13:354–356, 1969.

[19] M. Weiser. Programmers use slices when debugging. *Numer. Math.*, 7(25):446–452, 1982.

[20] B. Welch. *Practical Programming in Tcl and Tk.* Prentice Hall, Englewood Cliffs, NJ, 1994.

[21] M. V. Zelkowitz. Perspectives on software engineering. *Numer. Math.*, 2(10):197–216, 1978.

17

The TAMPR Program Transformation System: Simplifying the Development of Numerical Software

James M. Boyle[1]
Terence J. Harmer[2]
Victor L. Winter[3]

ABSTRACT
Writing correct numerical software is a complex, demanding, and, at times, even a boring, task. In this chapter, we describe an approach to constructing software—program specification and transformation—and allied tools that can help not only to ensure the correctness of numerical computations but also automate much of the drudge–work involved in preparing such software. This approach to software construction holds out the exciting prospect of enabling the numerical analyst or specialist in scientific computing to concentrate on correctly capturing the problem to be solved, while delegating the details of programming the software and adapting it to specialized computing environments to automated tools.

17.1 Introduction

Numerical computation presents ever–increasing demands for complex, correct, and efficient software. The numerical and scientific computations being attempted today are considerably more complex than in the past, and many of these computations are "virtual experiments", intended to replace, rather than to corroborate, physical experiments that are too difficult, too dangerous, or too expensive to perform. Computations intended for use in

---

[1]Technology Development Division, Argonne National Laboratory, Argonne, IL 60439, U.S.A. Email: `boyle@mcs.anl.gov`.
[2]The Queen's University of Belfast, Department of Computer Science, Belfast, BT7 1NN, U.K. Email: `t.harmer@qub.ac.uk`.
[3]Sandia National Laboratories, Intelligent Systems and Robotics Center, Dept 9622, MS–0660, P.O. Box 5800, Albuquerque, NM 87185–0660, U.S.A. Email: `vlwinte@sandia.gov`.

such applications require *trusted software*, software for which there is a high degree of assurance—possibly even a proof—of correctness.

Complexity and the need for correctness notwithstanding, numerical and scientific computations must be efficient: the problems being solved lie always at the limits of the capabilities of available hardware. Moreover, many of these computations, such as weather forecasting, must be done in real–time.

In addition, the field of numerical computation abounds in examples of programs that are needed in a number of closely related variants: versions for real and complex arithmetic, dense and sparse matrices, sequential and parallel computers are but a few examples. Preparing such variants by hand is not only boring, but prone to introduce error, even when the basic algorithm has already been correctly implemented.

In this chapter, we describe a software development technique—*program specification and transformation*—that not only can simplify and remove some of the drudgery from the development of mathematical software but also can ensure the correctness of the software and avoid the need for debugging. Moreover, the program specification and transformation methodology provides these advantages while producing programs that are as, or more, efficient than handwritten ones (see Section 17.2.4).

Two powerful ideas, both familiar to the numerical mathematician from the study of mathematics, lie behind program specification and transformation: *abstraction* and *algebraic manipulation.*

Program specification employs *abstraction* to enable the writing of a single, general representation of an algorithm that is independent of specific computational requirements and environments. Just as one can study certain properties of the integers by studying the abstract theory of groups, one can write an abstract specification for linear algebra software in terms of matrices over a division ring. Such a specification can then be used not only to create a program that solves the problem for matrices having real elements, but also (at least in principal) to create programs for matrices whose elements are polynomials, other matrices, or even symbols.

Program transformation employs *algebraic manipulation* to instantiate an abstract specification for a particular computational requirement (for example, sparse real matrices) or environment (for example, a shared-memory parallel machine). Just as one uses identities from ordinary algebra or trigonometry to change the variable of a polynomial (see Section 17.3.2) or to prove a trigonometric identity, so program transformation uses identities from various "program algebras" to manipulate abstract specifications into a form suitable for efficient execution.

At the heart of using program specification and transformation lies a major psychological shift in one's approach to programming: the shift from viewing the programs one writes as static, completed objects to viewing programs as *mutable objects*, which can be easily adapted to the circumstances of their intended use. When one views a program as a static object, efficiency inevitably enters one's thinking when writing the program, and the need for efficiency destroys the possibility of abstraction. With the shift to viewing programs as mutable, one can concentrate on writing an abstract specification that, while appearing inefficient, has the greatest possible adaptability and flexibility. Program transformations can then create any number of efficient variant implementations from this single specification, without any hand modification to that specification whatsoever.

Abstraction and algebraic manipulation also permit the program specification and transformation methodology to address the problem of program correctness. Abstraction permits specifications to be problem–oriented, employing concepts and even notation from the problem domain. To prove that such a specification correctly implements an algorithm is much easier than to prove that an efficient program does. Thus, one can have strong assurance that the specification is correct. Similarly, the identities from program algebra that form the basis for transformation permit one to prove that program transformations preserve correctness. Because the application of correctness–preserving transformations cannot destroy the correctness of a specification, the resulting implementation is as correct as the specification from which it is derived. In this way, program specification and transformation yields highly trusted software.

In software engineering terms, the specification–transformation methodology thus qualifies as what is called a *formal method*: the methodology requires a formal specification and provides a formal demonstration (proof) that a program implementing the specification is correct relative to that specification. An alternative formal method is to prepare a formal specification and then to write a program by hand, verifying that it implements the specification. In comparison to this approach, the specification–transformation methodology is attractive both because it saves the expensive labor involved in programming, and also because it greatly reduces the cost of proof. Instead of requiring a large and complex proof of correctness that is specific to a particular *implementation* of a program, the specification–transformation methodology breaks the proof into pieces (the proof of the high–level specification and the proofs of individual transformations) that are both of manageable size and sufficiently general that they can immediately be reused for other programs.

Such is the promise of program specification and transformation, a promise that has remained largely unfulfilled for a quarter century. How close is it to being realized? We provide evidence in this chapter that the TAMPR (Transformation Assisted Multiple Program Realization) System [2, 4] is capable of fulfilling much of this promise.

In the following sections, we discuss some of the applications that have been made of the TAMPR system, by way of demonstrating its usefulness and practicality. Then we discuss in general terms how program transformation can be automated using TAMPR. Finally, we give an example of the type of program specification we use and how such object–oriented specifications can be transformed into efficiently executable programs.

17.2 Some TAMPR Applications

As discussed in the preceding section, the program specification and transformation methodology starts from a high–level, abstract specification for a computation, to which program transformations are applied to produce executable programs. For most TAMPR applications, it is convenient to start with an object–oriented specification expressed in a functional programming language, usually a language similar to pure Lisp (Scheme) or SML. (Part of such a specification is presented in Section 17.4.1).

The derivation of an efficiently executable program from such a specification is a complex task. To manage this complexity, one breaks the derivation into a number of discrete steps by identifying *intermediate forms* of the program that lie between the specification language and the final implementation. For example, rather than making the transition directly from SML to Fortran, a specification can first be converted into the λ–calculus, and then into Fortran. Such major transitions may be further subdivided into other intermediate forms, to whatever level is necessary and convenient. (This technique of breaking the derivation into intermediate steps is discussed further in Section 17.3.2.) Correspondingly, the derivation to implement a specification is divided into *subderivations*; one subderivation being used to create each intermediate form.

Many of these subderivations perform identifiable, independently useful tasks (for example, *in–lining* of function definitions, also called *unfolding*) which may be "mixed and matched" to create derivations to produce programs having a variety of desired properties. A number of subderivations that we have created for use with TAMPR are shown in Figure 17.1. Some of the important subderivations illustrated are

Unfolded, which performs optimization at the specification level;

Evaluated, which partially evaluates and simplifies the unfolded definitions;

Sparse, which optimizes evaluation for (user–defined) sparse matrices;

Data Parallel, which targets a derivation toward data–parallel computers;

Vector, which targets a derivation toward vector computers;

MIMD, which targets a derivation toward MIMD computers.

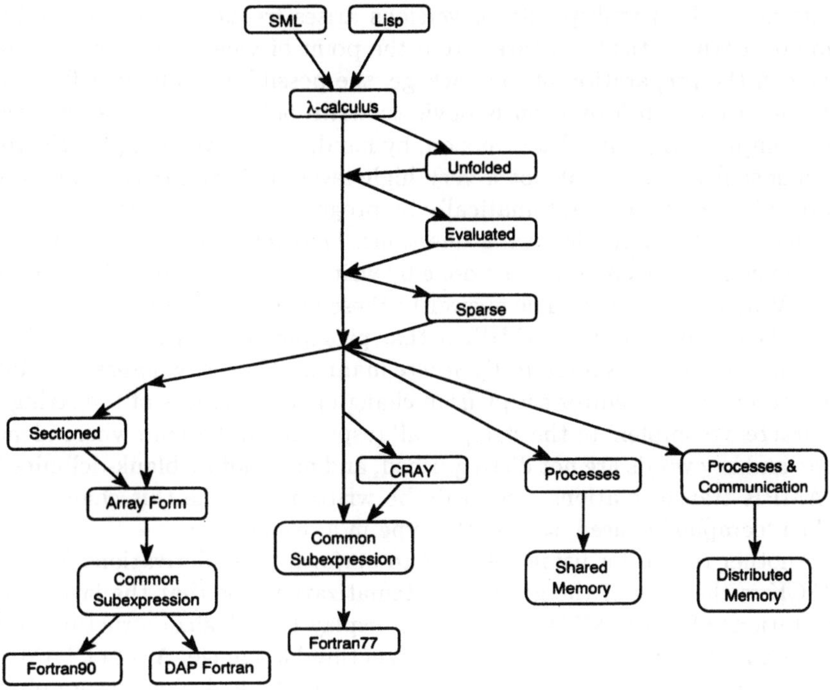

FIGURE 17.1. A family of derivations

Among the applications of the derivations of the type depicted in Figure 17.1, five particularly stand out, either because of their historical significance or their contribution to the demonstration that the program specification and transformations methodology is practical. We discuss them in the following subsections.

17.2.1 LINPACK

The first major application of TAMPR was to derive the programs included in the LINPACK package of subroutines for solving systems of linear equations [8]. The goal of the LINPACK project was to provide "certified" software—software that had been thoroughly systematized and tested—for solving systems of linear equations. The original LINPACK package consists of 48 Fortran subroutines plus additional test programs, and has been used at sites all over the world for more than 20 years.

Most LINPACK routines are available in four versions: complex single– and double–precision arithmetic, and real single– and double–precision arithmetic. This multiplicity of versions arises frequently in the production of mathematical software. From the point of view of avoiding drudge work in the preparation of the package, the possibility to benefit from using program transformation is obvious: write only one of these versions, the complex single–precision version, by hand, let this version play the role of a specification (albeit not a very high–level one), and derive the other three versions from it automatically by program transformation.

One might ask the following questions, however: Isn't complex–to–real conversion simple enough to be done by writing, say, a script for a text editor? Why use TAMPR? The answer to these questions is rooted in *correctness*: One reason to use TAMPR is that program transformations perform the conversion in a *syntactically secure* manner. That is, whereas it is difficult to prevent an editor script from changing all instances of the string of characters 'complex' to the string 'real' (especially in Fortran where identifiers and key words are not distinguished, and need not be blank-delimited), TAMPR transformations can easily be written to make this change only when 'complex' is used, say, as the type in a declaration.

Another reason for using TAMPR to perform these derivations for LINPACK is that one of the goals of systematization was that the layout (indentation) of the LINPACK programs display their logical structure, even though Fortran 66 is a notoriously unstructured language. A set of TAMPR transformations was therefore written to infer the logical structure of the programs, converting, for example, suitable sequences of Fortran "if" and "go to" statements and corresponding labels to "if-then-else" form. The TAMPR output component (the formatter) was programmed to indent statements within loops and within the "then" and "else" parts of "if" statements to indicate structure. Once these transformations had been applied to the complex versions (the specifications) of the LINPACK programs, it was trivial to apply or omit the complex–to–real transformations as a subderivation, to produce the indented complex version or indented real version of each LINPACK subroutine.

While the use of TAMPR to produce LINPACK did not start from a high–level functional specification nor directly make use of any of the subderivations shown in Figure 17.1, use of TAMPR as a Fortran–to–Fortran transformer did reduce labor and assured uniformity, improved readability, and enhanced correctness of the subroutines.

17.2.2 TAMPR Bootstrapping

A second major application of TAMPR (a non–numerical one) has been "bootstrapping" its own implementation. The specifications for the TAMPR transformer and other components are written in pure functional Lisp (functional Scheme). These specifications are object–oriented and highly flexible, and the full use of data abstraction permits the algorithms to be expressed in terms of structures such as trees, rather than the Lisp **cons**, **car**, and **cdr** operations. Whenever we wish to modify TAMPR, we modify the high-level specification, and use TAMPR to rederive its own implementations in C or Fortran. The derivation includes a number of transformational steps that improve the efficiency of the derived implementation, including the unfolded and evaluated subderivations shown in Figure 17.1. Unfolding in-lines data abstractions (object definitions), which can then often be simplified to reduce computation (see Section 17.4.2). The flexibility inherent in program specification and transformation is well illustrated by the fact that TAMPR can derive both sequential and parallel implementations of itself from the *same* specification [2].

17.2.3 MIMD Parallel Auction–ILINE Algorithm

A third application of TAMPR was the first in a sequence of attempts to compare the execution efficiency of transformationally derived programs with that of their handwritten counterparts. This application involved derivation of a parallel implementation (for MIMD shared–memory parallel computers) of a resource–assignment algorithm called Auction–ILINE [1]. An object–oriented specification for Auction–ILINE, similar to that discussed in Section 17.4.1, was written in pure Lisp.

An interesting feature of the use of functional–language specifications is that for many algorithms it is not necessary to indicate parallelism explicitly. Rather, parallelism appears implicitly in the specification when a function has two or more non–trivial arguments. Functions in functional programming languages are true mathematical functions; they cannot have "side–effects" (they cannot change the global state of the program without explicitly including the changed values as part of the function value). Thus, the values of two functions that are arguments to a third are guaranteed to be independent of one another, and the functions can therefore be evaluated in parallel. The same specification from which a sequential program can be derived thus enables the derivation of a parallel program. In fact, the parallel version of the TAMPR transformer was derived from a specification written with no thought of parallelism, long before practical parallel computers existed.

From the Auction–ILINE specification, TAMPR derived a parallel implementation in Fortran. Concurrently, the sequential Fortran implementation of Auction–ILINE was parallelized by hand. While the Auction–ILINE algorithm does not lend itself to highly parallel computation, the TAMPR–derived program achieved a somewhat higher parallel speedup than the handwritten parallel version, and with this speedup the overall execution speed of the TAMPR–derived program equalled that of the handwritten parallel version.

17.2.4 Vector Hyperbolic PDE Solver

The efficiency of the Auction–ILINE implementation encouraged us to explore whether TAMPR–derived programs could achieve, or even exceed, *parity* in execution speed compared to handwritten programs. For the fourth application, we chose a new cellular–automaton algorithm for solving hyperbolic partial differential equations (PDEs) [10]. Hyperbolic PDEs arise in fluid flow problems involving *shock*, such as supersonic flow, traffic flow, and the breaking of waves on a beach. We prepared a higher–order functional specification (discussed in Section 17.4) for this algorithm, taking care to write the specification in a way that made no concessions to efficiency and once again made full use of data abstraction. Indeed, one might even say that we wrote the specification to be as *inefficient* as possible—we always chose clarity of the specification over efficiency. Nevertheless, the TAMPR–derived implementations (in Fortran) are highly efficient: that tailored for the CRAY X–MP vector computer (using the unfolded, evaluated, and CRAY subderivations of Figure 17.1) is faster than the handwritten program prepared by the inventors of this algorithm, and that tailored for sequential computers (Sun) is again faster than the handwritten version on such machines [5, 6].

17.2.5 SIMD POT Eigenvalue Solver

Deriving an efficient implementation of a parallel eigenvalue solver (POT) is a fifth TAMPR application [3,9]. For this derivation, we expressed the specification in a variant of the SML functional language, which we transformed to pure Lisp. The derivation targets SIMD parallel machines, in particular the AMT DAP–510 array processor, and uses the unfolded, evaluated, and array–form subderivations in Figure 17.1. SIMD parallel programs are efficient only when most of the program can be executed using *array* operations; such operations apply the same arithmetic or logical operation to up to 1024 elements of a vector or a rectangular array simultaneously. The derivation uses an *array algebra* to generate efficient SIMD, and ultimately DAP, code from the specification [9]. Again, the TAMPR–derived implementation in DAP Fortran achieves parity with the handwritten program.

17.2.6 Application Summary

These examples illustrate that transformational programming's promise of flexibility and correctness can be realized, while preserving or even enhancing efficiency relative to hand coding. Moreover, even without formal proofs of correctness for the transformations involved, most of these derived programs worked correctly the first time they were run—they required *no debugging* whatsoever. (For parallel programs, avoiding debugging is an even greater advantage than for sequential ones!) Using the program specification and transformation methodology can thus save much of the time and effort traditionally invested not only in programming but also in debugging.

How is it possible to avoid debugging without formal proofs of correctness? First of all, the abstract specifications are problem–oriented and easy to understand; therefore, they are more likely to correctly express the solution to a problem than are traditional low–level programs. Second, TAMPR enables and encourages the user to construct sequences of transformations that consist of easily understood transformations—their correctness is often self–evident, without a formal proof. For the production of very high–assurance software, we are developing a methodology for formally proving the correctness of TAMPR transformations [11].

17.3 The TAMPR Approach to Program Transformation

TAMPR derives correct programs because TAMPR transformations are based on the syntax and semantics of the programming language being transformed (the subject language). TAMPR applies transformations not to a *string* representing the program, but rather to the syntax derivation *tree* (SDT) of the program, constructed according to the subject language grammar. A TAMPR transformation is a rewrite–rule consisting of a syntactic *pattern* and *replacement*. TAMPR searches the SDT of a program for a fragment that matchs the pattern of a transformation, and, if one is found, replaces that fragment by one constructed from the replacement of the transformation. The transformation

```
car ( cons ( <expr>"1" ) ) ==> <expr>"1"
```

expresses a simplification for lists discussed in Section 17.4.2.

The idea and basic notation for TAMPR transformations come from Chomsky's Transformational Grammar [7], developed in the late 1950s for use in describing natural languages. To make automated program transformation possible, TAMPR extends Chomsky's ideas in three important ways:

- TAMPR requires that transformations be *intra–grammatical*; that is, the result of applying a transformation to a sentence (program) in a particular language is guaranteed to be again a sentence (program) in that same language. Given a set of transformations that are intra–grammatical, it is easy and natural for them to apply repeatedly, the input of one transformation being the result of applying another.

- TAMPR added very simple *control notation* for describing how to apply transformations. This control notation makes possible the specification of sets of transformations that apply fully automatically.

- TAMPR provides a *computer implementation* of transformational grammar. Without a computer implementation, it is difficult to explore large applications of transformational grammar.

17.3.1 Automated Transformation

The following question naturally arises: How is it possible for TAMPR to apply program transformations automatically?

One might imagine that to automate the application of program transformations would require writing a set of program transformations and accompanying each with control directives that specify what transformations are to be tried next and where in the SDT, relative to the current application, they are to be applied. Such an approach is not very robust, however, because writing the control directives requires anticipating all possible combinations of program structures.

A much simpler, and highly robust, approach is to let the program being transformed guide the application of the transformations. In this approach, a set of transformations consists of several lists of transformations, each of which applies to some syntactic construct, such as `<expression>` or `<statement>`, in the subject language. Given such a set of transformations, the transformation system searches the program and applies a transformation wherever one from the set matches a syntactically suitable fragment of program. Because this strategy does not depend on any property of a particular program, it enables the application of a set of transformations to adapt itself to transform *any* program.

Applying transformations from a set wherever they match in a program, including to the results of earlier transformations, is called *application to exhaustion*, because the transformations in the set continue to apply until none matches anywhere in the program.

Application to exhaustion leads naturally to the concept of a set of transformations having a *goal*, the goal being the removal of all program fragments that match a pattern in the transformation set from the program. When transformation is considered from the point of view of *implementing* a specification into a program, the goal of many transformation sets is to *remove* specification–level constructs, ultimately leaving only implementation–level ones. Thus, complex, high–level language constructs can be implemented simply by writing transformations that match the high–level constructs and remove them from the program, replacing them with aggregations of lower level constructs having the same meaning. (Think, for example, of implementing Fortran in assembly language; Fortran–level constructs such as DO–loops and IF–statements must be replaced by assembly–level constructs such as tests and jumps.)

The goal of a set of transformations is thus to produce a program that lies in a restricted subset of the subject language, a subset in which specification–level constructs do not occur. This process can also be seen as carrying the program from one language level (for example, functional language) to another (for example, procedural, or even assembly, language). As discussed in the next subsection, when constructing TAMPR transformation sequences one tries to take advantage of properties of the intermediate language levels during the course of a derivation.

17.3.2 Structure of Derivations

The usual method of applying a set of TAMPR transformations, application to exhaustion, places a program (provided application terminates) in a canonical form. The canonical form is defined by the absence of all constructs that occur as patterns in the set of transformations. Of what use are canonical forms? They enable automatic transformation, because a *sequence* of canonical forms can be used to accomplish a complex transformational task.

An example of such a task from ordinary high–school algebra can help one to understand the connection between a sequence of canonical forms and a complex algebraic manipulation. Suppose that one is given a polynomial in the variable y

$$(x^2 + 2x + 1)y^2 + (x^2 - 9)y - (20x^2 + 18x - 18)$$

and asked to convert it into the corresponding polynomial in x:

$$(y^2 + y - 20)x^2 + (2y^2 - 18)x + (y^2 - 9y + 18).$$

To perform this conversion systematically, one intuitively uses a sequence of intermediate canonical forms. The choice of these intermediate canonical forms is dictated by the algebraic knowledge available to solve the problem. Specifically, one knows how to convert very simple polynomials, *monomials*, from y to x, using the commutative law of multiplication. For example, the monomial $2xy^2$ becomes $2y^2x$, which has the desired variable, x, on the right. Similarly, if one has a sum of monomials, one can apply the commutative and associative laws of addition to collect terms with like powers of the variable x. These laws are the only tools available to perform the change of variable.

Unfortunately, neither law can be applied directly to the example polynomial in y, because it is more complicated than a sum of monomials. However, other laws permit one to multiply out the multinomial coefficients to obtain the canonical form "sum of monomials in y":

$$x^2y^2 + 2xy^2 + y^2 + x^2y - 9y - 20x^2 - 18x + 18.$$

Then, one can apply commutativity of multiplication to obtain the canonical form "sum of monomials in x" (x commuted to the right):

$$y^2x^2 + 2y^2x + y^2 + yx^2 - 9y - 20x^2 - 18x + 18.$$

Application of commutativity of multiplication is followed by application of commutativity and associativity of addition to obtain "sum of monomials in x with like powers of x collected" canonical form:

$$y^2x^2 + yx^2 - 20x^2 + 2y^2x - 18x + y^2 - 9y + 18.$$

Finally, one can apply the *inverse* of the first manipulation to produce "multinomial coefficients of powers of x" form:

$$(y^2 + y - 20)x^2 + (2y^2 - 18)x + (y^2 - 9y + 18).$$

This is the desired form, which expresses the original polynomial in terms of x.

One can think of this sequence of algebraic manipulations as a function $\mathcal{T}_{cvx}$ "convert to x", which is itself a composition of functions, each of which carries out one step of the conversion:

$$\mathcal{T}_{cvx} = \mathcal{T}_{sm}^{-1} \circ \mathcal{T}_{lpc} \circ \mathcal{T}_{xctr} \circ \mathcal{T}_{sm}.$$

Here, $\mathcal{T}_{sm}$ is a function that produces "sum of monomials" canonical form, $\mathcal{T}_{xctr}$ produces "x commuted to right" canonical form, and $\mathcal{T}_{lpc}$ produces "like powers collected" canonical form.

This example illustrates four key points that are applicable to any algebraic manipulation, including transformational manipulations (derivations) in a "program algebra":

- Lengthy, complex algebraic manipulations (program derivations) can be carried out by using a sequence of canonical forms.

- Use of a *sequence* of canonical forms is important, because some of the manipulations (derivation steps) may be inverses of one another, and therefore cannot be applied simultaneously.

- At the heart of each manipulation (derivation) lies one or more *pivotal steps*, in which basic laws from the appropriate algebra are used to solve a simple version of the problem.

- The overall manipulation (derivation) consists of some preliminary steps that put the problem in the simple form required to apply the pivotal steps, the pivotal step or steps themselves, and some "postliminary" steps that produce the final form required of the manipulation.

These concepts form the basis for constructing program transformations that perform manipulations in the "program algebras" that underlie the languages in which specifications and programs are expressed. The use of a sequence of canonical forms enables complex optimizations to be performed during a derivation, including in–lining of function definitions, symbolic execution, and simplification. The effect of these optimizations is illustrated in the following section.

17.4 Example of the TAMPR Approach to Program Transformation

Against this background of program transformation, we discuss how program transformation can be used to produce highly efficient programs from object–oriented specifications.

When one employs an approach to programming or specification, such as object–oriented programming, that attempts to achieve clarity, modifiability, adaptability, and portability, the efficiency of the resulting program will suffer. In a general sense, inefficiency arises because information (such as the definition of a very simple operation on an object) must be localized for clarity; if executed in this form, a program will waste time referencing the local definition describing how to perform the operation. In contrast, efficiency is achieved by spreading out (copying) such information to the points in the program where it is used. Copying definitions of operations to their point of use (in–lining functions) has the further advantage of often enabling simplifications (removal of certain operations altogether), which further reduce the operational overhead of the program.

A number of techniques can be employed during a derivation to make both problem–domain–oriented and hardware–oriented optimizations to a program specification. These techniques include in–lining definitions, partial evaluation, simplification, finite differencing to reduce computation or storage, tail–recursion elimination, and converting arrays of structures to structures of arrays. Application of these techniques not only makes the program run with efficiency comparable to or exceeding that of traditional handwritten Fortran or C code but also makes the derived program *look* more like traditional code.

The hyperbolic PDE solver application (Section 17.2.4) provides a rich source of examples to illustrate the effect of these optimization techniques on an object–oriented specification. We discuss a number of such optimizations in the papers [5, 6]. Here we concentrate on the role of in–lining, partial evaluation, and simplification.

17.4.1 Specification for Part of the Hyperbolic PDE Solver

The hyperbolic PDE solver is based on a cellular–automaton algorithm invented by Garbey and Levine. This algorithm solves first–order, quasi-linear hyperbolic equations that arise from conservation laws. In the one–dimensional case, the problem is

$$\frac{\partial u}{\partial t} + \frac{\partial f(u)}{\partial x} = 0; \quad (x,t) \in \mathcal{R} \times [0,T]$$

$$u(x,0) = g(x); \quad x \in \mathcal{R}.$$

As discussed in [10], the algorithm computes the solution by the method of characteristics. This method computes the field of characteristics and derives the answer from it, rather than solving the problem directly.

The specification for the cellular–automaton algorithm is representative of the type of object–oriented specification that we typically employ. Space does not permit a detailed discussion of the specification, but we present a few of function definitions to give its flavor. Further details may be found in [5,6].

One solves a hyperbolic problem by applying a function `steptime` to an initial grid (for a Riemann problem in our tests) of a specified size, a set of boundary values, a current timestep number, and a number of time steps remaining to be performed:

```
timestep (initgrid (gridsize), bv, 1, maxsteps)
```

The result of taking a time step is the argument `grid` if the preceding time step was the last one; otherwise, the result is that obtained by taking another time step on an updated grid:

```
timestep (grid, bv, step, maxsteps) ==
    if step > maxsteps then
        grid
    else
        timestep (updategrid (grid, bv), bv, step+1, maxsteps)
```

Applying a local update rule for a cell to all the cells of the grid produces an updated grid:

```
updategrid (grid, bv) ==
    mapgrid (lambda((grid, loc) updatecell (grid, loc, bv)), grid)
```

The `mapgrid` higher–order function applies ("maps") a function (`mapgrid`'s first argument) to each cell in a grid (`mapgrid`'s second argument). For each cell in the grid, `mapgrid` applies the function to a pair of arguments—the grid and the location of the cell in the grid. Use of `mapgrid` in this way commits the algorithm to a Jacobi method, in which the update of a cell depends only on the values of the cells in the argument grid and not on any cells newly updated in this iteration.

Updating a cell happens in one of two ways, depending on whether the cell is a boundary cell or an interior cell:

```
updatecell (grid, loc, bv) ==
    if isonboundary (loc, grid) then
        updateboundarycell (
            cellat (loc, grid), whichboundary (loc, grid), bv
        )
    else
        updateinteriorcell (
            cellat (loc, grid), neighborsat (loc, grid)
        )
```

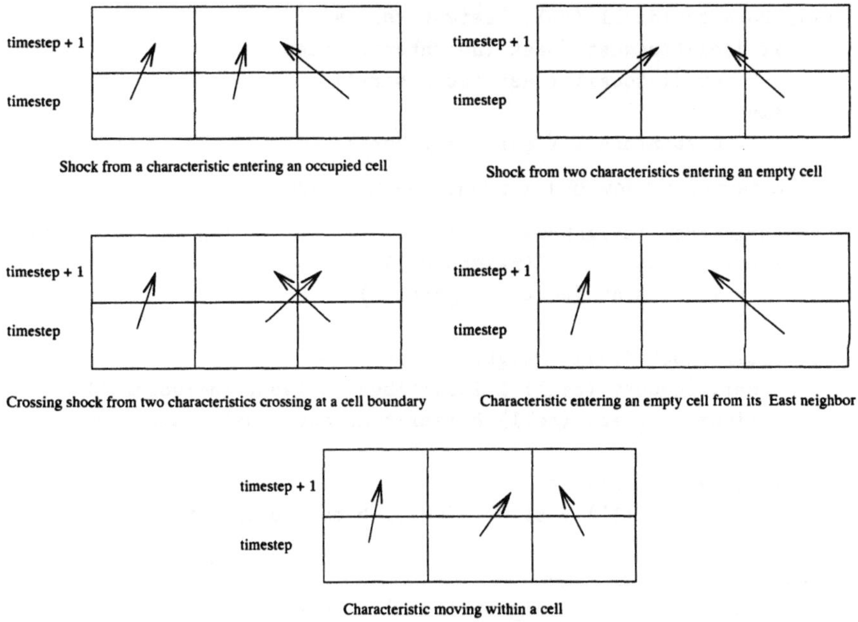

FIGURE 17.2. Movement of characterists

Updates of interior cells are of the most interest for this example; it is here that the cellular–automaton nature of the algorithm becomes apparent. There are five possible situations with regard to the movement of characteristics in the grid, illustrated in Figure 17.2. The cellular automaton treats the two shock case together; then crossing shock, then the movement of a characteristic from one cell into its neighbor, and finally the simple movement of a characteristic within a cell:

```
updateinteriorcell (cell, neighbors) ==
    if isshocked (cell, neighbors) then
        emptymarkedcell (shock ())
    else if iscrossingshock (cell, neighbors) then
        emptymarkedcell (crossingshock ())
    else if isenteringfrom (neighbors) then
        neighborenteredcell (neighbors)
    else
        timestepedcell (cell)
```

The movement of a particle into a cell from a neighboring cell is expressed as

```
neighborenteredcell (cell, neighbors) ==
    if isexitingwest (east (neighbors)) then
        movedintocell (east (neighbors))
    else
        movedintocell (west (neighbors))
```

Specifications for a few of the predicates used are

```
isenteringfrom (neighbors) ==
    isexitingwest (east (neighbors))
    | isexitingeast (west (neighbors))

iscrossingshock (cell, neighbors) ==
    (isexitingwest (cell) & isexitingeast (west (neighbors)))
    | (isexitingeast (cell) & isexitingwest (east (neighbors)))

isexitingeast (cell) ==
    newxvalue (cell) * signof (cell) > eastmostpoint ()

isexitingwest (cell) ==
    newxvalue (cell) * signof (cell) < westmostpoint ()
```

Finally, definitions associated with some of the data objects are

```
cellat (loc, grid) == arrayelement (grid, loc)

neighborsat (loc, grid) ==
    cons (arrayelement (grid, loc-1), arrayelement (grid, loc+1))

east (neighbors) == cdr (neighbors)

west (neighbors) == car (neighbors)

eastmostpoint () == 100

westmostpoint () == -100
```

One might wonder about the size of this specification; in full, it runs to slightly over five pages. This size results in part from the multiple layers of abstraction represented in the specification, which permit it to be understood at several levels of detail. The size is also increased by localization of the definitions of very simple operations on data, for which the function applications and definitions require more space than would repeating the operations where they are used. Of course, these abstraction and localization features make the specification highly flexible and understandable.

17.4.2 Strategies for Deriving Efficient Programs from Object-Oriented Specifications

The techniques of in-lining and simplification can greatly improve the efficiency of such a specification. Consider the definition of the function `updatecell`, which contains as its second alternative the function application:

```
updateinteriorcell (cellat (loc, grid), neighborsat (loc, grid))
```

In–lining the definition of `neighborsat` gives

```
updateinteriorcell (
    cellat (loc, grid),
    cons (elementof (grid, loc-1), elementof (grid, loc+1))
)
```

When the definition of `updateinteriorcell` is in–lined, the arguments are, of course, substituted into its definition. As a result, for example, the third predicate `isenteringfrom (neighbors)` becomes

```
isenteringfrom (
    cons (elementof (grid, loc-1), elementof (grid, loc+1))
)
```

Now, when the definitions of `isenteringfrom`, `east`, and `west` are in–lined, the code becomes

```
isexitingwest (
    cdr (cons (elementof (grid, loc-1), elementof (grid, loc+1)))
)
| isexitingeast (
    car (cons (elementof (grid, loc-1), elementof (grid, loc+1)))
)
```

Sufficient in–lining has now been done that symbolic execution can be performed on the code to apply simplifications based on laws for the list data type. Two of these laws are

```
car (cons (l1, l2)) = l1
cdr (cons (l1, l2)) = l2
```

Applying these laws simplifies this application of `isenteringfrom` to

```
isexitingwest (elementof (grid, loc+1))
| isexitingeast (elementof (grid, loc-1))
```

Further in–lining using the definitions of `isexitingeast`, `isexitingwest`, `eastmostpoint`, and `westmostpoint` produces

```
newxvalue (elementof (grid, loc+1))
        *signof (elementof (grid, loc+1))
    < -100
| newxvalue (elementof (grid, loc-1))
        *signof (elementof (grid, loc+1))
    > 100
```

At this point, further steps of the derivation take place, including one that converts the grid from array–of–structures to structure–of–arrays form. Finally, `newxvalue`, `signof`, and `elementof` simplify to become just array accesses, reducing the code for `isenteringfrom (neighborsat (loc, grid))` to

```
x(loc+1) * s(loc+1) < -100 | x(loc-1) * s(loc-1) < 100
```

which is the sort of code that would have been written in a typical Fortran program. While this code is much more efficient than that in the specification, it is much less clear to the human reader. This example shows how the transformational approach to program development enables one to "have his cake and eat it, too": to have both a clear specification and an efficient program.

In the transformation of the hyperbolic PDE solver, *all* of the uses of the `neighborsat` function simplify, so that it disappears entirely from the final Fortran program—there are no `conses` in the final program attributable to `neighborsat`. Thus, all of the overhead of using the `neighborsat` data abstraction is eliminated from the final program. In fact, this aspect of the example is typical; many of the accessors (except for array references) and constructors for data abstractions are eliminated by in–lining and simplification. That such simplification is possible should not be surprising, because one sees neither residues of data abstractions nor heroic means to avoid such residues in typical handwritten Fortran programs.

17.5 Conclusion

We have discussed how program specification and transformation in general, and the TAMPR program transformation system in particular, can be used to generate, from high–level specifications, efficiently executable code adapted to a wide variety of computational environments. By making use of the powerful mathematical ideas of abstraction and algebraic manipulation this approach to software creation frees the numerical analyst or specialist in scientific computation to concentrate on preparing a correct abstract description of an algorithm or problem, which can then be instantiated automatically to produce the required programs.

Why, then, is the program specification and transformation methodology not more widely used? Certainly one reason is that, while accustomed to think in terms of abstraction in mathematics, most potential users are not accustomed to applying these concepts in programming. The quest for efficiency is trained into every Fortran programmer; learning to pursue abstraction instead can be a difficult task. Moreover, program transformation is most easily applied to specifications expressed in declarative (functional or logical) form rather than in procedural form. Expressing programs declaratively and recursively also requires some reorganization of thinking.

In addition, while some applications can be handled by "canned" transformations, in many cases the user of program specification and transformation (or someone on a team using this methodology) may find it necessary to prepare new transformations that address issues unique to the particular problem being solved or to the particular hardware architecture being targeted. Finally, TAMPR is a research system, and so lacks, at the present time, a user–friendly interface and adequate training material.

Nevertheless, we believe that the ideas of abstraction and algebraic manipulation are powerful and attractive to the numerical community, and we are working to make the tools to use them more accessible.

Acknowledgments: The authors wish to acknowledge the U.S. Dept. of Energy, under Contract W–31–109–Eng–38 to Argonne National Laboratory and the BM/C3 directorate, Ballistic Missile Defense Organization, U.S. Dept. of Defense, which supported the work of J. M. Boyle, and the U.S. Dept. of Energy under Contract DE–AC04–94AL85000 to Sandia National Laboratory, which supported the work of V. L. Winter. The authors also wish to acknowledge the referees and one of the editors (HPL) for their suggestions and probing questions, which greatly improved the presentation of this chapter.

17.6 References

[1] D. P. Bertsekas and D. A. Castanon. Parallel synchronous and asynchronous implementations of the auction algorithm. *Parallel Computing*, 1990.

[2] J. M. Boyle. Abstract programming and program transformation—an approach to reusing programs. In T. J. Biggerstaff and A. J. Perlis, editors, *Software Reusability*, volume I, pages 361–413, New York, 1989. Addison–Wesley.

[3] J. M. Boyle, M. Clint, S. Fitzpatrick, and T. J. Harmer. The construction of numerical mathematical software for the AMT DAP by program transformation. In L. Bouge, M. Cosnard, Y. Robert, and D. Trystam, editors, *Parallel Processing: CONPAR 92—VAPP V, Second Joint International Conference on Vector and Parallel Processing, LNCS 634*, pages 761–767, Berlin, 1992. Springer–Verlag.

[4] J. M. Boyle, K. W. Dritz, M. M. Muralidharan, and R. Taylor. Deriving sequential and parallel programs from pure lisp specifications by program transformation. In L.G.L.T. Meertens, editor, *Proceedings of the IFIP TC2/WG2.1 Working Conference on Program Specification and Transformation, Bad–Toelz, Germany, 15–17 April, 1986*, pages 1–19, Amsterdam, 1987. North–Holland.

[5] J. M. Boyle and T. J. Harmer. A practical functional program for the Cray X–MP. *Journal of Functional Programming*, 2(1):81–126, January 1992.

[6] J. M. Boyle and T. J. Harmer. Practical transformation of functional programs for efficient execution. In B. Moeller, H. Partsch, and S. Schuman, editors, *Formal Program Development*, pages 62–88, Berlin, Heidelberg, New York, 1993. Springer–Verlag.

[7] N. Chomsky. Three models for the description of language. In E. Galanter R. D. Luce, R. Bush, editor, *Readings in Mathematical Psychology, Vol. II*, New York, 1965. John Wiley & Sons.

[8] J. J. Dongarra, C. B. Moler, J. R. Bunch, , and G. W. Stewart. *LINPACK User's Guide*. SIAM, Philadelphia, PA, 1979.

[9] S. Fitzpatrick, T. J. Harmer, A. Stewart, M. Clint, and J. M. Boyle. The automated transformation of abstract specifications of numerical algorithms into efficient array processor implementations. *Science of Computer Programming*, 1996. (to appear).

[10] M. Garbey and D. Levine. Massively parallel computation of conservation laws. *Parallel Computing*, 16:293–304, 1990.

[11] V. L. Winter and J. M. Boyle. Proving refinement transformations for deriving high–assurance software. In *Proceedings of the IEEE High–Assurance Systems Engineering Workshop*, 1996.

List of Contributors

Andersen, Svein Olav	Department of Informatics University of Bergen N-5020 Bergen, Norway Email: sveinoa@ii.uib.no
Arge, Erlend	SINTEF Applied Mathematics Box 124 Blindern, N-0314 Oslo, Norway Email: Erlend.Arge@math.sintef.no
Balay, Satish	Mathematics and Computer Science Division Argonne National Laboratory 9700 South Cass Avenue Argonne, IL 60439-4844, USA Email: balay@mcs.anl.gov
Bartels, Richard	Computer Science Department Computer Graphics Laboratory University of Waterloo Waterloo, Ontario N2L 3G1, Canada Email: rhbartel@uwaterloo.ca
Beck, Rudolf	ZIB Takustrasse 7, D-14195 Berlin, Germany Email: Beck@zib.de
Berzins, Martin	Computational PDEs Unit School of Computer Studies, University of Leeds Leeds LS2 9JT, UK Email: martin@dcre.leeds.ac.uk
Boyle, James M.	Technology Development Division Argonne National Laboratory Argonne, IL 60439, USA Email: boyle@mcs.anl.gov
Bruaset, Are Magnus	SINTEF Applied Mathematics Box 124 Blindern, N-0314 Oslo, Norway Email: Are.Magnus.Bruaset@math.sintef.no
Chudoba, Rostislav	Institute for Statics and Dynamics Faculty of Civil Engineering Ruhr-University Bochum Email: rch@mail.sd.bi.ruhr-uni-bochum.de
Erdmann, Bodo	ZIB Takustrasse 7, D-14195 Berlin, Germany Email: Erdmann@zib.de

Gropp, William D.	Mathematics and Computer Science Division Argonne National Laboratory 9700 South Cass Avenue Argonne, IL 60439-4844, USA Email: gropp@mcs.anl.gov
Harmer, Terence J.	Department of Computer Science The Queen's University of Belfast Belfast, BT7 1NN, UK Email: t.harmer@qub.ac.uk
Hjelle, Øyvind	SINTEF Applied Mathematics Box 124 Blindern, N-0314 Oslo, Norway Email: Oyvind.Hjelle@math.sintef.no
Holm, Erik Jarl	Institute for Energy Technology N-2007 Kjeller, Norway Email: erikh@ife.no
Hopkins, Timothy R.	Computing Laboratory University of Kent Canterbury, CT2 7NF, Kent, UK Email: trh@ukc.ac.uk
Houstis, Elias N.	Computer Science Department Purdue University W. Lafayette, IN 47907-1398, USA Email: enhcs.purdue.edu
Johnson, Christopher R.	Department of Computer Science University of Utah Salt Lake City, UT 84112, USA Email: crj@cs.utah.edu
Langtangen, Hans Petter	Department of Mathematics University of Oslo Box 1053 Blindern, N-0316 Oslo, Norway Email: hpl@math.uio.no
Manne, Fredrik	Department of Informatics University of Bergen N-5020 Bergen, Norway Email: fredrikm@ii.uib.no
McInnes, Lois Curfman	Mathematics and Computer Science Division Argonne National Laboratory 9700 South Cass Avenue Argonne, IL 60439-4844, USA Email: curfman@mcs.anl.gov

Mossberg, Eva
Department of Scientific Computing
Uppsala University
Box 120, S-751 04 Uppsala, Sweden
Email: evam@tdb.uu.se

Nelissen, Gert
Vrije Universiteit Brussel
Faculteit Toegepaste Wetenschappen
Dienst Electrotechniek en Industriële Electronica
Pleinlaan 2, 1050 Brussel, Belgium
Email: nelissen@etec4.vub.ac.be

Olsson, Hans
Department of Computer Science
Lund University
Box 118, S-221 00 Lund, Sweden
Email: Hans.Olsson@dna.lth.se

Olsson, Peter
Department of Scientific Computing
Uppsala University
Box 120, S-751 04 Uppsala, Sweden
Email: peter@tdb.uu.se

Otto, Kurt
Department of Scientific Computing
Uppsala University
Box 120, S-751 04 Uppsala, Sweden
Email: kurt@tdb.uu.se

Pantazopoulos, K. N.
Computer Science Department
Purdue University
W. Lafayette, IN 47907-1398, USA
Email: kp@cs.purdue.edu

Parker, Steven G.
Department of Computer Science
University of Utah
Salt Lake City, UT 84112, USA
Email: sparker@cs.utah.edu

Pennington, S. V.
Computational PDEs Unit
School of Computer Studies, University of Leeds
Leeds LS2 9JT, UK
Email: vp@dcre.leeds.ac.uk

Pratt, P. R.
PA Data Design
The Bishop's Mannor, Howden
East Yorkshire, DN14 7BL, UK
Email: paulp@padd.press.net

Rantakokko, Jarmo
Department of Scientific Computing
Uppsala University
Box 120, S-751 04 Uppsala, Sweden
Email: jarmo@tdb.uu.se

Roitzsch, Rainer	ZIB Takustrasse 7, D-14195 Berlin, Germany Email: `Roitzsch@zib.de`
Rüde, Ulrich	Institut für Mathematik Universität Augsburg 86135 Augsburg, Germany Email: `ruede@math.uni-augsburg.de`
Smith, Barry F.	Mathematics and Computer Science Division Argonne National Laboratory 9700 South Cass Avenue Argonne, IL 60439-4844, USA Email: `bsmith@mcs.anl.gov`
Thuné, Michael	Department of Scientific Computing Uppsala University Box 120, S-751 04 Uppsala, Sweden Email: `michael@tdb.uu.se`
Vankeirsbilck, Patrick	Vrije Universiteit Brussel Faculteit Toegepaste Wetenschappen Dienst Electrotechniek en Industriële Electronica Pleinlaan 2, 1050 Brussel, Belgium Email: `pvkeirsb@vnet3.vub.ac.be`
Ware, J. M.	Renaissance 22/25 Finsbury Square, London, UK Email: `justin@rs.com`
Weidmann, Matthias	Institut für Informatik Technische Universität München D-80290 München, Germany Email: `weidmann@informatik.tu-muenchen.de`
Weinstein, David M.	Department of Computer Science University of Utah Salt Lake City, UT 84112, USA Email: `dweinste@cs.utah.edu`
Winter, Victor L.	Intelligent Systems and Robotics Center Sandia National Laboratories Box 5800, Albuquerque, NM 87185–0660, USA Email: `vlwinte@sandia.gov`
Åhlander, Krister	Department of Scientific Computing Uppsala University Box 120, S-751 04 Uppsala, Sweden Email: `krister@tdb.uu.se`

Index

MIX
Papier aus verantwortungsvollen Quellen
Paper from responsible sources
FSC® C105338

If you have any concerns about our products,
you can contact us on
ProductSafety@springernature.com

In case Publisher is established outside the EU,
the EU authorized representative is:
Springer Nature Customer Service Center GmbH
Europaplatz 3, 69115 Heidelberg, Germany

Printed by Libri Plureos GmbH
in Hamburg, Germany